# Springer Complexity

Springer Complexity is an interdisciplinary program publishing the best research and academic-level teaching on both fundamental and applied aspects of complex systems - cutting across all traditional disciplines of the natural and life sciences, engineering, economics, medicine, neuroscience, social and computer science.

Complex Systems are systems that comprise many interacting parts with the ability to generate a new quality of macroscopic collective behavior the manifestations of which are the spontaneous formation of distinctive temporal, spatial or functional structures. Models of such systems can be successfully mapped onto quite diverse "real-life" situations like the climate, the coherent emission of light from lasers, chemical reaction-diffusion systems, biological cellular networks, the dynamics of stock markets and of the internet, earthquake statistics and prediction, freeway traffic, the human brain, or the formation of opinions in social systems, to name just some of the popular applications.

Although their scope and methodologies overlap somewhat, one can distinguish the following main concepts and tools: self-organization, nonlinear dynamics, synergetics, turbulence, dynamical systems, catastrophes, instabilities, stochastic processes, chaos, graphs and networks, cellular automata, adaptive systems, genetic algorithms and computational intelligence.

The two major book publication platforms of the Springer Complexity program are the monograph series "Understanding Complex Systems" focusing on the various applications of complexity, and the "Springer Series in Synergetics", which is devoted to the quantitative theoretical and methodological foundations. In addition to the books in these two core series, the program also incorporates individual titles ranging from textbooks to major reference works.

## Editorial and Programme Advisory Board

# Understanding Complex Systems

**Founding Editor: J.A. Scott Kelso**

Future scientific and technological developments in many fields will necessarily depend upon coming to grips with complex systems. Such systems are complex in both their composition - typically many different kinds of components interacting simultaneously and nonlinearly with each other and their environments on multiple levels - and in the rich diversity of behavior of which they are capable.

The Springer Series in Understanding Complex Systems series (UCS) promotes new strategies and paradigms for understanding and realizing applications of complex systems research in a wide variety of fields and endeavors. UCS is explicitly transdisciplinary. It has three main goals: First, to elaborate the concepts, methods and tools of complex systems at all levels of description and in all scientific fields, especially newly emerging areas within the life, social, behavioral, economic, neuroand cognitive sciences (and derivatives thereof); second, to encourage novel applications of these ideas in various fields of engineering and computation such as robotics, nano-technology and informatics; third, to provide a single forum within which commonalities and differences in the workings of complex systems may be discerned, hence leading to deeper insight and understanding.

UCS will publish monographs, lecture notes and selected edited contributions aimed at communicating new findings to a large multidisciplinary audience.

Alessandro Chiuso, Luigi Fortuna, Mattia Frasca,
Alessandro Rizzo, Luca Schenato,
and Sandro Zampieri (Eds.)

# Modelling, Estimation and Control of Networked Complex Systems

Springer

**Editors**

Alessandro Chiuso
Dipartimento di Tecnica e Gestione
dei Sistemi Industriali
Universitá di Padova
Stradella san Nicola, 3
36100 Vicenza
Italy
E-mail: CHIUSO@DEI.UNIPD.IT

Luigi Fortuna
Dipartimento di Ingegneria Elettrica
Elettronica e dei Sistemi
Universitá degli Studi di Catania
Viale A. Doria 6
95125 Catania CT
Italy
E-mail: LFORTUNA@DIEES.UNICT.IT

Mattia Frasca
Dipartimento di Ingegneria Elettrica
Elettronica e dei Sistemi
Universitá degli Studi di Catania
Viale A. Doria 6
95125 Catania CT
Italy
E-mail: MFRASCA@DIEES.UNICT.IT

Alessandro Rizzo
Dipartimento di Elettrotecnica ed Elettronica
Politecnico di Bari
Via E. Orabona 4
70125 Bari BA
Italy
E-mail: RIZZO@DEEMAIL.POLIBA.IT

Luca Schenato
Dipartimento di Ingegneria dell'Informazione
Università di Padova
Via Gradenigo, 6/b
35100 Padova
Italy
E-mail: SCHENATO@DEI.UNIPD.IT

Sandro Zampieri
Dipartimento di Ingegneria dell'Informazione
Università di Padova
Via Gradenigo, 6/b
35100 Padova
Italy
E-mail: ZAMPI@DEI.UNIPD.IT

ISBN 978-3-642-03198-4
e-ISBN 978-3-642-03199-1

DOI 10.1007/978-3-642-03199-1

Understanding Complex Systems
ISSN 1860-0832

Library of Congress Control Number: Applied for



*Typeset & Cover Design:* Scientific Publishing Services Pvt. Ltd., Chennai, India.

Printed in acid-free paper

9 8 7 6 5 4 3 2 1

springer.com

# Preface

The paradigm of complexity is pervading both science and engineering, leading to the emergence of novel approaches oriented at the development of a systemic view of the phenomena under study; the definition of powerful tools for modelling, estimation, and control; and the cross-fertilization of different disciplines and approaches. One of the most promising paradigms to cope with complexity is that of networked systems.

Complex, dynamical networks are powerful tools to model, estimate, and control many interesting phenomena, like agent coordination, synchronization, social and economics events, networks of critical infrastructures, resources allocation, information processing, control over communication networks, etc.

Advances in this field are highlighting approaches that are more and more often based on dynamical and time-varying networks, i.e. networks consisting of dynamical nodes with links that can change over time. Moreover, recent technological advances in wireless communication and decreasing cost and size of electronic devices are promoting the appearance of large inexpensive interconnected systems, each with computational, sensing and mobile capabilities. This is fostering the development of many engineering applications, which exploit the availability of these *systems of systems* to monitor and control very large-scale phenomena with fine resolution.

This book aims at presenting the state-of-the art of a wide range of research activities in terms of applications and advances on theoretical issues related to networked systems. The idea arose during the annual meeting of the Italian researchers in automation engineering (SIDRA Society) held in Vicenza (Italy) from 11th to 13th of September 2008, The contributions collected in this book will allow the reader to have a broad view of research topics at the cutting-edge of complex systems engineering, information processing, and control over complex networks.

The book is divided in four Parts. Part I highlights the aptitude of networked systems to model collective phenomena, such as synchronization, consensus, and formation control. Part II and III are devoted to important applications of networked system theory, in social phenomena, such as epidemic spreading and critical infrastructure networks (Part II), and in the design, management and optimization of sensor networks (Part III). Networked systems are also very important in different areas of system science, such as identification, estimation, optimization and control. These topics are dealt with in Part IV.

Bari, Catania, Padova, Vicenza,
May 2009

Alessandro Chiuso
Luigi Fortuna
Mattia Frasca
Alessandro Rizzo
Luca Schenato
Sandro Zampieri

# Contents

# List of Contributors

**Fabrizio Abrate**
Politecnico di Torino,
Corso Duca degli
Abruzzi 24, 10129
Torino, Italy,
fabrizio.abrate@polito.it

**Roberto Ambrosino**
Department of Technology,
University of Naples
Parthenope, Naples, Italy,
ambrosino@uniparthenope.it

**Giorgio Battistelli**
Dipartimento di Sistemi
e Informatica,
Università di
Firenze, 50139 Firenze,
Italy
battistelli@dsi.unifi.it

**Dario Bauso**
Universitá di Palermo,
viale delle Scienze,
90128 Palermo, Italy,
dario.bauso@unipa.it

**Alessio Benavoli**
Istituto "Dalle Molle" di Studi
sull'Intelligenza Artificiale,
6928 Manno-Lugano,
Switzerland
alessio@idsia.ch

**Francesco Bullo**
Center for Control,
Dynamical Systems and
Computation, University
of California at
Santa Barbara, Santa
Barbara, CA 93106, USA,
bullo@engineering.ucsb.edu

**Arturo Buscarino**
Scuola Superiore
di Catania, Laboratorio sui
Sistemi Complessi,
Università degli
Studi di Catania, via S.
Nullo, 5/i, 95125 Catania,
Italy,
arturo.buscarino@diees.unict.it

**Giuseppe C. Calafiore**
Politecnico di Torino,
Corso Duca degli
Abruzzi 24, 10129
Torino, Italy,
giuseppe.calafiore@polito.it

**Renato Casagrandi**
Dipartimento di

Elettronica e Informazione,
Politecnico di Milano,
Piazza Leonardo
da Vinci 32, I-20133
Milano, Italy,
casagran@elet.polimi.it

**Alessandro Casavola**
Dipartimento di Elettronica,
Informatica e Sistemistica,
Università degli
Studi della Calabria, Via
Pietro Bucci,
Cubo 42-c, Rende (CS),
87036, Italy,
casavola@deis.unical.it

**Luigi Chisci**
Dipartimento di Sistemi
e Informatica,
Università di Firenze,
50139 Firenze, Italy,
chisci@dsi.unifi.it

**Pietro De Lellis**
Department of Systems
and Computer Engineering,
University of Naples
Federico II, Via
Claudio 21,
80125 Napoli, Italy,
pietro.delellis@unina.it

**Stefano De Porcellinis**
Complex Systems and
Security Lab.,
Universitá CAMPUS
Bio-Medico, Roma, Italy,
s.deporcellinis@unicampus.it

**Francesco Delli Priscoli**
Università di Roma "La
Sapienza", Dipartimento
di Informatica e Sistemistica,
via Ariosto 25,
00162, Roma, Italy,
dellipriscoli@dis.uniroma1.it

**Mario di Bernardo**
Department of Systems
and Computer Engineering,
University of Naples
Federico II, Via
Claudio 21, 80125 Napoli, Italy,
mario.dibernardo@unina.it

**Cesare Fantuzzi**
DISMI - University of Modena
and Reggio Emilia,
via G. Amendola 2,
Morselli Building,
42100 Reggio Emilia, Italy,
cesare.fantuzzi@unimore.it

**Luigi Fortuna**
Dipartimento di Ingegneria
Elettrica Elettronica
e dei Sistemi,
Università degli
Studi di Catania,
viale A. Doria 6,
95125 Catania, Italy,
lfortuna@diees.unict.it

**Mattia Frasca**
Dipartimento di Ingegneria
Elettrica Elettronica
e dei Sistemi,
Università degli
Studi di Catania,
viale A. Doria 6,
95125 Catania, Italy,
mfrasca@diees.unict.it

**Francesco Garofalo**
Department of Systems
and Computer Engineering,
University of Naples
Federico II, Via Claudio 21,

80125 Napoli, Italy,
franco.garofalo@unina.it

**Emanuele Garone**
Dipartimento di Elettronica,
Informatica e Sistemistica,
Università degli
Studi della Calabria, Via
Pietro Bucci, Cubo 42-c,
Rende (CS), 87036, Italy,
egarone@deis.unical.it

**Andrea Gasparri**
University of "Roma Tre",
Via della Vasca
Navale 79, Roma, Italy,
gasparri@dia.uniroma3.it

**Laura Giarré**
Universitá di Palermo,
viale delle Scienze,
90128 Palermo, Italy,
giarre@unipa.it

**Giacomo Innocenti**
Università di Firenze,
Dipartimento di Sistemi
e Informatica, via
S. Marta 3, 50139
Firenze, Italy,
giacomo.innocenti@gmail.com

**Donatello Materassi**
University of Minnesota,
Department of Electrical
and Computer Engineering,
200 Union St SE, 55455,
Minneapolis (MN),
mater013@umn.edu

**Sandro Meloni**
University of "Roma Tre",
Via della Vasca
Navale 79, Roma, Italy,
sandro@dia.uniroma3.it

**Giuseppe Notarstefano**
Department of Engineering,
University of Lecce,
Via per Monteroni,
73100 Lecce, Italy,
giuseppe.notarstefano@unile.it

**Stefano Panzieri**
University of "Roma Tre",
Via della Vasca Navale
79, Roma, Italy,
panzieri@uniroma3.it

**Raffaele Pesenti**
Università di Venezia
Cà Foscari, Venezia,
Italy,
pesenti@unive.it

**Carlo Piccardi**
Dipartimento di Elettronica
e Informazione,
Politecnico di Milano,
Piazza Leonardo
da Vinci 32, I-20133
Milano, Italy,
carlo.piccardi@polimi.it

**Antonio Pietrabissa**
Università di Roma
"La Sapienza", Dipartimento
di Informatica e Sistemistica,
via Ariosto 25, 00162,
Roma, Italy,
pietrabissa@dis.uniroma1.it

**Kameshwar Poolla**
Department of Electrical
Engineering and Computer Science,
University of California
at Berkeley, Berkeley,
CA 94720, USA,
poolla@eecs.berkeley.edu

**Alessandro Rizzo**
Dipartimento di Elettrotecnica

ed Elettronica, Politecnico
di Bari, Via Re
David 200, 70125 Bari,
Italy,
rizzo@deemail.poliba.it

**Cristian Secchi**
DISMI - University of Modena
and Reggio Emilia,
via G. Amendola 2,
Morselli Building, 42100
Reggio Emilia, Italy,
cristian.secchi@unimore.it

**Roberto Setola**
Complex Systems and Security Lab.,
Universitá CAMPUS Bio-Medico,
Roma, Italy,
r.setola@unicampus.it

**Bruno Sinopoli**
Department of Electrical
and Computer Engineering,
Carnegie Mellon University,
Pittsburgh, PA 15213, USA,
brunos@ece.cmu.edu

# Part I
# Collective Phenomena

# Chapter 1
# Synchronization in Networks of Mobile Agents

Arturo Buscarino, Luigi Fortuna, Mattia Frasca, and Alessandro Rizzo

**Abstract.** In this Chapter we study synchronization issues in a system of mobile agents. Agents move as random walkers and interact with neighbouring units. Each agent carries a chaotic oscillator and coupling between oscillators occurs only when agents interact. Consequently, the interaction matrix is time-varying and appropriate synchronization criteria have to be defined.

## 1.1 Introduction

Studying synchronization in dynamical systems may give important insights for the understanding and modelling of phenomena which are ubiquitous in nature [40], and also for the development of devices in which a synchronous behaviour can lead to an enhancement of the performance (e.g., in the case of Josephson junctions arrays, higher output power values can be obtained when the junctions synchronously oscillate [20]). Hence, the study of synchronization in complex systems has a twofold appeal: from the theoretic point of view, it allows to better understand natural phenomena and, from the technological point of view, it is useful in the design of high performance devices and systems.

Synchronization is a process wherein two or more dynamical systems (either identical or non-identical), coupled through suitable configurations or driven by a

Arturo Buscarino
Scuola Superiore di Catania, Laboratorio sui Sistemi Complessi,
Università degli Studi di Catania, via S. Nullo, 5/i, 95125 Catania, Italy

Luigi Fortuna and Mattia Frasca
Dipartimento di Ingegneria Elettrica Elettronica e dei Sistemi,
Università degli Studi di Catania, viale A. Doria 6, 95125 Catania, Italy

Alessandro Rizzo
Dipartimento di Elettrotecnica ed Elettronica, Politecnico di Bari,
Via Re David 200, 70125 Bari, Italy

common forcing signal, can coordinate a particular dynamical property. Although the occurrence of synchronization has been initially studied among coupled periodic oscillators [39], more recently the attention has been moved to the synchronization of chaotic systems [34, 9, 7]. Such dynamical systems, characterized by a strong sensitivity to their initial conditions, can show a synchronous behaviour if suitably coupled. Moreover, several types of synchronization among chaotic systems can be revealed [9]: complete, phase, lag, generalized, intermittent, imperfect or almost synchronization. In the last decade, the emergence of synchronization in spatially extended or infinite dimension systems [2, 44], in experimental setups [12, 19, 3, 31] or in natural phenomena [40] has been investigated. The search for desynchronizing conditions has also been considered [27].

In the 90's the interest in chaotic synchronization focused on the definition of analytical conditions under which two coupled circuits could be synchronized and on the related techniques for the design of suitable coupling configurations, with applications especially in secure communications [1, 24, 25, 26, 33].

Progressively, the focus moved on synchronization of networks of dynamical units: at first, on regular structures like circuit arrays [43, 42]; then, on complex networks [10]. Most of the synchronization criteria that have been proposed for complex networks can be classified in two classes: those based on the Master Stability Function (MSF) and its generalizations, and those based on Lyapunov functions.

The MSF [35] allows to derive the analytical conditions under which a group of identical coupled dynamical systems can show a stable synchronous behaviour. More specifically, the MSF approach computes the maximum Lyapunov exponent of the variational equations, allowing the evaluation of the stability of the modes transverse to the synchronous manifold. The approach introduces a significant simplification of the problem of synchronization in complex networks since the study of the dynamical properties of the network node dynamics is made independent of that of the network topology. Thanks to this property, it is possible to characterize complex networks in terms of their aptitude to synchronization.

On the other hand, several synchronization criteria have been derived through approaches based on Lyapunov stability theory [27, 30, 29]. These studies provide slightly different results especially concerning the concept of aptitude to synchronization of a network. Nevertheless, it is a common finding that the ability of a given network to synchronize is strongly related to its topology.

The topology of a network also rules the onset of the globally synchronous state. For example, in random networks, isolated clusters of synchronous nodes tend to aggregate when the coupling strength is increased, whereas in scale-free networks a unique cluster grows up, including at each step more and more nodes of the network [23].

A less investigated aspect of synchronization in complex networks regards time-varying topologies. Many complex systems are made up by several simple units interacting through time-variant links [13, 21, 14, 22]. Such systems are characterized by dynamically changing coupling networks, whose links evolve in time according to either deterministic or stochastic rules.

Criteria for synchronization in time-varying networks have been derived under more stringent hypotheses than the case of static networks. Some studies conclude that the stability of the synchronization manifold can be evaluated by computing both the eigenvalues and the related eigenvectors of the time-varying coupling matrix [29, 28]. Other approaches are based on the definition of a matrix measure of complex matrices leading to criteria that can be applied to time-invariant, time-varying, and switching configurations, although the case of switching configuration requires the hypothesis of a periodically switching sequence [15]. Criteria that can be applied in a much simpler way can be derived under the hypothesis of *fast switching,* that is to say, that the switching among the possible coupling configurations occurs at a time scale which is much faster than that of the dynamics of the single node [36].

In this Chapter we consider a set of mobile agents in a two dimensional space, each one of them carrying a chaotic oscillator, and discuss the related synchronization issues under the framework of time-varying networks. In particular, under the constraint of fast switching [36, 37, 38] we derive synchronization conditions which relate synchronization to a scaled all-to-all coupling matrix and to the parameters describing the agent motion (for instance, the density of agents). We also study how the system scales as the number of agents is varied. Finally, we study the effect of considering a *topological interaction* scheme, that is to say, the case in which the generic agent interacts with a fixed number of agents rather than all the agents found within a fixed interaction radius [4].

The remainder of the Chapter is organized as follows: in Section 1.2 the MSF and fast switching theory will be reviewed; in Section 1.3 the model adopted in this Chapter will be presented. Simulation results will be then given both for the metric and for the topological interaction schemes, in Sections 1.4 and 1.5, respectively. Finally, conclusions are drawn in Section 4.6.

## 1.2 Synchronization of Time-Invariant and Time-Varying Networks

In this Section we introduce some of the most used techniques to study synchronization in networks. The first technique is the *Master Stability Function* (MSF), mainly used to characterize synchronization processes in networks with time-invariant links and identical dynamical nodes. Furthermore, we discuss some generalization of the MSF, including the case of networks with time-varying links. Finally, we illustrate the synchronization issues of networks with time-varying links under the assumption of fast switching.

The MSF approach derives the conditions under which $N$ identical oscillators (coupled by an arbitrary network configuration admitting an invariant synchronization manifold) can be synchronized [35].

The dynamics of each node is modelled as

$$\dot{\mathbf{x}}^i = \mathbf{F}(\mathbf{x}^i) - K\sum_j g_{ij}\mathbf{H}(\mathbf{x}^j) \tag{1.1}$$

where $i = 1,\ldots,N$, $\mathbf{x}^i$ is a $m$-dimensional vector of dynamical variables of the $i$-th node, $\dot{\mathbf{x}}^i = \mathbf{F}(\mathbf{x}^i)$ represents the dynamics of each isolated node, $K$ is the coupling strength, $\mathbf{H} : \mathbb{R}^m \to \mathbb{R}^m$ is the coupling function and $G = [g_{ij}]$ is a zero-row sum $N \times N$ matrix modelling network connections (i.e., the Laplacian of the network).

According to the analysis of Pecora and Carroll [35], the dynamics of the network is linearized around the synchronization manifold. Under the hypothesis that the network Laplacian is diagonalizable, a block diagonalized variational equation of the form

$$\dot{\xi}_h = [D\mathbf{F} - K\gamma_h D\mathbf{H}]\xi_h \tag{1.2}$$

is obtained. It represents the dynamics of the system around the synchronization manifold; where $\gamma_h$ is the $h$-th eigenvalue of $G$, $h = 1,\ldots,N$. $D\mathbf{F}$ and $D\mathbf{H}$ are the Jacobian matrices of $\mathbf{F}$ and $\mathbf{H}$ computed around the synchronous state, and are the same for each block. Therefore, the blocks of the diagonalized variational equation differ from each other only for the term $K\gamma_h$. If one wants to study synchronization properties with respect to different topologies, the variational equation must be studied as a function of a generic (complex) eigenvalue $\alpha + i\beta$. This leads to the definition of the Master Stability Equation (MSE):

$$\dot{\zeta} = [D\mathbf{F} - (\alpha + i\beta)D\mathbf{H}]\zeta. \tag{1.3}$$

The maximum (conditional) Lyapunov exponent $\lambda_{max}$ of the MSE is studied as a function of $\alpha$ and $\beta$, thus obtaining the Master Stability Function (MSF), i.e. $\lambda_{max} = \lambda_{max}(\alpha + i\beta)$. Then, the stability of the synchronization manifold in a given network can be evaluated by computing the eigenvalues $\gamma_h$ (with $h = 2,\ldots,N$) of the matrix $G$ and studying the sign of $\lambda_{max}$ at the points $\alpha + i\beta = K\gamma_h$. If all eigenmodes with $h = 2,\ldots,N$ are stable, then the synchronous state is stable at the given coupling strength.

If $G$ has real eigenvalues, the MSF can be computed only as function of $\alpha$, and the functional dependence of $\lambda_{max}$ on $\alpha$ can give rise to three different cases [10]. The first case (type I) is the case in which $\lambda_{max}$ is positive $\forall\alpha$ and, thus, the network nodes cannot be synchronized. In the second case (type II) $\lambda_{max}$ assumes negative values above a threshold value, and then this case always leads to a stable synchronous state for high enough coupling strength. In the third case (type III), $\lambda_{max}$ is negative only in a closed interval of values of $\alpha$, that is to say, network nodes can be synchronized only if $K\gamma_h$ for $h = 2,\ldots,N$ lies in this interval.

The MSF formalism allows to study how the overall topology of networks influences the propensity for synchronization. Specifically, it gives a necessary condition (the negativity of all Lyapunov exponents transverse to the synchronization manifold) for the stability of a complete synchronization process. With this approach, both heterogeneous and homogeneous networks, scale-free and small-world topologies, weighted and unweighted networks have been studied [10].

As stated above, the MSF approach is valid for diagonalizable networks. This assumption always holds for undirected networks, where the Laplacian is symmetric. For directed networks, this is not always the case. This case is dealt with in [32], where the Jordan form of the Laplacian is considered, instead of its diagonalization. Following this approach, the $N$ equations (1.2) are replaced by sets of variational equations, each associated with the generic $p \times p$ Jordan block:

$$\begin{aligned} \dot{\xi}_{h,1} &= [D\mathbf{F} - K\gamma_h D\mathbf{H}]\xi_1 \\ \dot{\xi}_{h,2} &= [D\mathbf{F} - K\gamma_h D\mathbf{H}]\xi_2 - KD\mathbf{H}\xi_1 \\ &\cdots \\ \dot{\xi}_{h,p} &= [D\mathbf{F} - K\gamma_h D\mathbf{H}]\xi_p - KD\mathbf{H}\xi_{p-1} \end{aligned} \tag{1.4}$$

where $\gamma_h$ is the generic Laplacian eigenvalue and $\xi_1, \xi_2, \ldots, \xi_p$ are the perturbation modes in the generalized eigenspace of eigenvalue $\gamma_h$. The condition under which the synchronization manifold is stable (i.e., the solution of equations (1.4) converges to zero) is the same as in the MSF approach ($\lambda_{max}(K\gamma_h) < 0$), where each mode of perturbation is decoupled from the others. The difference is that in the case of not diagonalizable networks the exponential convergence is no more simultaneous and decoupled; therefore the transient may be longer than in the diagonalizable case.

The MSF has also been extended to the case of networks with time-varying links [28, 29]. In this case, it is shown that synchronization properties not only depend on the Laplacian eigenvalues, but also on the corresponding eigenvectors. The variational equation (1.2) around the synchronization manifold now becomes:

$$\dot{\xi}_h = [D\mathbf{F} - K\gamma_h D\mathbf{H}(t) + K\beta_h(t)\mathbf{I}_m]\xi_h \tag{1.5}$$

where it is supposed that the time-varying Laplacian $G(t)$ can be diagonalized at each time instant by a matrix $\Phi(t)$ such that $\Phi^{-1}(t)G(t)\Phi(t) = \mathrm{diag}\{\gamma_1, \gamma_2, \ldots, \gamma_N\}$ and $\dot{\Phi}^{-1}(t)\Phi(t) = \mathrm{diag}\{\beta_1, \beta_2, \ldots, \beta_N\}$. This extension of the MSF is obviously of difficult applicability.

A simpler technique, introduced by Stilwell et al. in [38], relies on the hypothesis of fast switching. Their approach studies the synchronization properties of a time-varying topology of coupled chaotic oscillators by means of the time-average of the coupling matrix $G(t)$. The main result discussed in [38] can be expressed as follows: if the set of coupled oscillators defined by

$$\dot{\mathbf{x}}^i = \mathbf{F}(\mathbf{x}^i) - K\sum_{j=1}^{N} \bar{g}_{ij}\mathbf{E}\mathbf{x}^j \tag{1.6}$$

(with fixed topology $\bar{G} = [\bar{g}_{ij}]$) admits a stable synchronization manifold and if there exists a constant $T$ such that

$$\frac{1}{T}\int_t^{t+T} G(\tau)d\tau = \bar{G}, \tag{1.7}$$

then there exists $\varepsilon^*$ such that for all fixed $0 < \varepsilon < \varepsilon^*$ the set of coupled oscillators defined by

$$\dot{\mathbf{x}}^i = \mathbf{F}(\mathbf{x}^i) - K \sum_{j=1}^{N} g_{ij}(t/\varepsilon)\mathbf{E}\mathbf{x}^j \tag{1.8}$$

(i.e., coupled through a time-variant network $G(t/\varepsilon)$) also admits a stable synchronization manifold. This implies that, if the time-average of the coupling matrix $G(t)$, defined as in eq. (1.7), supports synchronization of the whole system and if the switching between all the possible network configurations is sufficiently fast, then the time-varying network will synchronize.

## 1.3 The Model

The model studied in this Chapter consists of mobile agents, each one associated with a chaotic oscillator coupled with those of the neighbouring agents. This situation, indeed, can be considered as a good representation of problems like clock synchronization in mobile robots, in which the communication is limited by the range of the communication system [13], or task coordination of swarming animals which are not only able to coordinate their motion in the plane, but also to react collectively and synchronously when subjected to external threats or attacks [16], or the appearance of synchronized bulk oscillations in a suspension of yeast cells, which has been experimentally observed for sufficiently high cell density [17, 18].

We consider $N$ moving individuals distributed in a planar space of size $L$ with periodic boundary conditions. Each individual moves with velocity $\mathbf{v}_i(t)$ and direction of motion $\theta_i(t)$ ($v$ being the modulus of the agent velocity, which is the same for all individuals). In our model, the agents are random walkers whose position and orientation are updated according to:

$$\begin{aligned} \mathbf{y}_i(t+\Delta t_M) &= \mathbf{y}_i(t) + \mathbf{v}_i(t)\Delta t_M \\ \theta_i(t+\Delta t_M) &= \eta_i(t+\Delta t_M) \ , \end{aligned} \tag{1.9}$$

where $\mathbf{y}_i(t)$ is the position of the $i$-th agent in the plane at time $t$, $\eta_i(t)$ are $N$ independent random variables chosen at each time step with uniform probability in the interval $[-\pi,\pi]$, and $\Delta t_M$ is the motion integration step size. The choice of random walk is motivated by the numerous examples of real systems which can be modelled through random walk (for example, living cells in a fluid [6], mobile robots performing exploration tasks [13], and social systems [11]).

In addition, to include the possibility that individuals can move through the bidimensional world with shorter time constants (as in some social system models [11]), we consider the case that individuals may perform long-distance jumps. This is accounted for by defining a parameter $p_j$ that quantifies the probability for an individual to perform a jump into a completely random position. In summary, at each time step, each agent evolves following Eqs. (1.9) (with $\mathbf{v}_i(t) = (v\cos\theta_i(t), v\sin\theta_i(t))$) with probability $1-p_j$ or performs a jump with probability $p_j$. In the latter case the

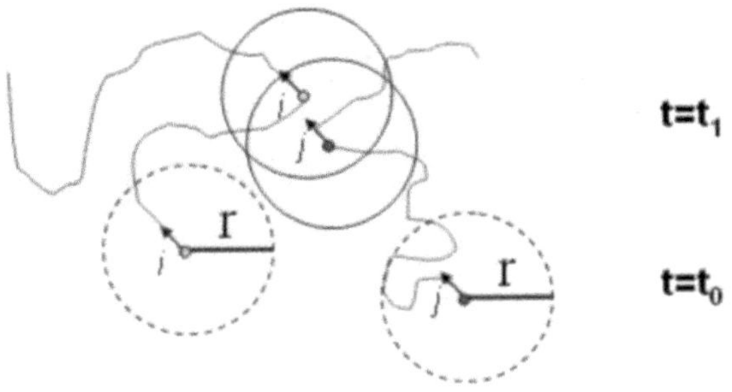


**Fig. 1.1** An example of the trajectories of two random walkers. The interaction radius is shown in dashed line for $t = t_0$ and in solid line for $t = t_1$. During their walk, agents can enter the interaction radius of other agents (as in $t = t_1$). In this case they are considered neighbours, otherwise the associated chaotic oscillators (for instance, for $t = t_0$) are not coupled.

agent position is updated into a new position chosen at random in the plane in which agents move.

Furthermore, a dynamical system, and in particular a chaotic one, is associated to each agent. Each agent is therefore characterized by a state variable vector $\mathbf{x}^i(t) \in \mathbb{R}^n$ which evolves according to a given chaotic law. In the following, without lack of generality, we consider the case of Rössler oscillators, where the state dynamics of each agent is described by:

$$\begin{cases} \dot{x}_1^i = -(x_2^i + x_3^i) \\ \dot{x}_2^i = x_1^i + ax_2^i \\ \dot{x}_3^i = b + x_3^i(x_1^i - c) \end{cases} \tag{1.10}$$

with $\mathbf{x}^i(t) = [x_1^i(t)\ x_2^i(t)\ x_3^i(t)]^T$. The following parameter values have been used: $a = 0.2$; $b = 0.2$; $c = 7$.

Each agent interacts at a given time with only those agents located within a neighbourhood of an interaction radius, defined as $r$ [13, 21, 41]. When two agents interact, the state equations of each agent are changed to include diffusive coupling with the neighbouring agent, so that, if for instance the $i$-th agent is coupled to the $j$-th agent, the diffusive term $K(x_1^j - x_1^i)$ is added to the first equation of (1.10). $K$ is the coupling parameter.

Based on these assumptions, the state dynamics of each agent can be described in terms of the following equations:

$$\dot{\mathbf{x}}^i = \mathbf{F}(\mathbf{x}^i) - K\sum_{j=1}^{N} g_{ij}(t)\mathbf{E}\mathbf{x}^j \tag{1.11}$$

for $i = 1,\ldots,N$ with $\mathbf{F}: \mathbb{R}^3 \to \mathbb{R}^3$ given by the Rössler dynamics (1.10),

$$E = \begin{bmatrix} 1\ 0\ 0 \\ 0\ 0\ 0 \\ 0\ 0\ 0 \end{bmatrix},$$

and $g_{ij}(t)$ are the elements of a time-varying matrix $G(t)$ which defines the neighbourhood of each agent at a given time $t$ and depends on the trajectory of each agent. More in detail, $g_{ij}(t) = g_{ji}(t) = -1$ if the $i$-th agent and the $j$-th agent are neighbour at time $t$; and $g_{ii}(t) = h$ where $h$ is the number of neighbours of the $i$-th agent at time $t$. An example is shown in Fig. 1.1, where two agent trajectories are reported: at $t = t_0$ agents are not neighbour (each one is not within the interaction radius of the other); at $t = t_1$ the two agents are neighbour.

Eqs. (1.11) are integrated with a fixed time step defined as $\Delta t_s$ (in all the simulations $\Delta t_s = 0.001$).

## 1.4 Simulation Results: Metric Interaction Scheme

The behaviour of the model described above is influenced by many parameters, and, in particular, it strongly depends on the agent velocity modulus $v$ and on the motion integration step size $\Delta t_M$. For sake of simplicity, let us first consider the case of two agents, i.e. $N = 2$. In this case, only two network configurations are possible at each time $t$, since the two agents can be neighbour or not.

Under the constraint of fast switching, we can analyze the synchronization properties inherited by the time average matrix $\bar{G}$ given by

$$\bar{G} = p_A G_A + p_0 G_0 \tag{1.12}$$

where $p_A$ is the probability that the two agents are neighbour (all-to-all coupling), while $p_0$ is the probability that the two agents are not neighbour, and

$$G_A = \begin{bmatrix} 1 & -1 \\ -1 & 1 \end{bmatrix}, G_0 = \begin{bmatrix} 0 & 0 \\ 0 & 0 \end{bmatrix} \tag{1.13}$$

are the corresponding $N \times N$ coupling matrices. Since $G_0 = 0$, $\bar{G}$ is given by $\bar{G} = p_A G_A$.

Thus, $p_A$ plays the role of a coupling parameter. The two eigenvalues of $\bar{G}$ are $\lambda_1 = 0$ (since $\bar{G}$ is defined as a zero-row sum matrix) and $\lambda_2 = 2p_A$. Let us consider a typical type III master stability function (i.e., a system for which stability of the synchronization manifold is only guaranteed in a given interval $[\alpha_1, \alpha_2]$) and let us suppose that $K\lambda_2 \in [\alpha_1, \alpha_2]$. Since $0 < p_A < 1$, this means that there is a critical value $p_c$ for $p_A$ so that if $p_A < p_c$ the chaotic dynamics of the two agents will not synchronize and, on the opposite, if $p_A > p_c$ the dynamics of the two agents will synchronize. Furthermore, this critical value can be estimated as

$$p_c = \frac{\alpha_1}{K\lambda_2}.$$

In the opposite case, i.e. if $K\lambda_2 > \alpha_2$, the system will behave in a different way; that is to say, two critical values of the parameters

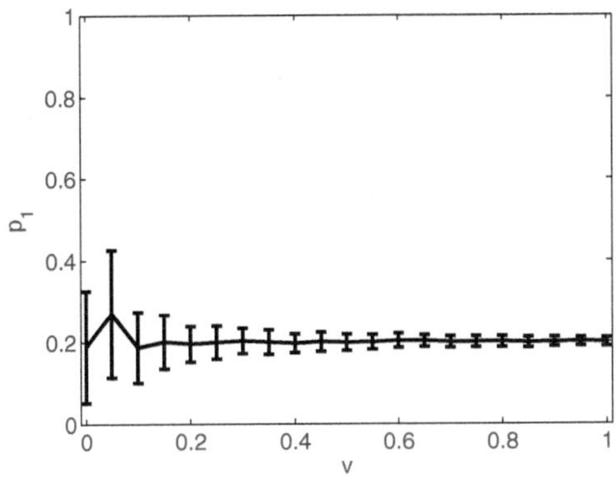


**Fig. 1.2** $p_A$ vs $v$. As $v$ increases, the actual value of $p_A$ tends to its estimation $p_A = \frac{\pi r^2}{L^2} = 0.2$. Simulation obtained for $\rho = 0.1273$, $N = 2$, $L = \sqrt{\frac{N}{\rho}}$, $r = 1$.

$$p_{c1} = \frac{\alpha_1}{K\lambda_2} \text{ and } p_{c2} = \frac{\alpha_2}{K\lambda_2}$$

can be defined so that the two agents will synchronize only if $p_A$ is within the range defined by these critical values.

Let us now discuss how the hypothesis of fast switching can be assumed in our model. This will also allow us to express the synchronization conditions with respect to the agent density $\rho$. The system has two time scales: the motion integration step size $\Delta t_M$ and the dynamics integration step size $\Delta t_s$. If the two time scales are chosen to be equal, then at each time step equations (1.11) will be integrated and a motion step will be performed according to equations (1.9). Otherwise, if, for instance, $\Delta t_M$ is chosen to be $\Delta t_M = \tau \Delta t_s$ ($\tau \in \mathbb{N}$), a motion step will be performed each $\tau$ integration step sizes of the dynamics equations (1.11). In the latter case, the neighbourhood of each agent will be updated each $\tau$ integration step sizes $\Delta t_s$, and thus the switching between the possible network configurations ($G_A$ and $G_0$) will be slower than in the case $\Delta t_M = \Delta t_s$. Fast switching can be therefore implemented by acting on the parameter $\Delta t_M$.

Another important parameter of the system is the motion velocity $v$. At the beginning of the simulation, the two agents have random initial positions. Thus, the probability that they are neighbour (i.e. agent 2 is within the interaction radius $r$ of agent 1) is given by

$$p_A = \frac{\pi r^2}{L^2}.$$

If $v$ is small, the two agents will perform small movements during the simulation, thus they will keep either close to each other or far from each other depending on their initial positions. However, when $v$ is large, one expects that the two agents will move more, and consequently $p_A = \frac{\pi r^2}{L^2}$ will be a good approximation of their probability of being neighbour. Thus, one expects that $v$ influences the value of the probability $p_A$ in terms of fluctuations: large values of $v$ lead to small fluctuations (as shown in Fig. 1.2).

Furthermore, for large $v$, by taking into account that $\rho = \frac{N}{L^2}$, it derives that $p_A$ depends on the density $\rho$ as

$$p_A = \frac{\pi r^2 \rho}{N}.$$

Since $p_A$ changes as $\rho$ changes, the existence of a critical $p_c$ (in the case of $\alpha_1 < K\lambda_2 < \alpha_2$) also implies the existence of a critical value of the density, defined as $\rho_c$ under which the two agents will not synchronize. The same argument applies when $K\lambda_2 > \alpha_2$.

The velocity $v$ has also effect on the switching between the network configurations. As $v$ increases, the distance covered by the agents at each motion step increases and the switching between possible network topologies becomes faster. This leads to the conclusion that for large values of $v$ and low values of $\Delta t_s$ we expect that the theoretical considerations discussed above can correctly predict the synchronization regions of the system.

Before generalizing our discussion to generic number of agents $N > 2$, let us now present numerical results for the case $N = 2$.

Let us start with $p_j = 0$, both in the cases of identical systems and in that of non-identical systems. In the latter case, we fixed $c = c_1 = 7$ (first agent) and $c = c_2 = 7.1$ (second agent) in equations (1.10).

For the analysis of the simulation results, we adopt the following synchronization error [8]:

$$\delta(t) = \frac{|x_1 - x_2| + |y_1 - y_2| + |z_1 - z_2|}{3} \tag{1.14}$$

and we define $\langle \delta \rangle = \langle \delta(t) \rangle$ as synchronization index, where the average is performed on the interval $[4/5T, T]$ on simulation of lenght $T = 500s$.

First of all, we discuss the behaviour with respect to the agent density $\rho$. The analysis of the MSF of the Rössler oscillator coupled through the $x_1$ variable leads to a type III MSF with $\alpha_1 \simeq 0.22$ and $\alpha_2 \simeq 4.08$. In the analysis that follows we define the synchronization region as that region for which $\langle \delta \rangle = 0$ holds. We use $\alpha_1$ and $\alpha_2$ to define the expected critical values of the parameters. For instance, if $K = 1$, we expect that the system has a single threshold parameter, so that the two agents synchronize if $\rho > \rho_c$ with

$$\rho_c = \frac{\alpha_1 N}{\lambda_2 K \pi r^2} = 0.07;$$

while, when $K = 10$, we expect that synchronization only occurs for $\rho_1 < \rho < \rho_2$ with

$$\rho_1 = \frac{\alpha_1 N}{\lambda_2 K \pi r^2} = 0.007 \text{ and } \rho_2 = \frac{\alpha_2 N}{\lambda_2 K \pi r^2} = 0.1299.$$

In Fig. 1.3 the behaviour of the system with $K = 10$ and with $K = 1$ (inset) is shown. As it will be clarified below, the behaviour of the system depends on

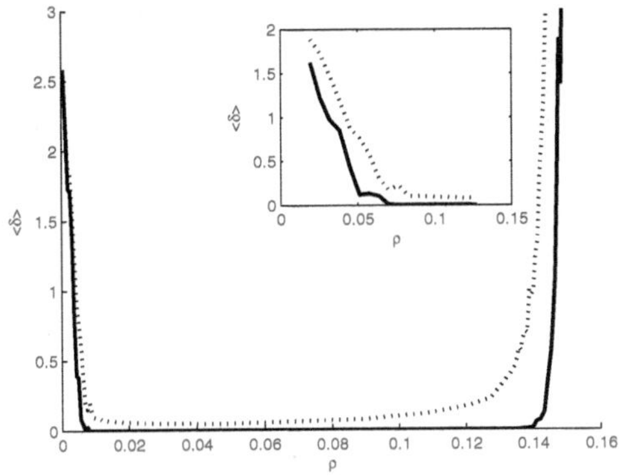


**Fig. 1.3** Synchronization index vs density $\rho$ of agents. When $K = 10$ the two agents synchronize if $0.008 = \rho_1 < \rho < \rho_2 = 0.1273$, while when $K = 1$ (inset) the two agents synchronize for $\rho > \rho_c = 0.07$. The other parameters have been chosen as follows: $\Delta t_M = 10^{-3}$, $v = 1$, $r = 1, N = 2$.

both $\Delta t_M$ and $v$: this graph has been obtained by fixing all the parameters of the simulation so that the fast switching hypothesis holds ($\Delta t_M = 10^{-3}$ and $v = 1$). From simulation results we estimate $\rho_c = 0.07$ for the case with $K = 1$, and $\rho_1 = 0.008$, $\rho_2 = 0.1273$ for the case with $K = 10$ , which agree with the theoretical predictions. Thus, depending on the choice of $K$, two qualitatively different behaviours arise: for $K = 10$ the two agents synchronize if $0.008 = \rho_1 < \rho < \rho_2 = 0.1273$, while when $K = 1$ the two agents synchronize for $\rho > \rho_c = 0.07$.

The influence of the fast switching hypothesis is studied by considering how the behaviour of the system changes when $\Delta t_M$ is varied. For sake of simplicity we focus on the case with $K = 1$, in which, as discussed above, $\rho_c = 0.07$. Two cases are shown in Fig. 1.4 and Fig. 1.5, which take into account both the cases of identical and non-identical agents. The two cases examined are $\rho = \bar{\rho} = 0.03$ and $\rho = \tilde{\rho} = 0.12$, which correspond to $\bar{\rho} < \rho_c$ and $\tilde{\rho} > \rho_c$. The analysis carried out in the previous Section predicts that the two agents will synchronize for sufficiently fast switching. As it can be observed in Fig. 1.4, this occurs only for low values of $\Delta t_M$. In the case with $\rho < \rho_c$, the two agents do not synchronize even if $\Delta t_M$ is decreased to its minimum value (i.e. $\Delta t_M = \Delta t_s$). In Fig. 1.4 and Fig. 1.5 the role of $v$ is also shown. It can be observed that, if $\rho > \rho_c$, for higher values of $v$ the synchronization index $\langle \delta \rangle = 0$ even if the integration step of the motion $\Delta t_M$ increases. Furthermore, in the case $\rho < \rho_c$, higher values of agent velocity lead to smaller values of $\langle \delta \rangle$. When non-identical systems are considered (Fig. 1.5), a similar effect on the synchronization index due to the higher velocity can be observed, though complete synchronization is never achieved.

We further investigate the dependence of the synchronization index on the velocity of the agents by considering two different cases: $\rho < \rho_c$ and $\rho > \rho_c$. When $\rho < \rho_c$, increasing $v$ has small effects on the synchronization properties of the

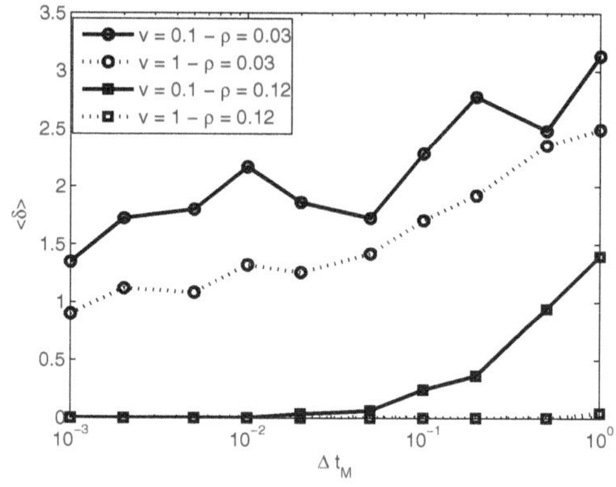


**Fig. 1.4** Synchronization index vs. $\Delta t_M$ for different values of $v$ and $\rho$. Curves are related to two identical systems.

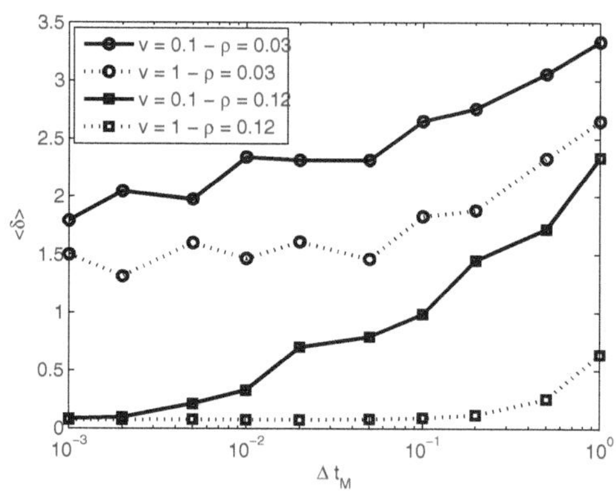


**Fig. 1.5** Synchronization index vs. $\Delta t_M$ for different values of $v$ and $\rho$. Curves are related to two non-identical systems.

system. The two agents do not synchronize even for relatively large values of $v$ as shown in Fig. 1.6.

On the opposite, if $\rho > \rho_c$, the velocity of the two agents plays an important role. If the switching between the two different networks ($G_A$ and $G_0$) is sufficiently fast, i.e. if $\Delta t_M$ is small, then synchronization occurs independently of the value of $v$. However, if $\Delta t_M$ is large, the agents synchronize only if their velocity is sufficiently high. This is shown in Fig. 1.7 for both the cases of identical and non-identical agents. It can be observed that when the two systems are non-identical, even if perfect synchronization is not achieved, increasing values of $v$ (when $\Delta t_M = 1$) lead to smaller values of the synchronization index.

When the agents also performs long-distance jumps (i.e. for $p_j \neq 0$), the behaviour of the system changes. Long-distance jumps may help synchronization, as

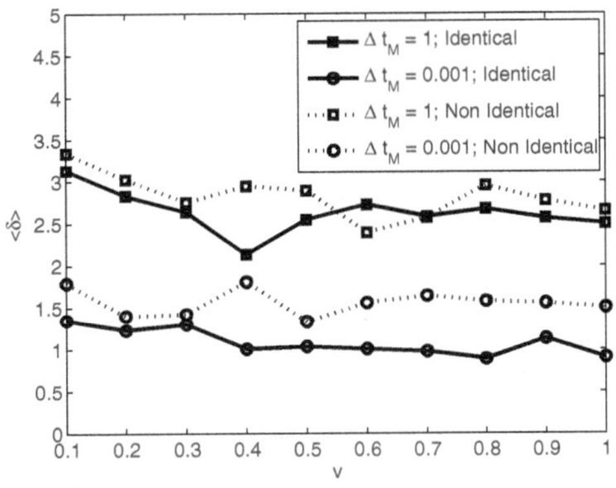


**Fig. 1.6** Synchronization index vs. $v$ for different values of $\Delta t_m$ and $\rho < \rho_c$, $K = 1$.

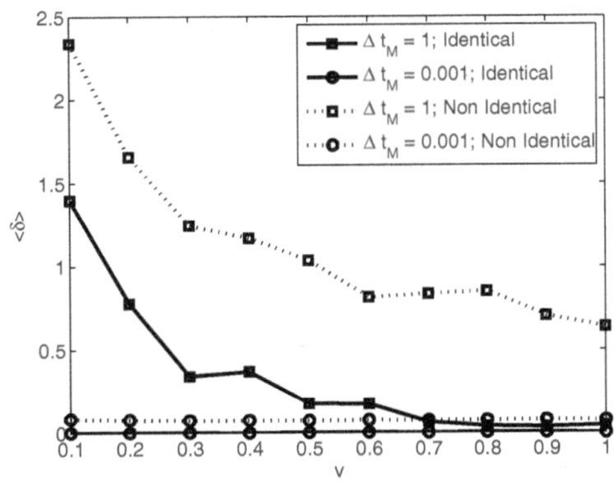


**Fig. 1.7** Synchronization index vs. $v$ for different values of $\Delta t_m$ and $\rho > \rho_c$, $K = 1$.

shown in Fig. 1.8 which reports the synchronization index $\langle \delta \rangle$ vs $\Delta t_M$ for different values of $p_j$. In the considered case the theoretical considerations predict synchronization. As discussed above, this occurs under the hypothesis of fast switching and thus only for low values of $\Delta t_M$. As it can be observed, when $p_j = 0.1$ synchronization may be achieved at higher values of $\Delta t_M$ with respect to the case of $p_j = 0$. This is more evident for $v = 0.1$ than for $v = 1$, since the switching rate also depends on $v$. This is an effect similar to that observed in other studies, where the presence of either long-distance jumps [21] or long-distance communications [13] improves communication between agents.

Let us now consider the case of $N > 2$ agents. We will show that a result similar to the previous case (i.e. $N = 2$), namely $\bar{G} = pG_A$, also holds for $N > 2$, where $G_A$ is the $N \times N$ all-to-all coupling matrix. In fact, in general, the possible configurations of the neighbourhood of $N$ agents, are obviously much more than two. For instance, for $N = 3$ there will be eight possible configurations (since the agents are numbered)

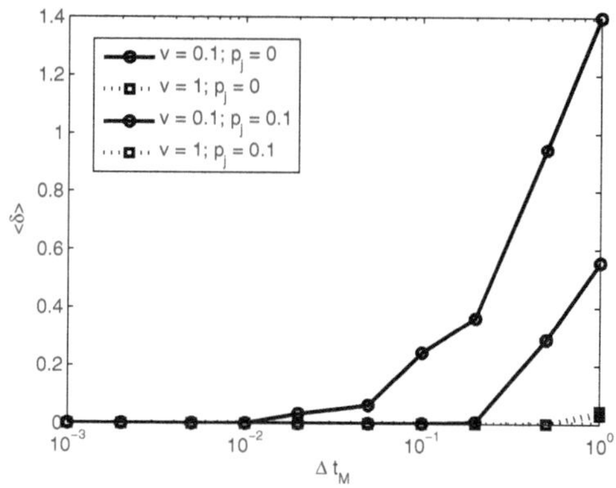


**Fig. 1.8** Synchronization index vs. $\Delta t_M$ for different values of $p_j$ and $v$ (identical systems).

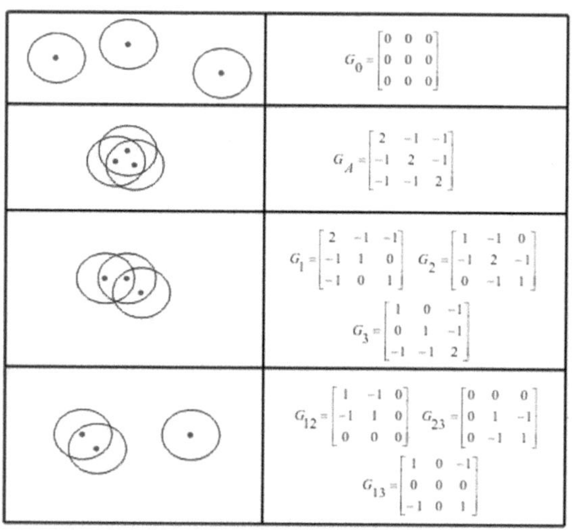

**Fig. 1.9** Possible neighbourhood configurations for $N = 3$ agents and corresponding Laplacian matrices.

which, as shown in Fig. 1.9, can be classified into four different cases. Therefore, we have

$$\bar{G} = p_A G_A + p_0 G_0 + p_{12} G_{12} + p_{23} G_{23} + p_{13} G_{13} + p_1 G_1 + p_2 G_2 + p_3 G_3 \quad (1.15)$$

where all the matrices in the sum are $N \times N$ matrices ($N = 3$), $p_A$ is the probability that all the agents are neighbour (and $G_A$ is the all-to-all coupling matrix), $p_{12}$ is the probability that agent 1 is close to agent 2, but agent 3 is not close neither to agent 1 nor to agent 2 (and $G_{12}$ is the related coupling matrix), $p_1$ is the probability that agent 2 is close to agent 1 and agent 3 is close to agent 1, but agent 2 and agent 3 are not neighbour (and $G_1$ is the related coupling matrix), and so on. By direct calculation it can be observed that

$$\begin{aligned} G_{12}+G_{23}+G_{13} &= G_A \\ G_1+G_2+G_3 &= 2G_A \end{aligned} \tag{1.16}$$

Since $p_{12}=p_{13}=p_{23}$ and $p_1=p_2=p_3$, it follows that

$$\bar{G}=p_AG_A+p_{12}(G_{12}+G_{23}+G_{13})+p_1(G_1+G_2+G_3)=(p_A+p_{12}+2p_1)G_A \tag{1.17}$$

and, consequently,

$$\bar{G}=pG_A. \tag{1.18}$$

The same considerations can be repeated for $N>3$. In a certain sense $\bar{G}$ is a kind of rescaled all-to-all matrix. In analogy with the case of blinking networks [5], it can be shown that $p$ is the probability that a link is activated (i.e., that two agents are neighbour) and thus $p$ is given by

$$p=\frac{\pi r^2}{L^2}. \tag{1.19}$$

By taking into account that $\rho=N/L^2$, it can be derived that $p$, and consequently $\bar{G}$, depend on the density $\rho$ as

$$p=\frac{\pi r^2\rho}{N},\ \bar{G}=\frac{\pi r^2\rho}{N}G_A \tag{1.20}$$

These relationships and the MSF allow us to derive the conditions under which the agents can be synchronized. To do this, the $N$ eigenvalues of the coupling matrix $\bar{G}$ are calculated. They are $\lambda_1=0$ (since $\bar{G}$ is zero-row sum) and $\lambda_i=pN$ for $i=2,\ldots,N$. Let us define $\lambda$ as $\lambda=pN$.

Let us suppose that $K\lambda\in[\alpha_1,\alpha_2]$. Since $0<p<1$, this means that there is a critical value of $p$

$$p_c=\frac{\alpha_1}{KN} \tag{1.21}$$

so that if $p<p_c$ the two agents will not synchronize and, on the opposite, if $p>p_c$ the two agents will synchronize. In terms of the density of agents, we can conclude that there exists a critical density threshold such that for

$$\rho>\rho_c=\frac{\alpha_1}{\pi r^2K} \tag{1.22}$$

agents do synchronize.

In the opposite case, i.e. if $K\lambda>\alpha_2$, the system will behave in a different way: in this case, the agents will synchronize at densities such that

$$\frac{\alpha_1}{\pi r^2K}=\rho_{c1}<\rho<\rho_{c2}=\frac{\alpha_2}{\pi r^2K}. \tag{1.23}$$

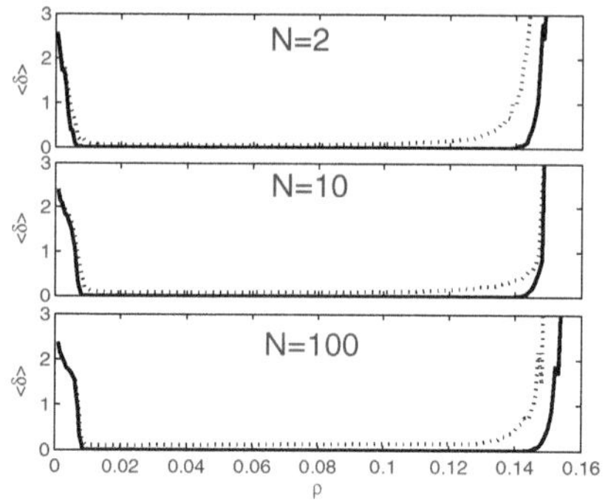


**Fig. 1.10** Synchronization index $\langle\delta\rangle$ vs density $\rho$ for identical (continuous line) and non-identical (dotted line) systems with $N=2$, $N=10$ and $N=100$. The coupling is fixed to $K=10$. The other parameters have been chosen as follows: $\Delta t_M = \Delta t_s = 10^{-3}$, $v=1$, $r=1$. Results are averaged over 50 realizations.

It can be observed that $p_c$ depends on the number of agents $N$, but $\rho_c$, $\rho_{c1}$ and $\rho_{c2}$ do not depend on $N$.

To verify that the critical value of the density does not depend on the number of agents, and to support the validity of the approximations made in the above analytical treatment, we carried out simulations of the full system at different densities and with different values of $N$ ($N=2$, $N=10$ and $N=100$). The other parameters have been fixed so that fast switching is achieved ($v=1$, $p_j=0$, $\Delta t_M = \Delta t_s = 10^{-3}$). We considered both identical and non-identical agents and we averaged the results of the simulations over 50 different realizations. In the case of non-identical agents, for each agent we fixed a different value of the parameter $c$ in the interval $[6.9, 7.1]$, so that each system has a chaotic behaviour.

For the analysis of the simulation results, we define the following synchronization error:

$$\delta(t) = \sum_{i=2}^{N} \frac{\left(|x_1^i - x_1^1| + |x_2^i - x_2^1| + |x_2^i - x_3^1|\right)}{3(N-1)} \tag{1.24}$$

and we also define

$$\langle\delta\rangle = \langle\delta(t)\rangle \tag{1.25}$$

as synchronization index, where the average is performed on the interval $[4T/5, T]$ ($T = 500$s is the total length of the simulation).

Fig. 1.10 reports the results for $K=10$ (which, for Rössler oscillators, is such that $K\lambda > \alpha_2$). In the case of identical systems, as expected, for

$$\frac{\alpha_1}{\pi r^2 K} = \rho_{c1} < \rho < \rho_{c2} = \frac{\alpha_2}{\pi r^2 K} \tag{1.26}$$

the index $\langle\delta\rangle$ is zero; this corresponds to complete synchronization on a synchronization manifold. In the case of non-identical systems, a synchronization manifold cannot be formally defined, and therefore the MSF approach does not rigorously apply. We expect however that, for $\rho_{c1} < \rho < \rho_{c2}$, a synchronization motion will be established where the difference between the states of the systems will only slightly oscillate around zero. This is confirmed in Fig. 1.10 where the corresponding values of $\langle \delta \rangle$ are very close to zero. The transition from a synchronized behaviour to a not synchronized one is sharper in the case of identical systems. It can be noticed that the critical values of the density are quite independent of the number of agents. This suggests that the real size of the system is measured not by $N$ but by the density of the agents. Furthermore, it is interesting to note that the effect of increasing the density is such that first the system synchronizes and then, for larger values of density, the synchronization is lost. This is related to the assumption of a type III MSF, whereas for type II MSF a single threshold is expected. The latter case may be pertinent to the experimental evidence in yeast cell populations where sustained oscillations depend on a sufficiently high cell density [18].

## 1.5 Simulation Results: Topological Interaction Scheme

Models based on moving particles which interact within a suitable radius have been often used to describe the behaviour of animal and human groups [16]. In a recent paper, Ballerini *et al.* [4] show that the interaction between birds in airborne flocks is governed by a topological distance rather than a metric one. In particular, through an accurate study based on image processing performed on bird flocks they show that each bird interacts on average with a fixed number of neighbours (six-seven), rather than with all neighbours within a fixed metric distance.

Based on this idea, in this Section we modify our model described in Section 1.4 by establishing interactions between agents which are considered neighbours according to a topological criterion (i.e., each agent interacts exactly with a fixed number of nearest neighbouring agents). In Fig. 1.11 an example of metric neighbourhood (a), topological neighbourhood with two (b) and four (c) neighbours, respectively, is illustrated.

In the following, we will show that results similar to those presented in Section 1.4 are achieved, with the difference that the role played by the agent density in the metric interaction scheme is now accomplished by the number of agents selected as neighbours.

Similarly to the previous Section, we adopt the paradigm of time-varying dynamical network of interactions to deal with the problem. The difference in the interaction scheme is that each agent now interacts with a fixed number of neighbours $N_n$.

We have performed numerous simulations with respect to different parameters of the system and for both identical and non-identical oscillators. The synchronization error (1.24) and index (1.25) have been adopted also in this case.

In Fig.1.12 the synchronization index with respect to the number of neighbours $N_n$ is shown. It is clear that for $N_n > 4$ synchronization is achieved. It is important to note that in the metric interaction a similar behaviour occurs, but the threshold

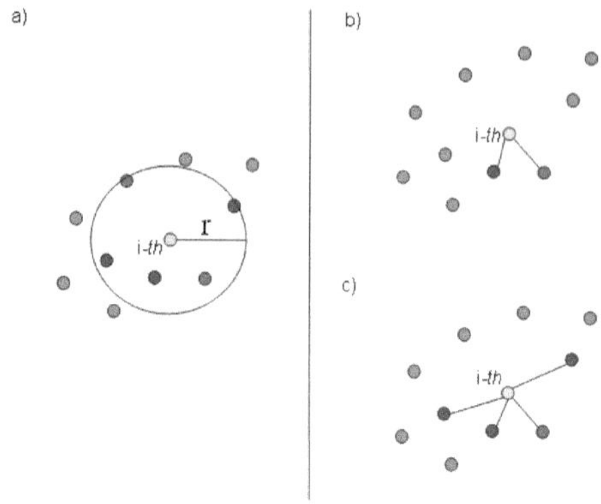


**Fig. 1.11** Examples of metric neighboourhood (a), topological neighbourhood with two (b) and four (c) neighbours.

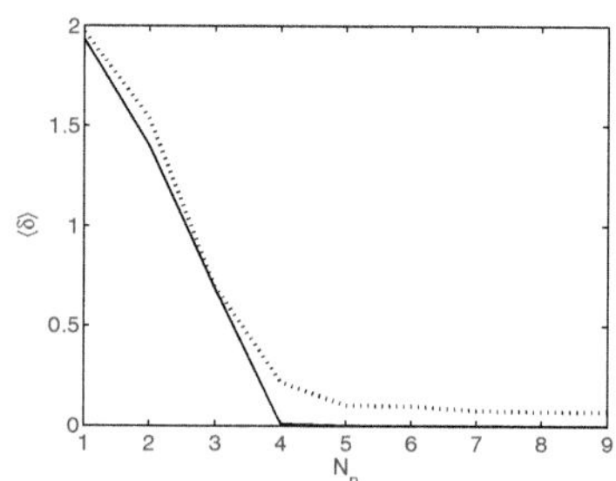


**Fig. 1.12** Synchronization index vs the number of neighbours $N_n$. The other parameters are $N = 10$, $v = 1$, $k = 0.05$, $\rho = 0.1$, $\Delta t_M = 0.001$. Results are average of 50 realizations. The dashed line refers to non-identical systems, the continuous one refers to identical ones.

is found in terms of agent density. In the topological interaction scheme, on the contrary, the agent density does not play any role.

Another important difference between the topological and metric case arises when the role of the other parameters of the system (such as the agent velocity $v$ or the motion step size $\Delta t_M$) is investigated. In fact, while in the metric case an increase in the velocity or a decrease in $\Delta t_M$ favour the switching between different interaction networks (and thus, according to the fast switching theorem, helps synchronization), the same consideration does not apply to the topological case, where varying these parameters has little effect on the synchronization index, as illustrated in Fig.1.13. This phenomenon is justified by the fact that an interaction radius is not considered, and an agent communicates with a fixed number of neighbours, irrespectively of their metric distance.

Even though the MSF approach holds only for networks with time-invariant links, as a preliminary study we monitor the location of the eigenvalues of the coupling matrix $G(t)$ at each time step. Recent literature shows that, under particular

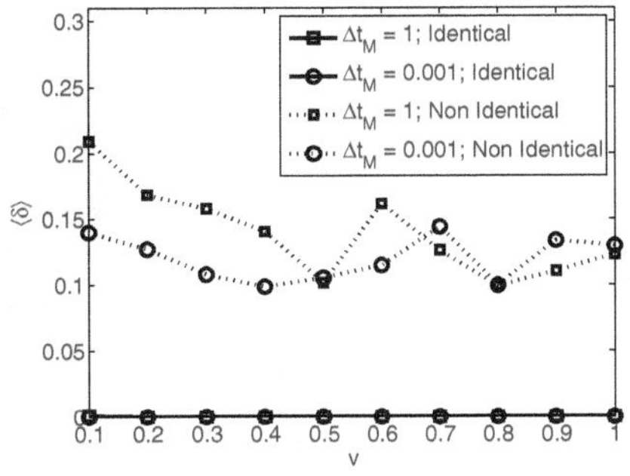


(a)

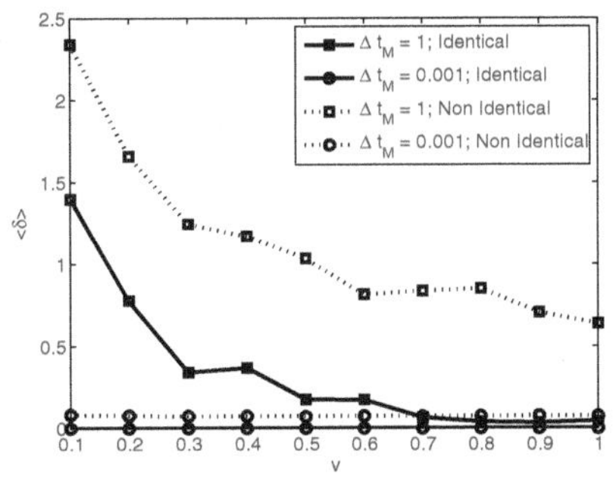


(b)

**Fig. 1.13** Effects of $v$ and $\Delta t_M$ on the synchronization index: comparison between topological (a) and metric case (b).

assumptions, synchronous motion can be observed even if some of the possible coupling configurations assumed over time do not guarantee synchronization under static conditions [5, 8]. However, it is worth to investigate the cases in which the synchronization in the time-varying case occurs in correspondence of configurations which, at any time instant, are statically synchronizable. By monitoring the eigenvalues of the coupling matrix at each time instant, we observe that this is indeed our case. In Fig. 1.14(a), the position of the eigenvalues is illustrated for $N_n = 3$ (which is the case of non synchronized systems), together with the MSF of the Rössler system with the given coupling strength $K$. In particular, the zero-level contour surrounding the region where $\lambda_{max}(\alpha + j\beta) < 0$ is shown. Fig. 1.14(b) is a zoom of the region of interest, showing that for $N_n = 3$ there exists configuration matrices whose eigenvalues do not lie in the synchronization region. In Fig. 1.14(c-d) the case with $N_n = 5$ (which is the case of synchronized systems) is shown. In this case, all the eigenvalues lie inside the region $\lambda_{max}(\alpha + j\beta) < 0$.

Summarizing, in this preliminary study about topological interaction we observe that:

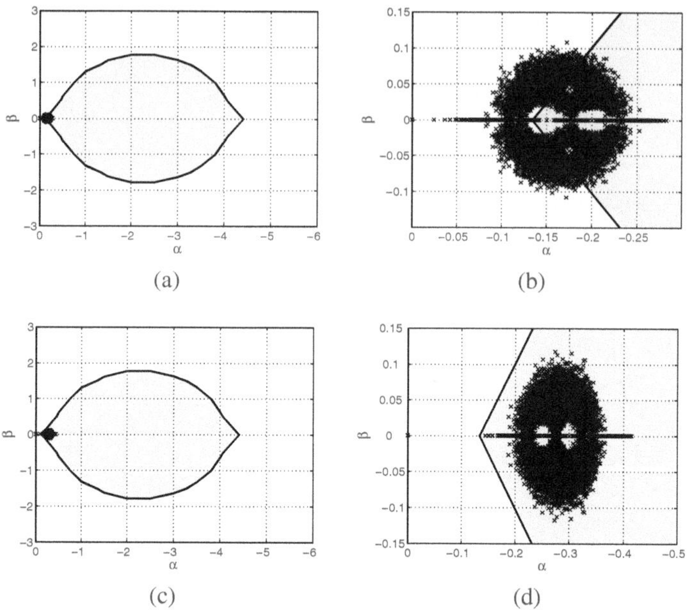


**Fig. 1.14** (a) Master Stability Function of the Rössler system as in Eq. (1.10) and eigenvalues of the Laplacian $G(t)$ for $N_n = 3$. (b) Zoom of the region of interest for $N_n = 3$. (c) Master Stability Function of the Rössler system as in Eq. (1.10) and eigenvalues of the Laplacian $G(t)$ for $N_n = 5$. (d) Zoom of the region of interest for $N_n = 5$.

- in the case of synchronized system the eigenvalues are all and at any time contained in the synchronization region individuated in the complex plane by the MSF approach;
- in the case of non-synchronized system some of the eigenvalues are located, during time, out of the synchronization region individuated in the complex plane by the MSF approach.

These observations provide interesting hints for further developments. In particular, main questions to answer are:

- Under which hypotheses the condition that at every time step the eigenvalues of the interaction matrix $G$ are confined in the region in which the MSF is negative ($\lambda_{max}(\alpha + j\beta) < 0$) is sufficient for synchronization?
- Is it possible to find a sharp bound for the nonzero eigenvalues of a generic time-varying interaction matrix with fixed number of neighbours (i.e., versus the number of agents $N$ and the number of neighbours $N_n$)?

Positive answers to these questions would make the topological interaction scheme a very powerful one to solve problems of collective behaviours (e.g., synchronization, consensus, coordinated motion, etc.), as the number of neighbours $N_n$ may become the only design parameter to achieve a coordinated behaviour.

## 1.6 Conclusions

In this Chapter we have studied the synchronization properties of a system of mobile agents, each carrying a chaotic oscillator. This is an interesting case of synchronization in time-varying networks, a topic of great interest yet not fully investigated. We drew conclusions on network synchronization by combining a tool for static networks, namely the Master Stability Function, and the hypothesis of fast switching. Two interaction schemes, metric and topological, have been considered. We found that a key role is played in the former case by agent density and in the latter case by the number of neighbours.

## References


1. Amitkar, R.E., Gupte, N.: Synchronization of chaotic orbits: The effect of a finite time step. Physical Review E 47, 3889–3895 (1993)
2. Arena, P., Bucolo, M., Fazzino, S., Fortuna, L., Frasca, M.: The cnn paradigm: Shapes and complexity. International Journal of Bifurcation and Chaos 15(7), 2063–2090 (2005)
3. Arena, P., Buscarino, A., Fortuna, L., Frasca, M.: separation and synchronization of piecewise linear chaotic systems. Physical Review E 74, 026, 212 (2006)
4. Ballerini, M., Cabibbo, N., Candelier, R., Cavagna, A., Cisbani, E., Giardina, I., Lecomte, V., Orlandi, A., Parisi, G., Procaccini, A., Viale, M., Zdravkovic, V.: Interaction ruling animal collective behavior depends on topological rather than metric distance: Evidence from a field study. Proceedings of the National Academy of Sciences 105, 1232–1237 (2008)
5. Belykh, I., Belykh, V., Hasler, M.: Blinking model and synchronization in small-world networks with a time-varying coupling. Physica D 195(1-2), 188–206 (2004)
6. Ben-Jacob, E., Cohen, I., Golding, I., Kozlovsky, Y.: Modeling branching and chiral colonial patterning of lubricating bacteria (1999), `http://arxiv.org/abs/cond-mat/9903382`
7. Boccaletti, S.: The synchronized dynamics of complex systems. Elsevier, Amsterdam (2008)
8. Boccaletti, S., Hwang, D.U., Chavez, M., Amann, A., Kurths, J., Pecora, L.M.: Synchronization in dynamical networks: Evolution along commutative graphs. Physical Review E 74, 016,102 (2006)
9. Boccaletti, S., Kurths, J., Valladares, D.L., Osipov, G., Zhou, C.: The synchronization of chaotic systems. Physics Reports 366, 1–101 (2002)
10. Boccaletti, S., Latora, V., Moreno, Y., Chavez, M., Hwang, D.U.: Complex networks: Structure and dynamics. Physics Reports 424, 175–308 (2006)
11. Brockmann, D., Hufnagel, L., Geisel, T.: The scaling laws of human travel. Nature 439, 462–465 (2006)
12. Buscarino, A., Fortuna, L., Frasca, M.: Experimental separation of chaotic signals through synchronization. Philosophical Transactions of the Royal Society A 366(1865), 569–577 (2008)
13. Buscarino, A., Fortuna, L., Frasca, M., Rizzo, A.: Dynamical network interactions in distributed control of robots. Chaos 16(1), 015, 116 (2006)
14. Buscarino, A., Fortuna, L., Rizzo, A.: Effects of long-range connections in distributed control of collective motion. International Journal of Bifurcation and Chaos 17(7), 2411–2417 (2007)

15. Chen, M.: Chaos synchronization in complex networks. IEEE Transactions on Circuits and Systems I 55(5), 1335–1346 (2008)
16. Couzin, I.D., Krause, J., Franks, N.R., Levin, S.A.: Effective leadership and decision-making in animal groups on the move. Nature 433, 513–516 (2005)
17. Danø, S., Hynne, F., Monte, S.D., d'Ovidio, F., Sørensen, P.G., Westerhoff, H.: Synchronization of glycolytic oscillations in a yeast cell population. Faraday Discussions 120, 261–275 (2002)
18. Danø, S., Sørensen, P.G., Hynne, F.: Sustained oscillations in living cells. Nature 402, 320–322 (1999)
19. Fortuna, L., Frasca, M.: Experimental synchronization of single-transistor-based chaotic circuits. Chaos 17, 043, 118 (2007)
20. Fortuna, L., Frasca, M., Rizzo, A.: Self-organising behavior of arrays of nonidentical josephson junctions. In: Proceedings of IEEE Int. Sym. on Circ. and Sys, ISCAS 2002, vol. 5, pp. 213–216 (2002)
21. Frasca, M., Buscarino, A., Rizzo, A., Fortuna, L., Boccaletti, S.: Dynamical network model of infective mobile agents. Physical Review E 74, 036, 110 (2006)
22. Frasca, M., Buscarino, A., Rizzo, A., Fortuna, L., Boccaletti, S.: Synchronization of moving chaotic agents. Physical Review Letters 100(4), 044, 102 (2008)
23. Gomez-Gardenes, J., Moreno, Y., Arenas, A.: Paths to synchronization on complex networks. Physical Review Letters 98, 034,101 (2007)
24. Guemez, J., Matias, M.A.: Modified method for synchronizing and cascading chaotic systems. Physical Review E 52, R2145–R2148 (1995)
25. Kapitaniak, T.: Synchronization of chaos using continuous control. Physical Review E 50, 1642–1644 (1994)
26. Kocarev, L., Parlitz, U.: General approach for chaotic synchronization with applications to communication. Physical Review Letters 74, 5028–5031 (1995)
27. Li, X., Chen, G.: Synchronization and desynchronization of complex dynamical networks: an engineering viewpoint. IEEE Transactions on Circuits and Systems I 50(7), 1381–1390 (2003)
28. Lü, J., Chen, G.: A time-varying complex dynamical network model and its controlled synchronization criteria. IEEE Transactions on Automatic Control 50(6), 841–846 (2005)
29. Lü, J., Yu, X., Chen, G.: Chaos synchronization of general complex dynamical networks. Physica A 334, 281–302 (2004)
30. Lü, J., Yu, X., Chen, G., Cheng, D.: Characterizing the synchronizability of small-world dynamical networks. IEEE Transactions on Circuits and Systems I 51(4), 787–796 (2004)
31. Meucci, R., Salvadori, F., Ivachenko, M.V., Naimee, K.A., Zhou, C., Arecchi, F.T., Boccaletti, S., Kurths, J.: Synchronization of spontaneous bursting in co2 lasers. Physical Review E 74, 066, 207 (2006)
32. Nishikawa, T., Motter, A.E.: Synchronization is optimal in nondiagonalizable networks. Physical Review E 73, 065, 106(R) (2006)
33. Parlitz, U., Kocarev, L., Stojanovski, T., Preckel, H.: Encoding messages using chaotic synchronization. Physical Review E 53, 4351–4361 (1996)
34. Pecora, L.M., Carroll, T.L.: Synchronization in chaotic systems. Physical Review Letters 64, 821–824 (1990)
35. Pecora, L.M., Carroll, T.L.: Master stability functions for synchronized coupled systems. Physical Review Letters, 2109–2112 (1998)
36. Porfiri, M., Stilwell, D.J., Bollt, E.M., Skufca, J.D.: Random talk: Random walk and synchronizability in a moving neighborhood network. Physica D 224(1-2), 102–113 (2006)

37. Skufca, J.D., Bollt, E.M.: Communication and synchronization in disconnected networks with dynamic topology: Moving neighborhood networks. Math. Biosci. Eng. 1(2), 347–359 (2004)
38. Stilwell, D.J., Bollt, E.M., Roberson, D.G.: Sufficient conditions for fast switching synchronization in time-varying network topologies. SIAM J. Appl. Dyn. Syst. 5(1), 140–156 (2006)
39. Strogatz, S.H.: From Kuramoto to Crawford: exploring the onset of synchronization in populations of coupled oscillators. Physica D 143, 1–20 (2000)
40. Strogatz, S.H.: SYNC: the emerging science of spontaneous order. Hyperion, New York (2003)
41. Vicsek, T., Czirók, A., Ben-Jacob, E., Cohen, I., Shochet, O.: Novel type of phase transitions in a system of self-driven particles. Physical Review Letters 75(6), 1226–1229 (1995)
42. Wu, C.W.: Synchronization in arrays of coupled nonlinear systems: passivity, circle criterion, and observer design. IEEE Transactions on Circuits and Systems I 48(12), 1257–1261 (2001)
43. Wu, C.W., Chua, L.O.: Synchronization in an array of linearly coupled dynamical systems. IEEE Transactions on Circuits and Systems I 42(4), 430–447 (1995)
44. Zanette, D.H.: Dynamics of globally coupled bistable elements. Physical Review E 55, 5315–5320 (1997)

# Chapter 2
# Decentralized Adaptive Control for Synchronization and Consensus of Complex Networks

Pietro De Lellis, Mario di Bernardo, and Francesco Garofalo

**Abstract.** In this Chapter we discuss a novel decentralized adaptive strategy to synchronize complex networks. We present two alternative strategies. A vertex-based method where each node assigns an adaptive coupling strength to all its incoming links and an edge-based one where mutually coupled nodes negotiates their strengths according to the mismatch between their output functions. Proof of asymptotic stability is given using an appropriate Lyapunov function. The theoretical results are validated on a set of representative examples: the synchronization of a network of Chua's circuits and the consensus of a network of integrators.

## 2.1 Emerging Dynamics in Complex Networks

Networked systems abound in nature and in applied science. Examples of networks in applications are the Internet, electric power grids and neural networks, but also network of subways or networks of scientific collaborations (see [3], [28]). The

Pietro De Lellis
Department of Systems and Computer Engineering, University of Naples Federico II,
Via Claudio 21, 80125 Napoli, Italy
e-mail: pietro.delellis@unina.it

Mario di Bernardo
Department of Systems and Computer Engineering, University of Naples Federico II,
Via Claudio 21, 80125 Napoli, Italy
e-mail: mario.dibernardo@unina.it

Francesco Garofalo
Department of Systems and Computer Engineering, University of Naples Federico II,
Via Claudio 21, 80125 Napoli, Italy
e-mail: franco.garofalo@unina.it

study of networks is the subject of a branch of mathematics, *graph theory*, born in 1736 when the mathematician Leonard Eulero published the solution to the so-called *problem of the Königsberg bridges*. During the last 15 years the study of networks has become of interest to other areas of sciences, as physics, biology and control system engineering. Recently, attention has been focused on large networks of interconnected systems, such as the internet or networks of proteins. To study these large networks, the standard graph-theory approach was not enough: in these kind of network, in fact named *complex networks*, there is a combination of the effects of the individual dynamics of the nodes with the topology of the interconnections. This combination can lead to the emergence of collective behaviour which could not be predicted if we look only at the individual dynamics. Two of the most fascinating and interesting types of collective behaviour are synchronization and consensus.

*Synchronization* is a common phenomenon in nature [3], [28], [21]. As examples, we can think about the lightning synchronization of fireflies and the clapping synchronization in a concert hall. The first scientific observation of this phenomenon was made by Huygens in 1665 [9]. He observed that coupling through a common beam a pair of asynchronous clocks, they finally achieve synchronization due to their weak interaction. Nowadays, the application of synchronization has moved from mechanical engineering to information engineering, chemistry and biology. In fact, we can think at clock synchronization in computer networks [17], or also networks of genes or cells which needs to achieve (or to avoid) synchronization in order guarantee a normal functioning of living beings (see [32], [16]).

A particular case of synchronization is *consensus* [22], in which all the nodes of the network converge to a fixed value. The *consensus problem* has many application in the fields of computer science and automata theory. In fact, in many applications of multiagent systems, a group of agents needs to agree upon a certain quantity. In technology, flocking is a typical consensus problem, in which a group of mobile agents has to align their velocity vectors and stabilize their inter-agent distances using decentralized nearest-neighbour interaction rules. Another common example of consensus is the rendezvous problem, in which a set of agents meet each other at a given point of the space using only information on the position of the nearest neighbours.

### 2.1.1 Mathematical Modeling

A classical mathematical scheme to study the behaviour of a complex network is to assume the network consists of $N$ identical nonlinear dynamical systems coupled through the edges of the network itself ([3], [21]). Each uncoupled system is described by a nonlinear set of ordinary differential equations (ODEs) of the form $\dot{x} = f(x)$, where $x \in R^n$ is the state vector and $f : R^n \mapsto R^n$ is a sufficiently smooth nonlinear vector field describing the system dynamics. Because of the coupling with the neighboring nodes in the network, the dynamics of each oscillator is affected by an input representing the interaction of all neighboring nodes with the oscillator itself. Hence, the equations of motion for the generic $i$-th system in the network become:

$$\frac{dx_i}{dt} = f(x_i) - \sigma \sum_{j=1}^{N} \mathscr{L}_{ij} h(x_j), \qquad i = 1,2,...N, \tag{2.1}$$

where $x_i$ represents the state vector of the $i$-th oscillator, $\sigma$ the overall strength of the coupling, $h(x) : R^n \mapsto R^n$ the output function[1] through which the systems in the network are coupled and $\mathscr{L}_{ij}$ the elements of the Laplacian matrix $\mathscr{L}$ describing the network topology. In particular, $\mathscr{L}$ is such that its entries, $\mathscr{L}_{ij}$, are zero if node $i$ is not connected to node $j \neq i$, while are negative if node $i$ is connected to node $j$, with $| \mathscr{L}_{ij} |$ giving a measure of the strength of the interaction. We say that a network is synchronized if and only if

$$\lim_{t \to \infty} (x_i(t) - x_j(t)) = 0$$

for all pairs $(i, j)$ of nodes such that $i \neq j$.

*Remark 2.1.* It is worth remarking that consensus, as defined in [20] is a particular case of synchronization in which $f(x) = 0$. That is why we can generally speak about synchronization.

In the recent literature, much effort has been devoted to characterize the topology of the interactions (for a survey on this subject see [28]) and later to understand its effects on the emergence of collective behaviours like synchronization and consensus. An approach based on the so-called Master Stability Function (MSF) has been established to evaluate the range of $\sigma$ for which synchronization is locally attained [3]. Basically, it has been shown that the width of the range of values of the coupling gain associated with the transversal stability of the synchronization manifold is related to the ratio between the largest and smallest non-zero eigenvalue of the network Laplacian $\mathscr{L}$. Such a ratio is often referred to as the network eigenratio. The MSF gives an indication of the network synchronizability, in terms of the values of $\sigma$ for which synchronization can be achieved.

Different approaches, aiming to achieve global results, are based on the *Lyapunov stability theory* (for further details on this approach, see [27], [29], [20]), or uses *contraction theory* (see [26], [23], [24]).

### *2.1.2 Why Adaptive Synchronization?*

The network model described by equation (2.1) is sometimes inadequate to describe the dynamics of a real network. In fact, many real phenomena are characterized by the presence of adaptive mechanisms. For example, we can mention wireless networks of sensors that gather and communicate data to a central base station [19]. Adaptation is also necessary to control networks of robots when the conditions

[1] Notice that $h(x)$ can also have null components, representing the case that not all the state variables are coupled.

change unexpectedly (i.e. a robot loses a sensor) [25]. Moreover, examples of adaptive networks can be found in biology [10]. Social insect colonies, for instance, have many of the properties of adaptive networks [8]. In all these cases, it is realistic to think that the strength of the interactions among nodes, characterized mathematically by $\sigma$, is not identical from node to node and time-invariant. Real-world networks are often characterized instead by evolving, adapting coupling gains which vary in time according to different environmental conditions. This is the main reason for introducing adaptive coupling in equation (2.1). In the next section we will show some of the most interesting and new adaptive mechanism.

## 2.2 An Adaptive Approach

As a first simple way of introducing an adaptive mechanism in the synchronization of complex networks, we can think of making the gain $\sigma$ in (2.1) a function of time and then assigning a continuous-time differential adaptive law for $\sigma$. In particular, we have:

$$\frac{dx_i}{dt} = f(x_i) - \sigma(t)\sum_{j=1}^{N}\mathscr{L}_{ij}h(x_j), \qquad i = 1,2,...N, \tag{2.2}$$

with $\sigma(t)$ given by a differential equation of the form:

$$\dot{\sigma}(t) = \phi(x_1,x_2,\ldots,x_N), \qquad \sigma(0) = \sigma_0 \geq 0, \tag{2.3}$$

where $\phi(\cdot)$ is some appropriately chosen scalar function of the state vectors $x_1,x_2$, $\ldots,x_N$ of all oscillators in the network and $\sigma_0$ is some initial condition on the evolution of $\sigma(t)$.

The strategy described by (2.2)-(2.3) is a *global, centralized adaptive strategy* where the adaptation mechanism is selected globally for all oscillators in the network. Specifically, we assume that unique adaptive gain, equal for all oscillators, is adapted in a centralized manner on the basis of information on all the oscillators in the network. A possible centralized adaptation strategy was proposed in [5], in which $\Phi$ is a quadratic function of the distances between the states of the nodes.

Anyway, it is surely more interesting and realistic, for the reasons explained in the previous section, to explore an alternative approach. Namely, to consider *local, decentralized strategies* [15], [14], [13]. In this case, the network can be described as

$$\frac{dx_i}{dt} = f(x_i) - \sum_{j=1}^{N}\sigma_{ij}(t)\mathscr{L}_{ij}h(x_j), \qquad i = 1,2,...N, \tag{2.4}$$

with $\sigma_{ij}(t)$ being the adaptive gains associated with each of the edges between different nodes $i$ and $j$. Each of the adaptive gains is then associated with an adaptive law of the form:

$$\dot{\sigma}_{ij}(t) = \phi_{ij}(I(x_i,x_j)), \qquad \sigma_{ij}(0) = \sigma_{ij}^0 \geq 0, \tag{2.5}$$

where $\phi_{ij}(\cdot)$ is now some appropriately chosen local function, $I(x_i, x_j)$ represents the subset of nodes in the neighborhood of the nodes $i$ and $j$ and $\sigma_{ij}^0$ are the initial conditions on each of the adaptive gains in the network.

In what follows, we shall focus our attention on such decentralized strategies, and we will show different adaptation mechanisms which involve different choices for the function $\phi_{ij}$ in (2.5).

## 2.2.1 Local Adaptive Strategies

We isolate two possible alternative local adaptive strategies for networks synchronization: (i) *vertex-based*, where the same adaptive gain is associated with all edges outcoming from a given node; and (ii) *edge-based* where an adaptive gain is associated with every link in the network.

Namely, we have the two following approaches.

1. **Vertex-based:** an adaptive gain $\sigma_i(t)$ is associated with every vertex $i = 1, \ldots$. Namely, the adaptation law is set to:

$$\dot{\sigma}_i(t) = \mu \left\| \sum_{j \in V_i} (h(x_j) - h(x_i)) \right\| := \Phi_i(t), \tag{2.6}$$

where $V_i$ is the set of neighbours of the node $i$.

Another simple vertex-based strategy is the one proposed by Kurths and Zhou in [33], given by:

$$\dot{\sigma}_i(t) = \frac{\mu \delta_i}{(1 + \delta_i)}, \qquad \mu > 0, \tag{2.7}$$

where

$$\delta_i = \left\| h(x_i) - (1/k_i) \sum_{j=1}^{N} A_{ij} h(x_j) \right\|. \tag{2.8}$$

2. **Edge-based:** an adaptive gain $\sigma_{ij}(t)$ is associated with each link. The dynamic of each $\sigma_{ij}(t)$ is described by the following differential equation:

$$\dot{\sigma}_{ij}(t) = \alpha \left\| (h(x_j) - h(x_i)) \right\| := \Phi_{ij}(t). \tag{2.9}$$

Another edge-based approach, proposed in [13], can be described by the following equations:

$$\dot{\sigma}_{ij}(t) = \frac{\alpha \delta_{ij}}{1 + \delta_{ij}}, \qquad \alpha > 0, \tag{2.10}$$

where

$$\delta_{ij} = \left\| h(x_i) - h(x_j) \right\|. \tag{2.11}$$

The idea behind the vertex-based approach is for each node to negotiate an appropriate coupling strength with all of its neighbours by comparing its output with that of all neighbouring nodes. Hence, an adaptive gain $\sigma_i$ is associated with each node. The edge-based approach is instead based on the negotiation between each pair of nodes in the network. Basically, two neighboring nodes compare their outputs and select their mutual coupling gain according to the relative distance between their outputs. In so doing, a different adaptive gain is associated with each link in the network.

## 2.3 Analytical Results

### 2.3.1 Mathematical Preliminaries

We introduce, now, some notation and give some definitions that will be used throughout the rest of the paper.

**Definition 2.1.** As in [5], we say that a function $f : \Re^n \times \Re^+ \mapsto \Re^n$ is QUAD iff, for any $x, y \in \Re^n$:

$$(x-y)^T[f(x,t)-f(y,t)]-(x-y)^T\Delta(x-y) \leq -\bar{\omega}\cdot \\ \cdot(x-y)^T(x-y), \tag{2.12}$$

where $\Delta$ is an arbitrary diagonal matrix of order $n$ and $\bar{\omega}$ is a positive scalar.

### 2.3.2 Main Results

In this section we expound the main stability result for decentralized adaptive strategies presented in this chapter, leaving the sketch of the proof to the appendix. In what follows, we consider that the following assumption is always verified:

*Assumption 1.* All the components of the output function $h(x_i)$ are affine transformations of the corresponding state variables. Mathematically:

$$h(x_i) = \begin{bmatrix} \alpha_1 x_{i1} + \beta_1 \\ \alpha_2 x_{i2} + \beta_2 \\ \vdots \\ \alpha_n x_{in} + \beta_n \end{bmatrix} \tag{2.13}$$

Moreover, we assume $\alpha_i \geq 0$, $\beta_i \in \Re$, $\forall i = 1,\ldots,n$ and $\sum_{i=1}^{n} \alpha_i > 0$.

Let us denote $\alpha_{max} = \max_i \alpha_i$. Now we can state the following theorem.

**Theorem 2.1.** *If $f$ is QUAD$(\varepsilon I, \bar{\omega})$, with $\varepsilon \leq \bar{\omega}/\alpha_{max}$ and the output function $h(\cdot)$ satisfies Assumption 1, then all the* vertex-based *and the* edge-based *strategies described respectively by equations* (2.6), (2.7) *and* (2.9), (2.10) *guarantee global asymptotic stability of the synchronization manifold of network* (2.4).

*Remark 2.2.* Assumption 1 means that each component of the output function is an increasing linear function of the state components plus a possible constant term. It is worth remarking that, from the assumptions on the $\alpha_i$ coefficients, it is possible also that only one $\alpha_i$ is different from 0. This means that Theorem 2.1 ensure that the above presented decentralized adaptive strategies guarantee synchronization also if the networked systems are coupled only on one state variable.

*Remark 2.3.* Despite its formal correctness, theorem 2.1 is based on the *QUAD* assumption on the $f(\cdot)$ that might appear difficult to verify. Notice that, as stated in [5], [4], many chaotic systems are QUAD and, moreover, we can choose $\Delta = 0$ (see [4]) (and thus $\Delta_{ii} = 0 < \bar{\omega}/\alpha_{max}$). For instance, in [13] was proven that the Chua's circuit is $QUAD(0,\bar{\omega})$. Thus, the next corollary follows.

**Corollary 2.1.** *If $f$ is $QUAD(0,\bar{\omega})$ and the output function $h(\cdot)$ satisfies Assumption 1, then all the* vertex-based *and the* edge-based *strategies described respectively by equations* (2.6), (2.7) *and* (2.9), (2.10) *guarantee global asymptotic stability of the synchronization manifold of network* (2.4).

## 2.4 Numerical Validation

In this section we use appropriate numerical simulations to validate the results presented in the previous section. We will consider two representative examples. In the first one, we will show that both the edge-based and the vertex-based strategies seem to guarantee network consensus, while in the second we test the synchronizability on a network of $N$ chaotic circuits. For the sake of brevity, in the first case we are going to show the simulations regarding strategies (2.6) and (2.9), otherwise in the second we will focus on strategies (2.7) and (2.10). In all the simulations we choose $h(x_i) = x_i$.

### 2.4.1 Adaptive Consensus

In our simulations we consider $N = 200$ integrators connected through a scale-free like network, constructed using the Barabàsi-Albert model with $N_0 = 5$ starting nodes (for further details on this topology, see [3], [21], [28]). The initial conditions are chosen randomly from a standard normal distribution. Adopting the vertex-based strategy (2.6), the equations of the network can be rewritten as follows:

$$\frac{dx_i}{dt} = -\sigma_i(t) \sum_{j=1}^{N} \mathcal{L}_{ij} x_j, \qquad i = 1,2,...N, \tag{2.14}$$

$$\dot{\sigma}_i = \mu ||\mathcal{L}_i x||, \qquad i = 1,\ldots,N. \tag{2.15}$$

where $\mu$ is chosen equal to 0.1, $\mathcal{L}_i$ is the $i-th$ row of the Laplacian matrix and we set $\sigma_i(0) = 0, \forall i = 1,\ldots,N$. In this case we have $\alpha_{max} = 1$ and it is trivial to show that $f = 0$ is $QUAD(\varepsilon I,\bar{\omega})$, with $\varepsilon = \bar{\omega}$. Thus, as predicted by Theorem 2.1,

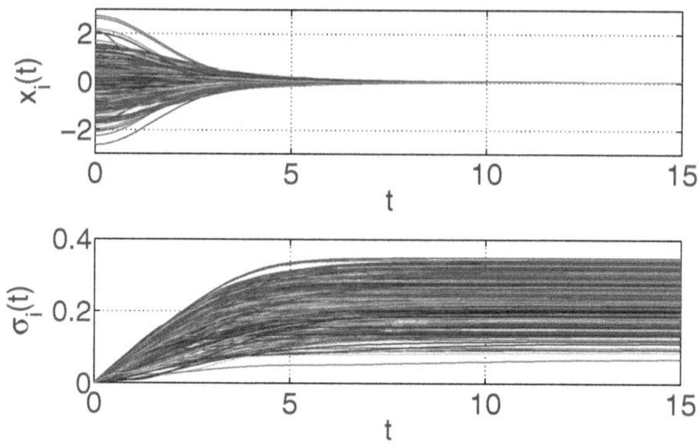


**Fig. 2.1** Evolution of the state variables (top) and coupling gains $\sigma_i$ (bottom) for a network of 200 integrators under the vertex-based adaptive strategy (2.6).

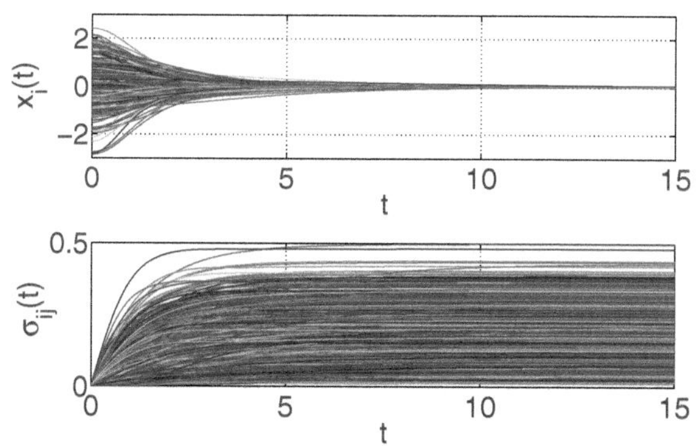


**Fig. 2.2** Evolution of the state variables (top) and coupling gains $\sigma_i$ (bottom) for a network of 200 integrators under the edge-based adaptive strategy (2.9).

synchronization is asymptotically achieved (see Fig. 2.1) with the various $\sigma_i$ settling to constant values.

Adopting the edge-based strategy (2.9), the equations of the network become:

$$\frac{dx_i}{dt} = -\sum_{j=1}^{N} \sigma_{ij}(t)\mathscr{L}_{ij}x_j, \qquad i = 1, 2, \ldots N, \tag{2.16}$$

$$\dot{\sigma_{ij}} = \alpha \left|\left|x_j - x_i\right|\right|, \qquad (i, j) \in \mathscr{E}. \tag{2.17}$$

where $\alpha$ is chosen equal to 0.1. The simulation shown in Fig. 2.2 confirms the effectiveness of such strategy in achieving synchronization while guaranteeing that all coupling gains converge towards some constant values.

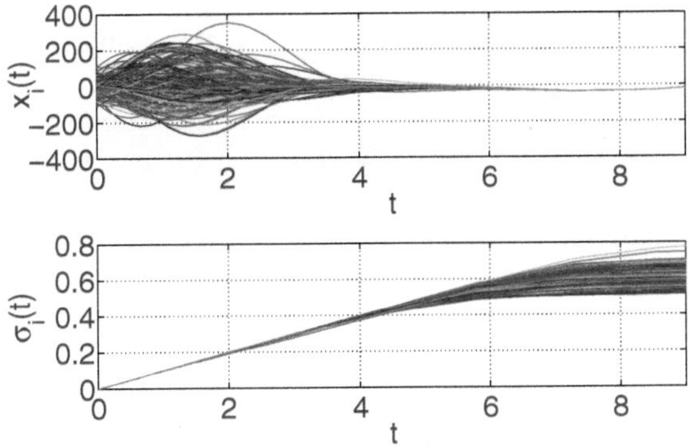


**Fig. 2.3** Synchronization of a scale-free like network of 200 Chua's circuits using the vertex-based adaptive strategy (2.7): evolution of $x_i$ (top) and $\sigma_i$ (bottom).

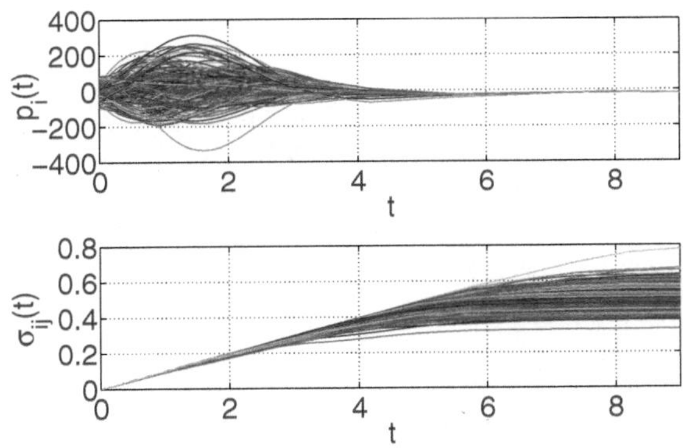


**Fig. 2.4** Synchronization of a scale-free like network of 100 Chua's circuits using the edge-based adaptive strategy (2.10): evolution of $x_i$ (top) and $\sigma_i$ (bottom).

## 2.4.2 Network of N Identical Chua's Circuits

The synchronization of networks of Chua's circuits [18] has been widely investigated in the existing literature on synchronization (see for instance [6, 7, 2, 31, 30] and references therein). Thus, it is an interesting testbed to further validate our approach to analyze the case of local adaptive gains. In what follows, we consider a network of $N$ Chua's chaotic circuits coupled through all the state variables. Each circuit is described by the vector field:

$$f(x_1,x_2,x_3) = (\beta(-x_2 - \phi(x_1)), x_1 - x_2 + x_3, \xi x_2)^T$$

where $\phi(x_1) = m_0 x_1 + \frac{1}{2}(m_1 - m_0)(|x_1 + 1| - |x_1 - 1|)$. The parameters are chosen as in [1] and correspond to chaotic behaviour of each circuit in the network; specifically, $\beta = 0.59/0.12$, $\xi = 0.59/0.162$, $m_0 = -0.07$, $m_1 = 1.5$.

The topology of the network describing the coupling between oscillators is a scale-free like network of $N = 200$ nodes, constructed using the Barabàsi-Albert model with $N_0 = 5$ starting nodes. Moreover, we select the initial conditions of the chaotic oscillators randomly from a normal distribution with mean equal to 0 and standard deviation equal to 40. To complement previous subsection, we consider now strategies (2.7) and (2.10), choosing $\mu = 0.1$ in the vertex-based strategy and $\alpha = 0.1$ in the edge-based one. As shown in [13], the Chua's circuit is $QUAD(0, \bar{\omega})$ and so, Corollary 2.1 ensure that the synchronous state is globally asymptotically stable for both adaptive strategies. The simulations shown in Figs. 2.3 and 2.4 confirm that the synchronization manifold is asymptotically stable and that, for both strategies, the adaptive coupling gains settle to constant values.

## 2.5 Robustness Analysis

In this section, we want to analyze the robustness of the presented adaptive schemes. In fact, in real applications the node dynamics are not exactly the same. Moreover, the network topology can change unexpectedly, for instance there could be a link failure (as an example, you can think of a robot loosing a sensor). Thus we will simulate firstly the behaviour of the presented strategy in presence of parameter mismatches, then we are going to investigate the robustness of the scheme to topological variations.

### *2.5.1 Robustness to Parameter Mismatch*

Before proceeding with our discussion, it is worth pointing out that, in the case of non-identical node dynamics, we cannot reach complete synchronization. All the trajectories, instead, will be observed to converge towards some common bounded neighbourhood in phase space. In this case, using any of the presented strategies we would have the coupling strength $\sigma_i$ indefinitely increasing. To solve this problem, we propose an hybrid modification of the adaptive strategy presented in this paper. Specifically, we consider a saturation of the coupling gains so that when the derivative of $\sigma_i$ becomes lower than a fixed threshold, say $l$, it is automatically set to 0.

In our simulations, we consider a network of 200 linearly coupled Kuramoto oscillators, characterized by natural frequencies distributed chosen from a standard normal distribution. Networks of Kuramoto oscillators were proposed in the literature as a viable paradigmatic example of synchronization of dynamical systems [12]. For the sake of brevity, we will show the robustness of this hybrid approach only for the vertex-based strategy (2.9). As shown in Figure 2.5, this hybrid scheme seem to be a viable approach for robust synchronization, as all the oscillators in

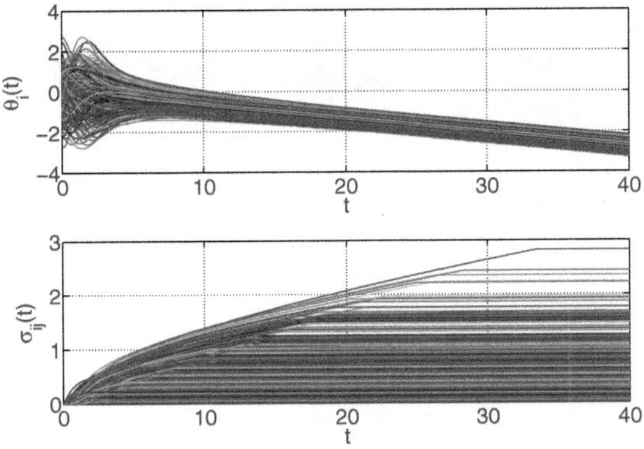


**Fig. 2.5** Evolution of phases (top), $\sigma$ (bottom) in a network of 200 Kuramoto oscillators with different natural frequencies.

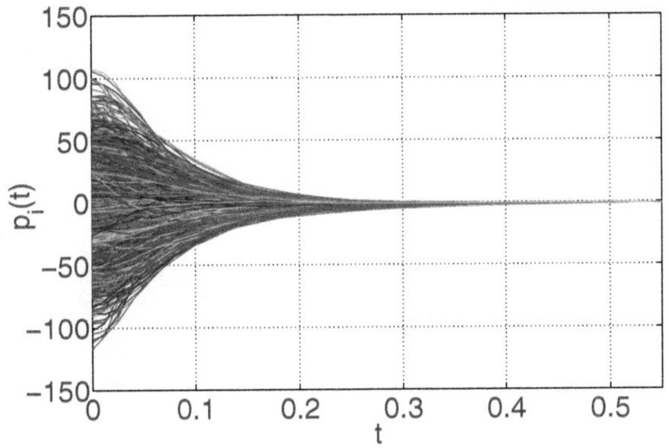


**Fig. 2.6** State evolution of all oscillators in case of periodical regeneration of the topology

the network reach at steady state a bounded neighbourhood of phase space. The adaptive gains saturate as expected.

## 2.5.2 Robustness to Topological Variations

In this section, we investigate whether the above presented strategies cope well with switches or changes in the network topology. In what follows, we will consider the vertex-based strategy (2.6). Similar results were obtained with the other strategies.

As a testbed, we have considered networks of 1000 Chua's circuits coupled at the beginning of the simulation through a scale-free topology, in which the network topology is randomly re-generated with a given frequency $\nu$. In other words, a

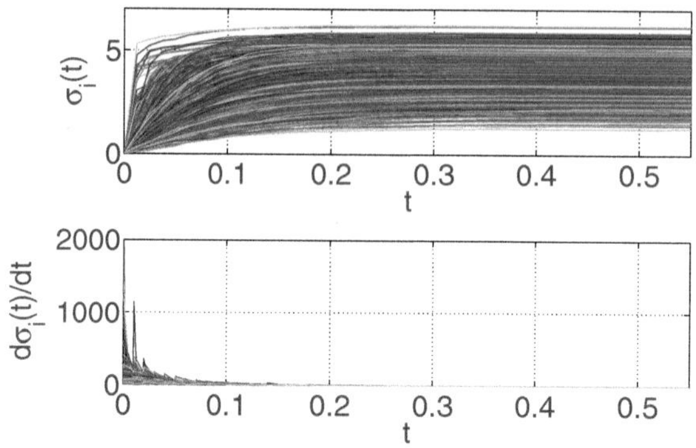


**Fig. 2.7** Evolution of $\sigma$ (top), $\dot{\sigma}$ (bottom) in case of periodical regeneration of the topology.

completely new scale-free topology is generated at each switching time. The switching frequency is set to be $\nu = 100Hz$.

As we can see from Fig. 2.6, synchronization is again attained with the various $\sigma_i$ settling onto constant values (see Fig. 2.7). This behaviour was still preserved when the switching frequency was increased (for the sake of brevity we omit the simulation results for this case). Thus, the numerical analysis seems to indicate that the presented adaptive strategies are strongly robust to topological variations. In fact, a sufficient condition for synchronization seems to be that the time-varying Laplacian always remains connected. The analytical proof of asymptotic stability in this case is currently under investigation.

## 2.6 Conclusions

In this chapter we discussed a novel decentralized adaptive strategy to synchronize complex networks. We presented two alternative strategies. A vertex-based method where each node assigns an adaptive coupling strength to all its incoming links and an edge-based one where mutually coupled nodes negotiates their strengths according to the mismatch between their output functions. After presenting different classes of adaptation laws for the gains, we investigated the asymptotic stability of the resulting network using Lyapunov methods. By choosing an appropriate Lyapunov function, we were able to show that all node trajectories converge asymptotically towards each other while the gains reach some bounded values at steady-state. The validation of the presented strategies was carried out on a set of representative examples confirming its effectiveness in making the network of interest achieve a synchronous state. A preliminary investigation of the robustness of the strategies discussed in the chapter was then performed, showing that both strategies perform well even in the presence of mismatches between the node vector fields or topological variations of the network.

## Appendix

In this appendix we show the procedure used to prove Theorem 2.1. To clarify this point, we go into the details of the proof in the case of the vertex-based adaptive strategy described by equation (2.7). An extension of the proof to the other cases can be found in [14], [13]. First of all, we need to introduce some notation that will be used through the derivation. Let us define $X = [x_1^T, \dots, x_N^T]^T$, $F(X) = [f(x_1)^T, \dots, f(x_N)^T]^T$, $H(X) = [h(x_1)^T, \dots, h(x_N)^T]^T$ and $S = diag\{\sigma_1, \dots, \sigma_N\}$. Moreover, let us define $\mathbf{L} = \mathscr{L} \otimes I_n$, $\mathbf{S} = S \otimes I_n$ and $\boldsymbol{\Delta} = I_n \otimes \Delta$, where $\otimes$ is the Kronecker product.

Using this definitions, we can recast equation (2.4) as:

$$\dot{X} = F(X) - \mathbf{SL}H(X). \tag{2.18}$$

To prove the theorem we use Lyapunov stability theory, so the first step of the proof is to introduce a suitable Lyapunov function:

$$V(e_{ij}, \sigma_{ij}) = \frac{1}{2}\eta X^T \mathbf{L} X + \frac{1}{2\gamma}(c-\sigma)^T(c-\sigma), \tag{2.19}$$

where $e_{ij} = x_i - x_j$.

Now, let us differentiate $V$. We have:

$$\begin{aligned}\dot{V} &= \eta X^T \mathbf{L}[F(X) - \mathbf{SL}H(X)] - \frac{1}{\gamma}\sum_{i=1}^{N}(c_i - \sigma_i)^T \dot{\sigma}^i = \\ &= \eta X^T \mathbf{L}[F(X) - \varepsilon X] + \eta\varepsilon X^T \mathbf{L} X - \eta X^T \mathbf{LSL}H(X) - \frac{1}{\gamma}\sum_{i=1}^{m}(c_i - \sigma_i)^T \dot{\sigma}_i\end{aligned} \tag{2.20}$$

Since $f$ is $QUAD(\varepsilon I, \bar{\omega})$, we can state that $X^T \mathbf{L}[F(X) - \varepsilon X] \leq -\bar{\omega} X^T \mathbf{L} X$.

Thus, we can write:

$$\dot{V} \leq -\eta\bar{\omega} X^T \mathbf{L} X + \eta\varepsilon X^T \mathbf{L} H(X) - X^T \mathbf{LSL} H(X) - \sum_{i=1}^{N}(c_i - \sigma_i)^T \frac{\delta_i}{(1+\delta_i)}. \tag{2.21}$$

Defining $X_i = \begin{bmatrix} x_{1i} \\ \vdots \\ x_{Ni} \end{bmatrix}$ and $H(X_i) = \begin{bmatrix} h(x_{1i}) \\ \vdots \\ h(x_{Ni}) \end{bmatrix}$ and using *Assumption 1*, we have:

$$\begin{aligned}\dot{V} \leq &-\eta\bar{\omega}\sum_{i=1}^{n} X_i^T L X_i + \eta\varepsilon\sum_{i=1}^{n}\alpha_i Z_i^T L X_i + \eta\varepsilon\sum_{i=1}^{n} X_i^T L \begin{bmatrix}\beta_i \\ \vdots \\ \beta_i\end{bmatrix} + \\ &- \eta\sum_{i=1}^{n}\alpha_i X_i^T LSLX_i + \eta\sum_{i=1}^{n} X_i LSL \begin{bmatrix}\beta_i \\ \vdots \\ \beta_i\end{bmatrix} - \sum_{i=1}^{N}(c_i - \sigma_i)^T \frac{\delta_i}{1+\delta_i}.\end{aligned} \tag{2.22}$$

Considering that both $L$ and $LSL$ have zero row-sums, we can state that the third and fifth term of the right-hand side of (2.22) are null, and so it follows that:

$$\dot{V} \leq -\eta \sum_{i=1}^{n} (\bar{\omega} - \alpha_i \varepsilon) X_i^T L X_i - \eta \sum_{i=1}^{n} \alpha_i X_i^T LSLX_i - \sum_{i=1}^{N} (c_i - \sigma_i)^T \frac{\delta_i}{1+\delta_i}. \tag{2.23}$$

Considering that $\varepsilon \leq \frac{\bar{\omega}}{\alpha_{max}}$, we have:

$$\dot{V} \leq -\eta \sum_{i=1}^{n} \alpha_i X_i^T LSLX_i - \sum_{i=1}^{N} (c_i - \sigma_i)^T \frac{\delta_i}{1+\delta_i}. \tag{2.24}$$

Let us denote respectively with $A$ and $B$ the first and the second term of the right-hand side of (2.24). Notice that both $A$ and $B$ are linear functions of each $\sigma_i$, given that at least one $\alpha_i$ is positive. Considering that $LSL$ is negative semidefinite (see [14] for further details), if each $\sigma_i$ were bounded we would clearly get $\dot{V} \leq 0$. We can now prove by contradiction that each $\sigma_i$ is bounded. Indeed, if $\sigma_i$ were unbounded, we would have $A$ and $B$ diverging linearly respectively to $-\infty$ and $+\infty$. So we can always find a suitable $\eta$ such that $\frac{|A|}{|B|} > 1$ and so $A + B \leq 0$. Hence $\dot{V} \leq 0$ against the assumption of $\sigma_i$ diverging. Being $\sigma_i$ monotone increasing ($\dot{\sigma} \geq 0$) then all the $\sigma_i$ converge to constant values that we call $c_i$. Thus, we finally get $\dot{V} \leq 0$. Moreover, given that the network is connected, $\dot{V} = 0$ only on the synchronization manifold ($e = X^T LX = 0$, $\sigma_i = c_i$). Thus, from the global Krasovskii-LaSalle principle [11], the synchronous trajectory is globally asymptotically stable and so $e(t) \to 0$.

Analogously it is possible to prove the effectiveness of strategies (2.6), (2.9) and (2.10).

## References

1. Bartissol, P., Chua, L.O.: The double hook. IEEE Transactions on Circuits and Systems 35(12), 1512–1522 (1988)
2. Belykh, V.N., Verichev, N.N., Kocarev, L.J., Chua, L.O.: On chaotic synchronization in a linear array of Chua's circuits. Journal of Circuits, Systems, and Computers 3(2), 579–589 (1993)
3. Boccaletti, S., Latora, V., Moreno, Y., Chavez, M., Hwang, D.U.: Complex networks: structure and dynamics. Physics Reports 424, 175–308 (2006)
4. Chen, T., Liu, X., Lu, W.: Pinning complex networks by a single controller. IEEE Transactions on Circuits and Systems 54, 1317–1326 (2007)
5. Chen, T., Liu, X.: Network synchronization with an adaptive coupling strenght (October 2006) arXiv:math/0610580
6. Chua, L.O., Itoh, M., Kocarev, L., Eckert, K.: Chaos synchronization in Chua's circuit. Technical Report UCB/ERL M92/111, EECS Department, University of California, Berkeley (1992)
7. Chua, L.O., Kocarev, L., Eckert, K., Itoh, M.: Experimental chaos synchronization in Chua's circuit. International Journal of Bifurcation and Chaos 2, 705–708 (1992)

8. Fewell, J.H.: Social insect networks. Science 301, 1867–1870 (2003)
9. Hugenii, C.: Horoloquium Oscilatorium. Apud F. Muguet (1673)
10. Kashiwagi, A., Urabe, I., Kaneko, K., Yomo, T.: Adaptive response of a gene network to environmental changes by fitness-induced attractor selection. PLoS ONE 1(e49) (2006)
11. Khalil, H.K.: Nonlinear systems, 3rd edn. Prentice-Hall, Englewood Cliffs (2001)
12. Kuramoto, Y.: Chemical oscillations, waves and turbolence. Springer, Heidelberg (1984)
13. De Lellis, P., di Bernardo, M., Garofalo, F.: Novel decentralized adaptive strategies for the synchronization of complex networks. Accepted for Publication on Automatica (2008)
14. De Lellis, P., di Bernardo, M., Garofalo, F.: Synchronization of complex networks through local adaptive coupling. Chaos 18, 037110 (2008)
15. De Lellis, P., di Bernardo, M., Sorrentino, F., Tierno, A.: Adaptive synchronization of complex networks. International Journal of Computer Mathematics 85(8), 1189–1218 (2008)
16. Li, C., Chen, L., Aihara, K.: Stochastic synchronization of genetic oscillator networks. BMC Systems Biology (2007) doi:10.1186/1752-0509-1-6
17. Li, Q., Rus, D.: Global clock synchronization in sensor networks. IEEE Transactions on Computers 55(2), 214–226 (2006)
18. Matsumoto, T.: A chaotic attractor from Chua's circuit. IEEE Transactions on Circuits and Systems 31(12), 1055–1058 (1984)
19. Moallemi, C.C., Van Roy, B.: Distributed optimization in adaptive networks. In: Thrun, S., Saul, L.K., Schölkopf, B. (eds.) Advances in Neural Information Processing Systems (NIPS), vol. 16 (2004)
20. Murray, R.M., Olfati-Saber, R.: Consensus problems in networks of agents with switching topology and time-delays. IEEE Transactions on Automatic Control 49(9), 1520–1533 (2004)
21. Newman, M.E.J., Barabàsi, A.L., Watts, D.J.: The structure and dynamics of complex networks. Princeton University Press, Princeton (2006)
22. Olfati-Saber, R., Fax, J.A., Murray, R.M.: Consensus and cooperation in networked multi-agent systems. Proceedings of the IEEE 95 (January 2007)
23. Pham, Q.-C., Slotine, J.-J.: Stable concurrent synchronization in dynamic system networks. Neural Networks 20(1), 62–77 (2007)
24. Russo, G., di Bernardo, M.: Contraction theory and the master stability function: linking two approaches to study synchronization. IEEE Transactions on Systems and Circuits II (in press, 2009)
25. Stanley, K.O., Bryant, B.D., Mikkulainen, R.: Evolving adaptive neural networks with and without adaptive synapses. In: The 2003 Congress on Evolutionary Computation. CEC 2003 (2003)
26. Wang, W., Slotine, J.-J.E.: On partial contraction analysis for coupled nonlinear oscillators. Biological Cybernetics 92(1), 38–53 (2005)
27. Wang, X.F., Chen, G.: Synchronization in scale-free dynamical networks: Robustness and fragility (May 2001) arXiv:cond-mat/0105014v2
28. Wang, X.F., Chen, G.: Complex networks: small-world, scale-free and beyond. IEEE Circuits and Systems Magazine 3(1), 6–20 (2003)
29. Wu, C.W., Chua, L.O.: Synchronization in an array of linearly coupled dynamical systems. IEEE Transactions on Circuits and Systems 42(8), 430–447 (1995)
30. Wu, X., Cai, J., Zhao, Y.: Some new algebraic criteria for chaos synchronization of Chua's circuits by linear state error feedback control. International Journal of Circuit Theory and Applications 34(3), 265–280 (2006)

31. Yalcin, M.E., Suykens, J.A.K., Vandewalle, J.P.L.: Cellular Neural Networks, Multi-Scroll Chaos and Synchronization, vol. 50. World Scientific Publishing Co. Pte Ltd., Singapore (2005)
32. Yamaguchi, S., Isejima, H., Matsuo, T., Okura, R., Yagita, K., Kobayashi, M., Okamura, H.: Synchronization of cellular clocks in the suprachiasmatic nucleus. Science 302(5649), 1408–1412 (2003)
33. Zhou, C., Kurths, J.: Dynamical weights and enhanced synchronization in adaptive complex networks. Physical Review Letters 96, 164102 (2006)

# Chapter 3
# Dealing with Uncertainty in Consensus Protocols

Dario Bauso, Laura Giarré, and Raffaele Pesenti

**Abstract.** Recent results on consensus protocols for networks are presented. The basic tools and the main contribution available in the literature are considered, together with some of the related challenging aspects: estimation in networks and how to deal with disturbances is considered. Motivated by applications to sensor, peer-to-peer, and ad hoc networks, many papers have considered the problem of estimation in a consensus fashion. Here, the Unknown But Bounded (UBB) noise affecting the network is addressed in details. Because of the presence of UBB disturbances convergence to equilibria with all equal components is, in general, not possible. The solution of the $\varepsilon$-consensus problem, where the states converge in a tube of ray $\varepsilon$ asymptotically or in finite time, is described. In solving the $\varepsilon$-consensus problem a focus on linear protocols and a rule for estimating the average from a compact set of candidate points, the lazy rule, is shown.

## 3.1 Introduction

*Consensus protocols* are distributed control policies based on neighbors' state feedback that allow the coordination of multi-agent systems. According to the usual meaning of consensus, the system state must converge to an equilibrium point with all equal components in finite time or asymptotically [3, 14, 15, 19, 21, 24, 26, 31, 32]. Two recent surveys on consensus [22], [25] report in details the main contribution of the past few years on consensus.

Consensus problems have a history in *computer science* where the word *consensus* instead of *control policy* or *agreement* was first used, [18]. Also, in the early

Dario Bauso and Laura Giarré
Universitá di Palermo, viale delle Scienze, 90128 Palermo, Italy
e-mail: dario.bauso@unipa.it, giarre@unipa.it

Raffaele Pesenti
Università di Venezia Cà Foscari, Venezia, Italy
e-mail: pesenti@unive.it

works [18], [29], [30] convergence of consensus problems was studied in a stochastic setting, but the exchange of random messages between the agents was assumed to be error-free. In particular, [30] obtained consensus results for a group of agents minimizing their common cost function via stochastic gradient based optimization.

The main applications of the cooperative control for multi-agent systems are of two kinds, [25]. The *formation control problems* with applications to mobile robots unmanned air vehicles (UAVs), autonomous underwater vehicles (AUVs), satellites, aircraft, spacecraft, and automated highway systems. The *nonformation control problems* such as task assignment, payload transport, role assignment, air traffic control, timing, and search, inventory problems, sensor networks, P2P. In particular,for the nonformation problems, a lot of work has been done to speed up convergence through estimation.

The main issues that are addressed are related to the *shared information*. In particular, the definition and management of *shared information* among a group of agents to facilitate the *coordination* among agents. The *shared information* may take the form of common objectives, common control algorithms, relative position information. Information is necessary for cooperation, and it may be shared in a variety of ways. Examples of information necessary for cooperation can be: the relative position sensors may enable vehicles to construct state information for other vehicles (UAV's formations); the joint knowledge might be pre-programmed into the vehicles before a mission begins (robotics team) and the knowledge may be communicated between vehicles using a wireless network (vehicle formations).

*Why Consensus Problems?*
For cooperative control strategies to be effective a team of agents must be able to respond to unanticipated situations sensed as a cooperative task. As the environment changes, the agents on the team must be in agreement as to what changes took place. A consequence that shared information is assumed necessary for coordination is that cooperation requires that the group of agents reach *consensus* on the coordination data.

*What is Consensus?*
Convergence to a common value is called *agreement* or *consensus* problems. We recall that a distributed averaging algorithm is a procedure using which the agents can exchange messages and update their values iteratively, so that eventually, each agent is able to compute the average of all initial values. Consensus protocols have the advantages of letting the policy converge to *the best for all agents* equilibrium, admissible Nash with a minimal exchanging of information among agents, [2].

## 3.2 Graph Theory Background and Consensus Problem

It is natural to model information exchange between agents in a cooperative team by a network An (undirected) network consists of a pair $(N, E)$, where $N$ is a finite nonempty set of *nodes* and $E$ is a set of pairs of nodes, called *edges*. Each node corresponds to an agent and each edge to a bidirectional communication link between two agents. A path of length $r$ between node $i$ and node $j$ is a sequence of edges of the form $(i_0,i_1),(i_1,i_2),\ldots,(i_{r-1},i_r)$, where $i_k \in N$, for $k=0,1,\ldots,r$, and $i_0 = i$ and $i_r = j$. The existence of a path between $i$ and $j$ guarantees that agent $i$ and $j$ can share information at least indirectly through some other agents. A network is called connected if there is a path between any distinct pair of nodes.

A network may be characterized by the adjacency matrix $A=[a_{ij}]$ of a weighted digraph is defined as $a_{ii}=0$ and $a_{ij}>0$ if$(j,i)\in E$ where $i\neq j$.

The Laplacian matrix of the weighted digraph is defined as $L=[\ell_{ij}]$, where $\ell_{ii} = \sum_j a_{ij}$ and $\ell_{ij} = -a_{ij}$ where $i \neq j$.

## *Example*

A team of 4 UAVs in longitudinal flight and initially at different heights. Each UAV controls the vertical rate without knowing the relative position of all UAVs but only of neighbors

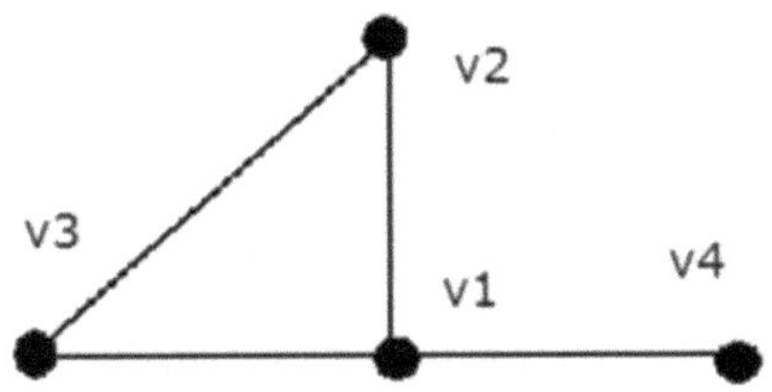


**Fig. 3.1** Topology of the information network of the uavs

Here:

$$A = \begin{bmatrix} 0 & 1 & 1 & 1 \\ 1 & 0 & 1 & 0 \\ 1 & 1 & 0 & 0 \\ 1 & 0 & 0 & 0 \end{bmatrix}$$

$$L = \begin{bmatrix} 3 & -1 & -1 & -1 \\ -1 & 2 & -1 & 0 \\ -1 & -1 & 2 & 0 \\ -1 & 0 & 0 & 1 \end{bmatrix}$$

## *Some Properties*

For a graph $G$ and its Laplacian matrix $L$ with eigenvalues $\lambda_0 \leq \lambda_1 \leq \ldots \leq \lambda_{n-1}$ $L$ is always positive-semidefinite ($\forall i, \lambda_i \geq 0$). The second smallest eigenvalue of

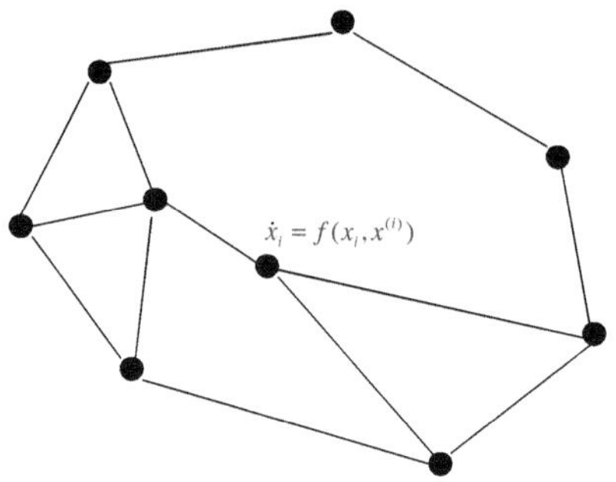


**Fig. 3.2** The information flow in a network of dynamic agents

graph Laplacian $\lambda_1$ is called the *algebraic connectivity*. In the previous example the eigenvalues are: $0,1,3,4$.

## *The Consensus Problem*

In this and in the following sections, if not differently specified, we assume that the communications links between the agents of the systems considered are bidirectional and can be modeled as undirected connected networks. We can generalize the results of this and the following sections to systems with unidirectional links by modeling the systems as directed networks, i.e., networks in which the edges are directed. However, it is interesting to note that in this case we should need not only that the information can flow between each pair of nodes through some directed paths but also that the flow of information is in some way well-distributed. Formally, we should require the oriented network to be fully connected (for each pair of nodes $i$ and $j$ there exists a directed path from $i$ to $j$ and a directed path form $j$ to $i$) and to be balanced (each node $i$ presents the same number of entering and exiting communication links/edges).

Consider a system of $n$ dynamic agents $\Gamma = \{1,\ldots,n\}$ Let us model the interaction through *connected* network $G = (\Gamma, E)$.

It is possible to define the consensus problem as follows.

*Consensus Problem:* Find a distributed and stationary control policy (*protocol* )

$$\dot{x}_i = u_i(x_i, x^{(i)}) \quad \forall i \in \Gamma.$$

such that, if $\hat{\chi} : \mathbb{R}^n \to \mathbb{R}$ is the *agreement function*, usually $\hat{\chi}(\xi) = ave(\xi_1,\ldots,\xi_n)$. Then the consensus is reaching

$$\|x_i - \hat{\chi}(x(0))\| \longrightarrow 0 \quad \text{for } t \longrightarrow \infty.$$

The system converges to

$$\hat{\chi}(x(0))\mathbf{1}.$$

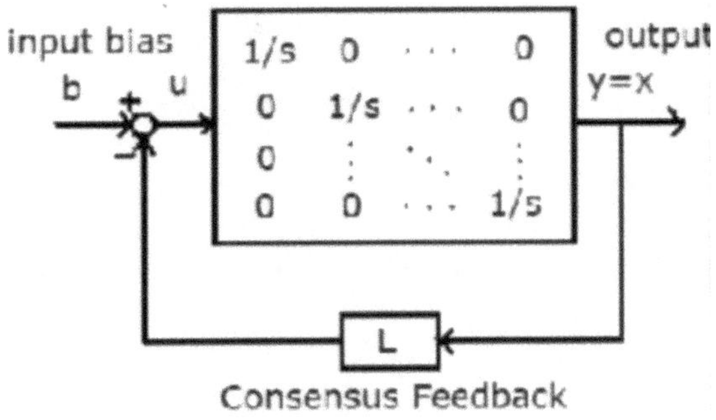


**Fig. 3.3** MIMO system with Laplacian feedback matrix.

*A Consensus Algorithm for Averaging:* ([22]) A simple consensus algorithm to reach an averaging agreement regarding the state of $n$ integrator agents with dynamics $\dot{x}_i = u_i$ can be expressed as an $n$th-order linear system on a graph:

$$\dot{x}_i(t) = \sum_{j \in N_i} (x_j(t) - x_i(t)), x_i(0) = z_i. \tag{3.1}$$

The *collective dynamics* of the group of agents following protocol can be written as

$$\dot{x} = -Lx \tag{3.2}$$

where $L = [l_{ij}]$ is the Laplacian of the network. For a connected network: the consensus value is equal to the average of the initial values. Irrespective of the initial value of the state of each agent, all agents reach an asymptotic consensus regarding the value of the function

$$f(z) = \frac{1}{n} \sum_i z_i$$

## Example

Consider dynamics 3.2 with Laplacian matrix:

$$L = \begin{bmatrix} 3 & -1 & -1 & -1 \\ -1 & 2 & -1 & 0 \\ -1 & -1 & 2 & 0 \\ -1 & 0 & 0 & 1 \end{bmatrix}$$

with eigenvalues: $0, 1, 3, 4$ Then the second eigenvalue of $-L$ is $-1$ and quantifies the speed of convergence of consensus algorithms. The network with the consensus rule can be summarized in the MIMO system (without disturbance) displayed in Fig. 3.3 where $L$ is the Laplacian matrix.

## 3.3 Challenging Aspects in Consensus

### *Comments on the Linear Algorithm*

We recall that the algorithm (3.1) is working for linear protocols. The linearity can be compromised by many factors: saturations, disturbances, noisy measurements, tec. The convergence is only guaranteed for arithmetic mean. Extension to nonlinear algorithm can be found in literature, in particular not only averaging algorithm, have been presented. In ([3]) it has been shown that consensus must be reached on a group decision value returned by a function of the agents' initial state values. The agents can reach consensus if the value of such a function computed over the agents' state trajectories is time invariant. This basic result has been used to introduce a protocol design rule allowing consensus on a quite general set of values. Such a set includes, e.g., any generalized mean of order $p$ of the agents' initial states.

### *Comments on the Decentralized Information*

How to deal with decentralized information, decentralized decision making and distributed objectives (less explored) is very challenging. In particular, agents may need to be in *agreement* as to what happens (information sharing); what to do (coordinated decision making) and/or what to aim at. The last aspect has been extensively treated in ([3]) where a mechanism design approach to consensus problems has been solved.

### *Comments on the Topology*

Switching topology: some link can be lost (e.g. a dropping in a transmission channel). If the network is still connected, under certain hypothesis on the switching time, it is possible to prove convergence. The topology is not connected at each instant, but it is on averaging. A vast literature has been devoted to the switching topology results on convergence. We recall hereafter the work of [17],[21],[15].

### *Comments on the Convergence in Presence of Delays*

Convergence properties and stability have been studied in many papers. Interesting is the problem of a possible delay in the channel communications. In particular, in [1] previously known results [10] and [19] have been extended by including the possibility of arbitrary bounded time-delays in the communication channels and relaxing the convexity of the allowed regions for the state transition map of each agent.

### *Comments on the Estimation Algorithms*

Motivated by applications to sensor, peer-to-peer, and ad hoc networks, many papers have considered the problem of estimation in a consensus fashion.

In [7] distributed algorithms, also known as gossip algorithms, for exchanging information and for computing in an arbitrarily connected network of nodes are studied. The relationship with consensus problem is shown. In particular, the topology of such networks changes continuously as new nodes join and old nodes leave the network and additionally, nodes in sensor networks operate under limited computational, communication, and energy resources. These constraints have motivated the design of "gossip" algorithms: schemes which distribute the computational burden and in which a node communicates with a randomly chosen neighbor.

In [23] three different algorithms for distributed Kalman filter are described. In particular, a continuous-time distributed Kalman filter that uses local aggregation of the sensor data but attempts to reach a consensus on estimates with other nodes in the network is derived. This is an interesting peer-to-peer distributed estimation method.

Distributed estimation for ad hoc wireless networks for noisy links is presented in [27]. Here, distributed algorithms for estimation of unknown deterministic signals using ad hoc WSNs based on successive refinement of local estimates is presented. Consensus algorithm on the local variable is introduced. The crux of their approach is to express the desired estimator, either MLE or BLUE, as the solution of judiciously formulated convex optimization problems.

In [6], the importance of the estimation in speeding up the convergence of consensus algorithms was considered. We recall that in networks, the information is not globally distributed, and nodes need to estimate it locally from aggregate information. Algorithms that improve their convergence by additive local estimation are presented. In particular the obvious statement: "estimating is better than not estimating" is supported and proved in the paper [6] where linear and non linear predictor are used to improve the consensus speed in distributed inventory noncooperative games.

### *Comments on the Quantized Consensus*

The problem of quantized consensus is a distributed averaging problem on arbitrary connected network, with the additional constraint that the value at each node is an integer, [16]. This discretized distributed averaging problem models several problems of interest, such as averaging in a network with finite capacity channels and load balancing in a processor network.

A similar problem has been studied in [20], where distributed iterative algorithms for the averaging problem over time varying topologies is considered. Here the focus is on the convergence time of such algorithms when complete (unquantized) information is available, and on the degradation of performance when only quantized information is available. They study a large and natural class of averaging algorithms, which includes the vast majority of algorithms proposed to date, and provide tight polynomial bounds on their convergence time.

### *Comments on Dealing with Noise*

Actually, despite the literature on consensus is now becoming extensive, only few approaches have considered a disturbance affecting the measurements. In the literature, most existing algorithms assume exact state exchange between the agents with only very few exceptions (see, e.g., [33],[5] and [13]). In ([33]), a least mean square optimization method is used to choose the constant coefficients in the averaging rule. The long term consensus error is then minimized. Recently, the other two papers, independently, were dealing with noise in a similar way. When the noise is affecting the network the consensus is in general a problem, because the convergence to an average value is not guaranteed. The two papers, ([5] and [13]), consider different a-priori knowledge assuumptions on the noise affecting the network that is, respectively, Unknown But Bounded and stochastic. But both papers adopt a similar solution to guarantee the convergence: a "lazy" consensus, that is to say, the introduction of a factor $a(t)$ in the consensus rule, that is a sort of *forgetting factor*. This factor, related to a different weighting of the information, guarantees an *almost sure* convergence in ([13]) and an $\varepsilon$-*consensus*, a convergence in a tube of ray $\varepsilon$ in ([5]).

## 3.4 Consensus with UBB Disturbances

We summarize hereafter the main results presented in ([4]) and ([5]). The novelty of this approach is in the presence of Unknown But Bounded (UBB) disturbances [9] in the neighbors' state feedback. An UBB noise has been assumed, because it requires the least amount of a-priori knowledge on the disturbance. Only the knowledge of a bound on the realization is assumed, and no statistical properties need to be satisfied. Moreover, we recall that starting from [9], the UBB framework has been used in many different fields and applications, such as, mobile robotics, vision, multi-inventory, data-fusion and UAV's and in estimation, filtering, identification and robust control theory.

Because of the presence of UBB disturbances convergence to equilibria with all equal components is, in general, not possible. The main contribution is then the introduction and solution of the $\varepsilon$-consensus problem, where the states converge to a tube of ray $\varepsilon$ asymptotically or in finite time. In solving the $\varepsilon$-consensus problem we focus on linear protocols and present a rule for estimating the average from a compact set of candidate points, say it lazy rule, such that the optimal estimate for the $i$th agent is the one which minimizes the distance from $x_i$.

We consider **Unknown But Bounded** (UBB) noise affecting a network of $n$ dynamic agents $\Gamma = \{1,\dots,n\}$ modeling the interaction through *connected* network $G = (\Gamma, E)$ that is a matrix of disturbances $d(t) = [d_{ij}(t)]_{i,j \in \Gamma} \in D$ affecting it.

### *Illustrative Example*

Consider the chain network displayed in Fig. 3.4 with the following dynamics:

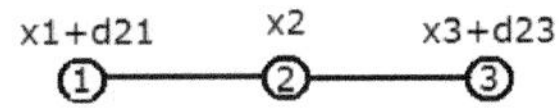


**Fig. 3.4** Chain network of three agents.

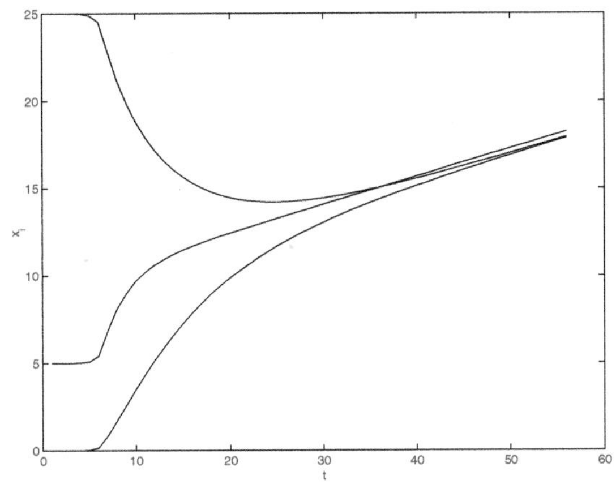


**Fig. 3.5** Time plot of the state variables when $d_{ij} = 1$ for all $i, j$.

$$
\begin{aligned}
\dot{x}_1 &= (x_2 + d_{12}) - x_1 \\
\dot{x}_2 &= [\underbrace{(x_1 + d_{21}) - x_2}_{y_{21}}] + [\underbrace{(x_3 + d_{23}) - x_2}_{y_{23}}] \\
\dot{x}_3 &= (x_2 + d_{32}) - x_3
\end{aligned}
$$

We can find equilibria by imposing $\dot{x} = 0$ and obtain

$$
\begin{aligned}
x_1 &= x_2 + d_{12}, \\
0 &= d_{12} + d_{21} + d_{23} + d_{32}, \\
x_3 &= x_2 + d_{32}
\end{aligned}
$$

We note that no equilibria exist unless $d_{12} + d_{21} + d_{23} + d_{32} = 0$.

If we consider $d_{ij} = 1$ for all $i, j$ then $\dot{x} = -Lx + (1,2,1)^T$ has a drift, if we take $d_{12} = d_{23} = 1$, $d_{32} = d_{21} = -1$, then $\dot{x} = -Lx + (1,0,-1)^T$ as depicted in figures 3.5 and 3.6.

## Problem Formulation: $\varepsilon$- Consensus

Let a network $G = (\Gamma, E)$ be given, find $\varepsilon$-consensus protocol $u_{\sigma(t)}(.)$ and study the dependence of $T$ on the topology $E$ and disturbances $D$.

The hypercube $D = \{d : -\xi \leq d_{ij} \leq \xi,\ \forall (i,j) \in \Gamma^2\}$ is a bounded tube of radius $\varepsilon$,

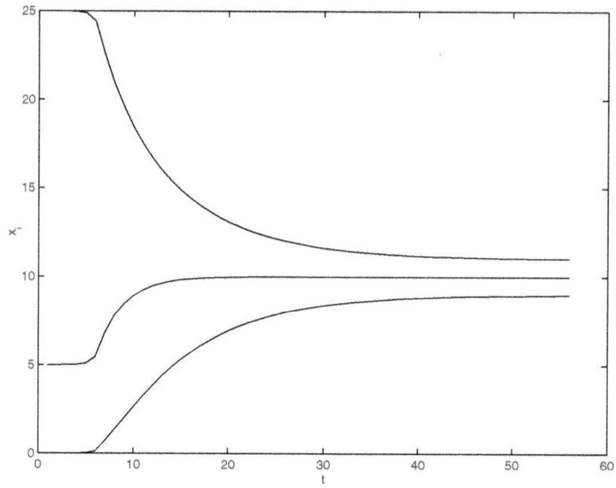


**Fig. 3.6** Time plot of the state variables when $d_{12} = d_{23} = 1, d_{32} = d_{21} = -1$.

$$
\begin{aligned}
T &= \{x \in \mathbb{R}^n : |x_i - x_j| \leq 2\varepsilon, \forall i, j \in \Gamma\}, \\
s.t. & \max_{i \in \Gamma, x \in T} x_i \leq \max_{j \in \Gamma} x_j(0) \\
& \min_{i \in \Gamma, x \in T} x_i \geq \min_{j \in \Gamma} x_j(0).
\end{aligned}
$$

Consider a network $G_{\sigma(t)} = (\Gamma, E_{\sigma(t)})$ that has a time variant edgeset $E_{\sigma(t)} \in \mathscr{E}$. We define an edgeset $E_k$ as recurrent for a given realization of $\sigma(t)$ if for all $t \geq 0$, there exists $t_k \geq t$ such that $\sigma(t_k) = k$. As $\mathscr{E}$ is finite there exists at least a recurrent edgeset for any realization of $\sigma(t)$. Throughout the rest of the paper we assume that all the edgesets in $\mathscr{E}$ are recurrent over time for any realization of $\sigma(t)$, more formally

**Assumption.** *Any realization $\sigma(t)$ is such that, for all $t \geq 0$, there exists $t_k \geq t$ such that $\sigma(t_k) = k$ for all $k \in \mathscr{I}$.*

When we say that the edgeset $E_{\sigma(t)}$ is time variant we understand that the Assumption holds and all the edgesets in $\mathscr{E}$ are recurrent. Note that there is no loss of generality in Assumption, if it were false, the results of this section would hold for the subset $\hat{\mathscr{E}} \subseteq \mathscr{E}$ of recurrent edgesets.

Let $\Psi$ be the set of switching times. For all $i \in \Gamma$, consider the family of *first-order* dynamical systems controlled by a *distributed* and *stationary* control policy

$$
\begin{aligned}
\dot{x}_i &= u_{i\sigma(t)}(x_i, y^{(i)}) \quad \forall t \geq 0,\ t \notin \Psi \\
x_i(t^+) &= x_i(t^-) \quad \forall t \in \Psi
\end{aligned}
\tag{3.3}
$$

where $y^{(i)}$ is the information vector from the agents in $N_{i\sigma(t)}$ with generic component $j$ defined as follows,

$$
y_j^{(i)} = \begin{cases} y_{ij} \text{ if } j \in N_{i\sigma(t)}, \\ 0 \quad \text{otherwise.} \end{cases}
$$

In the above equation, $y_{ij}$ is a disturbed measure of $x_j$ obtained by agent $i$ as

$$y_{ij} = x_j + d_{ij}$$

and $d_{ij}$ is an UBB disturbance, i.e., $-\xi \leq d_{ij} \leq \xi$ with a-priori known $\xi > 0$. All agents have a perfect measure of their own state, that is $d_{ii} = 0$ for all $i \in \Gamma$.

We have proved in [4] that the system trajectory $x(t)$ is bounded as $t \rightarrow \infty$ and also that the difference between the maximum and the minimum agent states may not increase over the time. More formally, denote by $\mathscr{V}(x(t)) = x_{i_1}(t) - x_{i_n}(t)$ then

$$\mathscr{V}(x(t)) \geq \mathscr{V}(x(t+\Delta t)) \quad \text{for any } t \geq 0 \text{ and } \Delta t > 0. \tag{3.4}$$

We use this last implication to introduce some additional results that will turn useful when dealing with switching topology systems.

A typical consensus problem is the average consensus one, i.e., the system state converges to the average of the initial state. It is well known that the average consensus can be solved by linear protocols, in absence of disturbances and under certain conditions on the network. With this in mind, let the following linear protocol be given as

$$u_{i\sigma(t)}(x_i, y^{(i)}) = \sum_{j \in N_{i\sigma(t)}} (\tilde{y}_{ij} - x_i), \forall i \in \Gamma \tag{3.5}$$

where $\tilde{y}_{ij}$ is the estimate of state $x_j$ on the part of agent $i$. For a given disturbed measure $y_{ij}$ the state $x_j$ and consequently its estimate $\tilde{y}_{ij}$ must belong to the interval

$$y_{ij} - \xi \leq \tilde{y}_{ij} \leq y_{ij} + \xi. \tag{3.6}$$

Among all possibile $\tilde{y}_{ij}$ we choose the one defining the following *lazy consensus* ([4],[5]): pick the solution of

$$\tilde{y}^{(i)} = \arg \min_{\tilde{y}_{ij} \in [y_{ij} - \xi, y_{ij} + \xi],\ j \in N_i} \Big| \sum_{j \in N_i} (\tilde{y}_{ij} - x_i) \Big|.$$

Denote by $\mathscr{V}_\infty = \lim_{t \rightarrow \infty} (x_{i_1}(t) - x_{i_n}(t))$ the final value of $\mathscr{V}(x(t))$. Observe that for some network $G(\Gamma, E)$ and initial state $x(0)$, there may exist some disturbance realizations $\{d(t) \in D : t \geq 0\}$ such that even if $\mathscr{V}(x(0)) > \mathscr{V}_\infty$, the value $\mathscr{V}(x(t))$ may be constant over some finite time interval before reaching its final value $\mathscr{V}_\infty$. More specifically, there may exist $t$ and $\Delta t$ such that $\mathscr{V}(x(t)) = \mathscr{V}(x(t+\Delta t)) > \mathscr{V}_\infty$.

Given the lazy rule (3.6), protocol (3.5) turns out to have a feedback structure

$$\begin{aligned} \dot{x}_i(t) = u_{i\sigma(t)}(x_i, y^{(i)}) = \mathscr{D}_{|N_{i\sigma(t)}|\xi}\left(\Sigma_{j \in N_{i\sigma(t)}} (y_{ij} - x_i)\right) = \\ \mathscr{D}_{|N_{i\sigma(t)}|\xi}\left(\Sigma_{j \in N_{i\sigma(t)}} (x_j + d_{ij} - x_i)\right) \end{aligned} \tag{3.7}$$

where $\mathscr{D}_{|N_{i\sigma(t)}|\xi}$ is a dead-zone function of type

$$\mathscr{D}_{|N_{i\sigma(t)}|\xi}(x) = \begin{cases} 0 & \text{if } |x| \leq |N_{i\sigma(t)}|\xi \\ x - |N_{i\sigma(t)}|\xi & \text{if } x \geq |N_{i\sigma(t)}|\xi \\ x + |N_{i\sigma(t)}|\xi & \text{if } x \leq -|N_{i\sigma(t)}|\xi \end{cases}.$$

Hereafter, we simply refer to system (3.7) when considering the linear protocol (3.5) with $\tilde{y}^{(i)}$ as in (3.6). First, we report results for the non switching network and then for the switching one.

## *Case* $\sigma(t) = const$ *Non Switching Network*

The lazy rule *maintains the sign* that is: for each $i \in \Gamma$, for each $t \geq 0$ either

$$\begin{aligned} \text{sign}(u_i(x_i, y^{(i)})) &= \text{sign}(\Sigma_{j \in N_i}(x_j - x_i)), \\ & \quad \textit{or} \\ \text{sign}(u_i(x_i, y^{(i)})) &= 0. \end{aligned}$$

Considering a Lyapunov function $V(x(t)) = \frac{1}{2}\Sigma_{(i,j)\in E}(x_j(t) - x_i(t))^2$, is it possible to prove that trivially, $V(x(t)) \geq 0$; $V(x(t)) = 0 \Leftrightarrow x(t) \in span\{\mathbf{1}\}$.

$$\begin{aligned} &\dot{V}(x) = \\ &\Sigma_{(i,j)\in E}(x_j - x_i)(u_j - u_i) = \\ &-\Sigma_{i\in\Gamma} u_i \Sigma_{j\in N_i}(x_j - x_i) = \\ &= -\Sigma_{i\in\Gamma} \text{sign}(u_i) \, \text{sign}\left(\Sigma_{j\in N_i}(x_j - x_i)\right) \\ &|u_i| \left|\Sigma_{j\in N_i}(x_j - x_i)\right| \\ &\leq 0 \end{aligned}$$

Is has been shown that equilibria exist. Moreover, all equilibria are such that, for all $i \in \Gamma$, for all $t \geq \bar{t}$ and for all $\Delta t > 0$ and finite

$$\int_t^{t+\Delta t} u_{i\sigma(\tau)}(x_i^*, y^{(i)}) d\tau = 0.$$

Assuming $d(t)$ unknown but constant, then all equilibria are in the polyhedron

$$\begin{aligned} P(d,E) = \Big\{ x : -\tfrac{\Sigma_{j\in N_i} d_{ij}}{|N_i|} - \xi \leq \tfrac{\Sigma_{j\in N_i} x_j}{|N_i|} - x_i \leq \\ \leq -\tfrac{\Sigma_{j\in N_i} d_{ij}}{|N_i|} + \xi, \; \forall i \in \Gamma \Big\}; \end{aligned}$$

The Polyhedron $P(0,E)$ is shown in Fig. 3.7

If we assume *bang-bang* disturbance $d \in \{\xi, -\xi\}$ with dwell time, then all equilibria are in

$$P(\xi, E) \cap P(-\xi, E) = \{\pi\mathbf{1}\}.$$

Same result can be obtained if $d$ takes on any value in $D$ with dwell time.

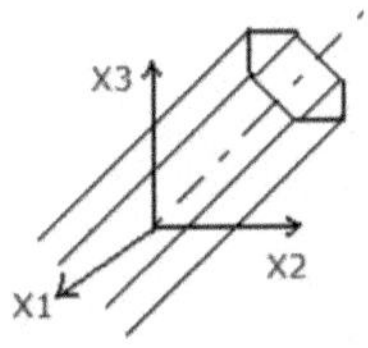


**Fig. 3.7** Polyhedron $P(0,E)$.

For a generic $\{d(t) \in D,\ t \geq 0\}$, all equilibria are in

$$P(D,E) := \bigcup_{d \in D} P(d,E) = \left\{ -2\xi \leq \frac{\sum_{j \in N_i} x_j}{|N_i|} - x_i \leq 2\xi, \forall i \in \Gamma \right\}.$$

### Stability

In [4] it has been proved the following results.

**Theorem.** *There is an equilibrium point $x^* \in P(D,E)$ such that $x(t) \rightarrow x^*$.*
**Corollary:** *Assume d bang-bang all system trajectories converges to $\{\pi \mathbf{1}\}$.*

*Sketch of the Proof of the Theorem:* As any $x^*$ is in $P(D,E)$ then we only prove $x(t) \rightarrow x^*$. The Lyapunov function $V(x(t)) = \frac{1}{2} \sum_{(i,j) \in E} (x_j(t) - x_i(t))^2$. Trivially, $V(x(t)) \geq 0$; $V(x(t)) = 0 \Leftrightarrow x(t) \in span\{\mathbf{1}\}$; remind $\dot{V} \leq 0$. For any sequence $\{t_k\}_0^\infty$ with $t_{k+1} \geq t_k$ for all $k$, $t_0 = 0$ and $\lim_{k \rightarrow \infty} t_k = \infty$, rewrite

$$\int_0^\infty \dot{V}(x(t))dt = \sum_{k=0}^{\infty} \int_{t_k}^{t_{k+1}} \dot{V}(x(\tau))d\tau. \tag{3.8}$$

RHS of (3.8) is bounded $\Rightarrow \lim_{k \rightarrow \infty} \int_{t_k}^{t_{k+1}} \dot{V}(x(\tau))d\tau = 0$ From $\dot{V}$ expression, $-\dot{V}(x(t)) \geq \|u(t)\|_2^2$, then $\lim_{k \rightarrow \infty} \int_{t_k}^{t_{k+1}} \|u(\tau)\|_2^2 d\tau = 0$.

## $\varepsilon$-Consensus with Switching Topology

A basic observation is that even in presence of switches $\mathscr{V}(x(t))$ is not increasing on $t$ as stated in (3.4). To see this note that the protocol $u_{i\sigma(t)}(x_i, y^{(i)})$, for all $i \in \Gamma$, induces a continuous and bounded state trajectory even for a network with switching topology (3.3) on $G_{\sigma(t)} = (\Gamma, E_{\sigma(t)})$. Then, let $t_k < t_{k+1}$ be two generic consecutive switching times. The arguments of the previous section applied at time $t = t_k$ instead of $t = 0$ guarantee that $x_{i_n}(t_k) \leq x_i(t) \leq x_{i_1}(t_k)$ for all $i \in \Gamma$ and for all $t_k < t < t_{k+1}$. As $x(t)$ is bounded, then $u_{i\sigma(t)}(x_i, y^{(i)})$, for all $i \in \Gamma$, is also bounded in the same time interval. Hence the first condition in (3.5) implies that $x(t)$ is continuous for $t_k < t < t_{k+1}$, whereas, the second conditions in (3.5) imposes the continuity of the state trajectory in $t_k$ as $x(t_k^-) = x(t_k^+)$ and in $t_{k+1}$ as $x(t_{k+1}^-) = x(t_{k+1}^+)$. As a consequence, for all $t \geq 0$, we also have that $x_{i_n}(0) \leq x_i(t) \leq x_{i_1}(0)$ for all $i \in \Gamma$ and $\mathscr{V}(x(t))$ is not increasing.

We also need to define the set $\mu(.)$, for given realizations $d(t)$ and $\sigma(t)$, and a subset $Q$ of $D$,

$$\mu(Q,E_k) = \{t \in \mathbb{R} : \; d(t) \in Q \; and \; \sigma(t) = k\}. \tag{3.9}$$

For each $Q \subseteq D$: $L(Q,E_k) = \max_{x \in P(Q,E_k)} \mathscr{V}(x)$ the maximum value of $\mathscr{V}(x)$ for points in $P(Q,E_k)$; $S(Q,E_k,\nu) = \{x \in \mathbb{R}^n : \mathscr{V}(x) \leq L(Q,E_k) + 2\nu\}$ the set of points whose maximum difference between two components does not exceed $L(Q,E_k)+2\nu$. It is worth to be noted that $S(Q,E_k,\nu)$ are tubes of radius less than or equal to $L(Q,E_k)+2\nu$ and then, $S(Q,E_k,\nu) \subseteq S(Q,E_{\hat{k}},\nu)$ whenever $L(Q,E_k) \leq L(Q,E_{\hat{k}})$. Also, observe that, by definition, it holds that $P(Q,E_k) \subseteq P(Q,E_k) + B(\nu) \subseteq S(Q,E_k,\nu)$.

Finally, we introduce a minimum dwell time $\tau(\nu,\gamma) = max_{k \in \mathscr{I}}\{q_k(\nu,\gamma)\}$. In other words, the minimum length of the switching intervals is equal to the maximal value over the different $E_k \in \mathscr{E}$ of the times $q(\nu,\gamma)$.

We can now prove the following Theorem.

**Theorem.** *Given the switched system (3.7) with initial state $x(0)$, with a minimum dwell time $\tau(\nu,\gamma)$, assume that two values $\nu > 0$ and $0 < \gamma < 1$ are also given. The state is driven in finite time to the target set*

$$T = \bigcap_{E_k \in \mathscr{E}} S(D,E_k,\nu). \tag{3.10}$$

Note that convergence to $T$, as in (3.10), means that the agents have reached $\varepsilon$-consensus with $\varepsilon = \min\{\varepsilon_k\}$ where $\varepsilon_k$ is the radius of tubes $S(D,E_k,\nu)$. The above theorem does not guarantee the convergence to an equilibrium point. Actually, switching systems may oscillate as shown by the following example.

*Example 3.1.* Consider a family of chain networks $G$ on the set of agents $\Gamma = \{1,2,3\}$ and edgsets $E_1 = \{(1,2),(2,3)\}$ and $E_2 = \{(1,2),(1,3)\}$ (see Fig. **??**). Let $x_1(0) = 2$, $x_2(0) = 0$, and $x_3(0) = 1$, and $\{d(t) = d = \text{const} \in D : t \geq 0\}$, with in particular $d_{12} = d_{13} = d_{31} = 1$ and $d_{21} = d_{23} = d_{32} = -1$, $d_{ij} = 0$ otherwise. If the switching time intervals are sufficiently long the systems trajectory oscillates in $\mathbb{R}^3$ along the segment delimited by points $[2,0,0]$ and $[2,0,2]$. Note that only the state of agent 3 changes over time.

We guarantee that the agents may reach an $\varepsilon$-consensus even within networks with switching topologies, provided the switching intervals (dwell times) have a sufficient length. However, the above mentioned theorems do not indicate any constructive ways to estimate a minimum dwell time $\tau(\nu,\gamma)$. This is a non trivial task and and open problem as reported in [28] and it is out of the scope of this paper. However, to our best knowledge, the first result available in literature to determine an upper bound of the dwell time for discrete linear systems is [12].

## 3.5 Final Remarks

In the present paper a tutorial on consensus problems has been presented. Challenging aspects arisen recently and motivated also by applications to sensor, peer-to-peer, and ad hoc networks, are described with reference to existing literature. In particular, we have considered some aspects related to how to deal with noise, with decentralized information, with estimation, with quantized information and with switching topology.

## References

1. Angeli, D., Bliman, P.-A.: Stability of leaderless discrete-time multi-agent systems. Mathematics of Control, Signals & Systems 18(4), 293–322 (2006)
2. Arslan, G., Marden, J.R., Shamma, J.S.: Autonomous vehicle-target assignment: A game theoretical formulation. ASME Journal of Dynamic Systems, Measurement, and Control, special issue on Analysis and Control of Multi-Agent Dynamic Systems, 584–596 (2007)
3. Bauso, D., Giarré, L., Pesenti, R.: Nonlinear Protocols for the Optimal Distributed Consensus in Networks of Dynamic Agents. Systems and Control Letters 55(11), 918–928 (2006)
4. Bauso, D., Giarrè, L., Pesenti, R.: Consensus for switched networks with unknown but bounded disturbances. SIAM Journal on Control and Optimization (2006) arXiv: math.OC/0612834v1
5. Bauso, D., Giarrè, L., Pesenti, R.: Lazy consensus for networks with unknown but bounded disturbances. In: Proceedings of IEEE CDC, New Orleans, pp. 2283–2288 (2007)
6. Bauso, D., Giarré, L., Pesenti, R.: Distributed consensus in noncooperative inventory games. European Journal of Operational Research (in press) (corrected proof, 17 October 2007)
7. Boyd, S., Ghosh, A., Prabhakar, B., Shah, D.: Randomized Gossip Algorithms. IEEE Trans. on Information Theory 52(6), 2508–2530 (2006)
8. Borkar, V., Varaiya, P.: Asymptotic agreement in distributed estimation. IEEE Trans. Automat. Control 27, 650–655 (1982)
9. Bertsekas, D.P., Rhodes, I.: Recursive state estimation for a set-membership description of uncertainty. IEEE Trans. on Automatic Control 16(2), 117–128 (1971)
10. Bertsekas, D.P., Tsitsiklis, J.N.: Parallel and distributed computation. Prentice-Hall International, Englewood Cliffs (1989); republished by Athena Scientific (1997), https://dspace.mit.edu/handle/1721.1/3719
11. Broy, M.: Software engineering — from auxiliary to key technologies. In: Broy, M., Dener, E. (eds.) Software Pioneers, pp. 10–13. Springer, Heidelberg (2002)
12. Geromel, J., Colaneri, P.: RMS gain with dwell time for discrete-time switched linear systems. In: Proc. of IEEE Mediterrean Conference, Ajaccio (July 2008)
13. Huang, M., Manton, J.H.: Stochasting approximation for consensus seeking: mean square and almost sure convergence. In: Proceedings of IEEE CDC, New Orleans, pp. 306–311 (2007)
14. Fax, A., Murray, R.M.: Information flow and cooperative control of vehicle formations. IEEE Trans. on Automatic Control 49(9), 1565–1576 (2004)
15. Jadbabaie, A., Lin, J., Morse, A.: Coordination of Groups of mobile autonomous agents using nearest neighbor rules. IEEE Trans. on Automatic Control 48(6), 988–1001 (2003)

16. Kashyap, A., Basar, T., Srikant, R.: Quantized Consensus. Automatica 43, 1192–1203 (2007)
17. Liberzon, D.: Switching in Systems and Control. Volume in series Systems and Control: Foundations and Applications. Birkhauser, Boston (2003)
18. Lynch, N.A.: Distributed Algorithms. Morgan Kaufmann Publishers, Inc., San Francisco (1996)
19. Moreau, L.: Leaderless coordination via bidirectional and unidirectional time-dependent communication. In: Proc. of the 42nd IEEE Conference on Decision and Control, Maui, Hawaii, pp. 3070–3075 (2003)
20. Nedić, A., Olshevsky, A., Ozdaglar, A., Tsitsiklis, J.N.: On Distributed Averaging Algorithms and Quantization Effects?, LIDS Report 2778 (November 2007)
21. Olfati-Saber, R., Murray, R.: Consensus problems in networks of agents with switching topology and time-delays. IEEE Trans. on Automatic Control 49(9), 1520–1533 (2004)
22. Olfati-Saber, R., Fax, J.A., Murray, R.M.: Consensus and Cooperation in Networked Multi-Agent Systems. Proceedings of the IEEE 95(1), 215–233 (2007)
23. Olfati-Saber, R.: Distributed Kalman Filtering for Sensor Networks. In: Proc. of IEEE CDC, New Orleans, pp. 5492–5498 (2007)
24. Ren, W., Beard, R., Atkins, E.M.: A survey of consensus problems in multi-agent coordination. In: Proc. of the American Control Conference, Portland, OR, USA, pp. 1859–1864 (2005)
25. Ren, W., Beard, R.W., Atkins, E.: Information Consensus in Multivehicle Cooperative Control: Collective Group Behavior through Local Interaction. IEEE Control Systems Magazine 27(2), 71–82 (2005)
26. Ren, W., Beard, R.: Consensus seeking in multi-agent systems under dynamically changing interaction topologies. IEEE Trans. on Automatic Control 50(5), 655–661 (2005)
27. Schizas, I.D., Ribeiro, A., Giannakis, G.B.: Consensus in Ad Hoc WSNs With Noisy Links- Part I: Distributed Estimation of Deterministic Signals. IEEE Trans. on Signal Processing 56(1) (January 2008)
28. Shorten, R., Wirth, F., Mason, O., Wulff, K., King, C.: Stability Criteria for Switched and Hybrid Systems? SIAM Review 49(4), 545–592 (2007)
29. Tsitsiklis, J.N., Athans, M.: Convergence and asymptotic agreement in distributed decision problems. IEEE Trans. Automat. Control 29(1), 42–50 (1984)
30. Tsitsiklis, J.N., Bertsekas, D.P., Athans, M.: Distributed asynchronous deterministic and stochastic gradient optimization algorithms. IEEE Trans. Automat. Contr. 31(9), 803–812 (1986)
31. Tanner, H.G., Jadbabaie, A., Pappas, G.J.: Stable flocking of mobile agents, part ii: Dynamic topology. In: Proc. of the 42nd IEEE Conference on Decision and Control, Maui, Hawaii, pp. 2016–2021 (2003)
32. Xiao, L., Boyd, S.: Fast linear iterations for distributed averaging. Systems and Control Letters 53(1), 65–78 (2004)
33. Xiao, L., Boyd, S., Kim, S.-J.: Distributed average consensus with least-mean-square deviation. Journal of Parallel and Distributed Computing 67, 33–46 (2007)

# Chapter 4
# Formation Control over Delayed Communication Network

Cristian Secchi and Cesare Fantuzzi

**Abstract.** In this Chapter we address the problem of formation control of a group of robots that exchange information over a communication network characterized by a non negligible delay. We consider the Virtual Body Artificial Potential approach for stabilizing a group of robots at a desired formation. We show that it is possible to model the controlled group of robots as a port-Hamiltonian system and we exploit the scattering framework to achieve a passive behavior of the controlled system and to stabilize the robots in the desired formation independently of any communication delay.

## 4.1 Introduction

A central problem in the coordination of multi-robot systems is the formation control. An example is in the framework of mobile sensor networks where it is desirable to control the disposition of the sensors in such a way to maximize the amount of information that they are able to detect. Several approaches for solving the formation control problem have been proposed in the literature [2, 3, 10, 7, 11, 9].

In most of the current approaches for formation control, it is assumed that the information is instantaneously exchanged among the robots and this can be quite restrictive. In fact, in real applications, it is necessary to choose a medium over which the robots exchange information. Packets switched networks are very good candidate to play this role, as noted in [7]. Nevertheless, the use of this kind of medium introduces some problems that need to be addressed. First of all the information is exchanged among the robots with a certain communication delay which tends

Cristian Secchi and Cesare Fantuzzi
DISMI - University of Modena and Reggio Emilia,
via G. Amendola 2, Morselli Building, 42100 Reggio Emilia, Italy
e-mail: {cristian.secchi,cesare.fantuzzi}@unimore.it

to destabilize the controlled system. Moreover, some packets can get lost and the communication delay can be variable.

In [9, 4] the virtual body artificial potential (VBAP) approach has been introduced. Artificial potentials are exploited for reproducing the social forces, a concept deduced from biological studies of animal aggregations [1], in a group of mobile robots. The goal of this work is to extend the VBAP approach in order to take into account a possible non negligible delay in the communication among the robots and between the robots and the virtual leaders. We will exploit the scattering framework in the VBAP formation control strategy in order to stabilize a group of robots at a desired formation when the information is exchanged over a delayed communication channel. First of all we will formulate the VBAP strategy within the port-Hamiltonian framework [12, 14] in order to put into evidence the interconnection structure through which energy is exchanged, namely, the communication protocol through which the robots and the elements associated to the artificial potentials exchange the information for obtaining the desired formation. We will then reformulate this communication protocol in terms of scattering variables and we will prove that the new communication strategy allows to stabilize the robots in the desired formation independently of any communication delay.

The paper is organized as follows: in Sec. 4.2 we will give some background on the VBAP approach, on port-Hamiltonian systems and on the scattering framework. In Sec. 4.3 we will interpret the VBAP control strategy within the port-Hamiltonian framework. In Sec. 4.4 we will exploit the scattering framework for implementing the VBAP formation control over a delayed network. In Sec. 4.5 we will provide some simulations to validate the results proposed in the paper and, finally, in Sec. 4.6 we will give some concluding remarks and we will address some future work.

## 4.2 Background

### *4.2.1 The VBAP Approach for Formation Control*

Consider a group of $N$ agents (that can be mobile robots, UAV, etc.) that we want to take in a desired formation. We will indicate with $x_{a_i} \in \mathbb{R}^3$, the position of the $i^{th}$ agent with respect to an inertial frame. Furthermore, we will indicate with $u_i \in \mathbb{R}^3$ the control input of the $i^{th}$ vehicle. We consider fully actuated, point mass agent models whose dynamics is represented by:

$$\ddot{x}_{a_i}(t) = u_i(t) \qquad i = 1,\dots,N \tag{4.1}$$

In the space where the agents are moving, a set of $L$ reference points, called *virtual leaders*, are introduced. The position of the $k^{th}$ virtual leader is denoted by $x_{l_k} \in \mathbb{R}^3$. In this paper we consider the case in which all the leaders are at rest. This means that we are focusing on the problem of taking all the agents in the desired formation and not on the problem of moving the formation.

Let $x_{a_{ij}} = x_{a_i} - x_{a_j}$ be the relative position of agent $i$ with respect to agent $j$. Between each pair of agents, an artificial potential $V_I(x_{a_{ij}})$, that we call *interagent potential*, is defined. This potential depends on the distance between the $i^{th}$ and the $j^{th}$ agents and its role is to produce a radial force $f_I(x_{a_{ij}})$ that regulates the distance between the agents. Similarly let $x_{a_il_k} = x_{a_i} - x_{l_k}$ be the relative position of agent $i$ with respect to leader $k$. Between each agent and each virtual leader a potential $V_h(x_{a_il_k})$, that we call *leader potential*, is defined. This potential depends on the distance between the $i^{th}$ agent and the $k^{th}$ leader and its role is to produce a radial force $f_h(x_{a_il_k})$ that regulates the distance between the agent and the leader.

The control of each agent is defined as the sum of the forces induced by the potentials and of a damping term, namely as

$$u_i = -\sum_{i=1}^{N} \frac{\partial V_I(x_{a_{ij}})}{\partial x_{a_{ij}}} - \sum_{k=1}^{L} \frac{\partial V_h(x_{a_il_k})}{\partial x_{a_il_k}} - R_i\dot{x}_{a_i} =$$
$$= -\sum_{i=1}^{N} \frac{f_I(x_{a_{ij}})}{\|x_{a_{ij}}\|} x_{a_{ij}} - \sum_{k=1}^{L} \frac{f_h(x_{a_il_k})}{\|x_{a_il_k}\|} x_{a_il_k} - R_i\dot{x}_{a_i} \quad (4.2)$$

where $R_i$ is a positive definite matrix. The shape of the potential $V_I$ has to be such to produce a repulsive force when two agents are too close, namely when $\|x_{a_{ij}}\| < d_0$, an attractive force when two agents are far, namely when $\|x_{a_{ij}}\| > d_0$ and a null force when the agents are too far, namely when $\|x_{a_{ij}}\| > d_1 > d_0$; $d_1$ and $d_0$ are design parameters. The shape of the potential $V_h$ is similar and it depends on possibly different design parameters $h_0$ and $h_1$ which play the same role as $d_0$ and $d_1$.

The group of agents is immersed in a potential field which is given by the sum of all the artificial potentials. The minimum of this potential is given by a configuration where all the agents are at a distance $d_0$ from the neighboring agents and at a distance $h_0$ from the neighboring leaders. Choosing properly the position of the virtual leaders and the parameters $h_0$, $d_0$, $h_1$ and $d_1$ it is possible to shape the overall potential in such a way that its minimum corresponds to a desired formation.

Defining the state of the agents group as $x = (x_{a_1}, \dots, x_{a_N}, \dot{x}_{a_1}, \dots, \dot{x}_{a_N})^T$ and using as a candidate Lyapunov function

$$V(x) = \frac{1}{2}\sum_{i=1}^{N}\left(\dot{x}_{a_i}^T\dot{x}_{a_i} + \sum_{j\neq i}^{N} V_I(x_{a_{ij}}) + 2\sum_{k=1}^{L} V_h(x_{a_il_k})\right) \quad (4.3)$$

it is possible to prove that the configuration $x = (\bar{x}, 0)$, where $\bar{x}$ is the minimum of the sum of the artificial potentials, is asymptotically stable. For further information see [6, 4]

### 4.2.2 Port-Hamiltonian Systems and the Scattering Framework

The port-Hamiltonian modeling framework is the mathematical formalization of the bond-graph strategy for representing physical systems. Loosely speaking, a port-Hamiltonian system is made up of a set of energy processing elements that exchange energy through a power preserving interconnection. More formally, port-Hamiltonian systems are defined on the state manifold of energy variables $\mathscr{X}$ and they are characterized by a lower bounded Hamiltonian energy function $H : \mathscr{X} \mapsto \mathbb{R}$, expressing the amount of stored energy, and by a Dirac structure $\mathscr{D}$, representing the internal energetic interconnections. Using coordinates, in their simplest form, they are represented by the following equations

$$\begin{cases} \dot{x}(t) = (J(x) - R(x))\frac{\partial H}{\partial x} + g(x)u(t) \\ y(t) = g^T(x)\frac{\partial H}{\partial x} \end{cases} \tag{4.4}$$

where $x \in \mathbb{R}^n$ is the coordinate vector of the energy variables $J(x)$ is a skew-symmetric matrix representing the Dirac structure, $R(x)$ is a positive semidefinite function representing the energy dissipated by the system, $H$ is the Hamiltonian function, $u$ is the time dependent input vector, $g(x)$ is the input matrix and $y$ is time dependent the conjugated output variable. The dependence on time of the state has been omitted in order to keep the notation simple. The pair of dual power variables $(u, y)$ forms the power port through which the port-Hamiltonian system exchanges energy with the rest of the world. The power provided to the system at the instant $t$ is given by $P(t) = u^T(t)y(t)$. It can be easily seen that the following power balance holds:

$$\dot{H}(t) = y^T(t)u(t) - \frac{\partial^T H}{\partial x}R(x)\frac{\partial H}{\partial x} \leq y^T(t)u(t) \tag{4.5}$$

The scattering framework allows to interpret the power flowing through a power port as the difference of incoming power wave and outgoing power wave rather than as a product of power variables. More formally, given a power port $(e(t), f(t))$, the power flowing through it can be decomposed into an incoming power wave $s^+(t)$ and an outgoing power wave $s^-(t)$ in such a way that

$$e^T(t)f(t) = \frac{1}{2}\|s^+(t)\|^2 - \frac{1}{2}\|s^-(t)\|^2 \tag{4.6}$$

where $\|\cdot\|$ is the standard Euclidean norm and the pair of scattering variables $(s^+(t), s^-(t))$ is defined as

$$\begin{cases} s^+(t) = \frac{1}{\sqrt{2b}}(e(t) + bf(t)) \\ s^-(t) = \frac{1}{\sqrt{2b}}(e(t) - bf(t)) \end{cases} \tag{4.7}$$

where $b > 0$ is the impedance of the scattering transformation. The relation in Eq.(4.7) is bijective and, consequently, a power port can be equivalently represented both as a pair of conjugated power variables or as a pair of scattering variables. For further information see [14, 12].

## 4.3 A Port-Hamiltonian Interpretation of VBAP Approach

In this section we will show that a group of agents whose formation is controlled by means of the VBAP strategy can be interpreted as a set of energy processing elements exchanging energy along a power preserving interconnection, that is as a port-Hamiltonian system.

First of all, let us consider the main actors in the VBAP approach: the agents, the interagent potentials and leader potentials. Each agent can be modeled as a kinetic energy storing element, namely as a mass characterized by a certain inertia. Thus, the energy variable describing the state of the agent $i$ is the momentum $p_{a_i}$ and the kinetic energy function associated to agent $i$ is given by

$$H_{a_i} = \frac{1}{2} M_i^{-1} p_{a_i}^T p_{a_i} \tag{4.8}$$

where $M_i$ is the inertia matrix of the agent. In order to take into account the damping effect introduced by the VBAP strategy for asymptotically stabilizing the formation, we embed a dissipative effect to the dynamics of each agent. Thus, indicating with $(e_{a_i}, f_{a_i})$ the power port associated to agent $i$ we have that a generic agent can be described by

$$\begin{cases} \dot{p}_{a_i} = -R_i M_i^{-1} p_i + e_{a_i} \\ \\ f_{a_i} = M_i^{-1} p_i \end{cases} \tag{4.9}$$

where $p_{a_i} = M_i \dot{x}_{a_i} \in \mathbb{R}^3$ is the momentum of the agent and $R_i$ is a symmetric positive definite matrix representing the damping imposed by the VBAP control strategy on the $i^{th}$ agent. The effort $e_{a_i}$ corresponds to the input $u_i$ in Eq.(4.1). We can easily see that

$$e_{a_i}^T f_{a_i} = \dot{H}_{a_i} + p_{a_i}^T M^{-T} R_i M^{-1} p_{a_i} \tag{4.10}$$

namely that the power incoming through the power port is either stored in form of kinetic energy or dissipated.

Both the interagent potentials and the leader potentials necessary for regulating the relative distance among the agents and among the agents and the leaders can be modeled as a set of potential energy storing elements which can be interpreted as virtual springs. The energy variables representing the state of the virtual spring between agent $i$ and agent $j$ is given by $x_{a_{ij}} \in \mathbb{R}^3$, namely by the relative position of agent $i$ with respect to agent $j$. The energy variable representing the state of the virtual spring between agent $i$ and leader $k$ is given by $x_{a_i l_k} \in \mathbb{R}^3$, namely by the relative position of agent $i$ with respect to leader $k$. The potential energy function representing the amount of energy stored in a given configuration for the elements associated to the interagent potentials and to the leader potentials are $V_I(\cdot)$ and $V_h(\cdot)$ respectively. The elements representing the interagent potential between agent $i$ and agent $j$ and the potential between agent $i$ and leader $k$ are described by

$$\begin{cases} \dot{x}_{a_{ij}} = f_{a_{ij}} \\ e_{a_{ij}} = \dfrac{\partial V_I(x_{a_{ij}})}{\partial x_{a_{ij}}} \end{cases} \qquad \begin{cases} \dot{x}_{a_i l_k} = f_{a_i l_k} \\ e_{a_i l_k} = \dfrac{\partial V_h(x_{a_i l_k})}{\partial x_{a_i l_k}} \end{cases} \tag{4.11}$$

where $(e_{a_{ij}}, f_{a_{ij}})$ and $(e_{a_i l_k}, f_{a_i l_k})$ are the the power ports associated to these elements. The efforts $e_{a_{ij}}$ and $e_{a_i l_k}$ represent the forces that have to be applied to agent $i$ because of the interagent potential regulating the distance from agent $j$ and of the leader potential regulating the distance from leader $k$ respectively. It follows directly from Eq.(4.11) that $e^T_{a_{ij}} f_{a_{ij}} = \dot{V}_I$ and that $e^T_{a_i l_k} f_{a_i l_k} = \dot{V}_h$, namely that the power incoming through the power port is stored in terms of potential energy.

Thus, all the main actors in the VBAP framework can be modeled as energy storing elements. The dynamics of the state variables and, consequently, the evolution of the agents, depend on the way in which the energy processing elements are interconnected, namely, by the way in which they exchange energy. The VBAP control strategy induces an interconnection structure along which all the elements that we have just defined exchange energy through their power ports. The interconnection structure represents a relationship between the power variables and it expresses the way in which the output variables have to be combined to form the input variables. In other words, it represents a communication protocol that defines how input/output information is exchanged among the energy processing elements. In order to make the notation compact, let us define

$$f_a = \begin{pmatrix} f_{a_1} \\ \vdots \\ f_{a_N} \end{pmatrix} \quad f_{aa} = \begin{pmatrix} f_{a_{12}} \\ \vdots \\ f_{a_{N-1N}} \end{pmatrix} \quad f_{al} = \begin{pmatrix} f_{a_1 l_1} \\ \vdots \\ f_{a_N l_L} \end{pmatrix}$$
$$e_a = \begin{pmatrix} e_{a_1} \\ \vdots \\ e_{a_N} \end{pmatrix} \quad e_{aa} = \begin{pmatrix} e_{a_{12}} \\ \vdots \\ e_{a_{N-1N}} \end{pmatrix} \quad e_{al} = \begin{pmatrix} e_{a_1 l_1} \\ \vdots \\ e_{a_N l_L} \end{pmatrix} \tag{4.12}$$

Let us consider the interconnection between the agents and the elements representing interagent potentials. Given $N$ agents, there are $N(N-1)/2$ interagent potentials elements, one per each pair of agents. The effort generated by each interagent potential element is applied to the pair of agents that it interconnects with opposite signs, as it happens for two masses connected by a spring. The effort acting on each agent is given by the sum of all the efforts generated by the interagent potential elements that interconnect it with the other agents. A possible interconnection between the agents and the interagent potential elements leads to the following efforts:

$$\begin{aligned} e_{a_1} &= e_{a_{12}} + \cdots + e_{a_{1N}} \\ e_{a_2} &= -e_{a_{12}} + e_{a_{23}} + \cdots + e_{a_{2N}} \\ &\cdots \\ e_{a_N} &= -e_{a_{1N}} - e_{a_{2N}} + \cdots - e_{a_{N-1N}} \end{aligned} \tag{4.13}$$

The flow to be given as an input to each interagent potential element is given by the difference between the flows of the agents that the elements interconnect. The sign convention chosen for composing the efforts produced by the interagent potential elements to form the efforts to apply to the agents and the power continuity of the interconnection between the agents and the interagent potential elements, induces the sign convention through which composing the flows of the agents in order to form those of the interagent potentials.

In summary, the interconnection structure that joins the agents and the interagent potential elements is given by

$$\begin{pmatrix} e_a \\ f_{aa} \end{pmatrix} = \begin{pmatrix} O_1 & J_1 \\ -J_1^T & O_2 \end{pmatrix} \begin{pmatrix} f_a \\ e_{aa} \end{pmatrix} \tag{4.14}$$

where $O_1$ and $O_2$ represent square null matrices of order $3N$ and $3N(N-1)/2$ respectively and $J_1$ is a $3N \times 3N(N-1)/2$ matrix given by

$$J_1 = \left( \begin{array}{ccccc|ccccc|c|c} \multicolumn{5}{c|}{\text{N-1 columns}} & \multicolumn{5}{c|}{\text{N-2 columns}} & & \\ \mathbf{1} & \mathbf{1} & \mathbf{1} & \dots & \mathbf{1} & \mathbf{0} & \mathbf{0} & \mathbf{0} & \dots & \mathbf{0} & & \mathbf{0} \\ -\mathbf{1} & \mathbf{0} & \mathbf{0} & \dots & \mathbf{0} & \mathbf{1} & \mathbf{1} & \mathbf{1} & \dots & \mathbf{1} & & \mathbf{0} \\ \mathbf{0} & -\mathbf{1} & \mathbf{0} & \dots & \mathbf{0} & -\mathbf{1} & \mathbf{0} & \mathbf{0} & \dots & \mathbf{0} & \dots & \vdots \\ \vdots & \vdots & \vdots & \vdots & \vdots & \vdots & \vdots & \vdots & \vdots & \vdots & & \mathbf{1} \\ \mathbf{0} & \mathbf{0} & \mathbf{0} & \dots & -\mathbf{1} & \mathbf{0} & \mathbf{0} & \mathbf{0} & \dots & -\mathbf{1} & & -\mathbf{1} \end{array} \right) \tag{4.15}$$

where $\mathbf{1}$ and $\mathbf{0}$ represent the $3 \times 3$ identity and null matrix respectively.

Let us now consider the energetic interconnection between the agents and the leaders. Each leader influences all the agents and, therefore, for each agent there are $L$ energy storing elements yielding the action that each leader exerts on the agent. Thus, in a system with $N$ agents and $L$ leaders, there are $NL$ potential energy storing elements acting of the agents.

The effort acting on each agent because of the interaction with the virtual leaders is given by

$$e_{a_i} = -\sum_{k=1}^{L} \frac{\partial V_h(x_{a_i l_k})}{\partial x_{a_i l_k}} = -\sum_{k=1}^{L} e_{ail_k} \tag{4.16}$$

The flow of the element associated to the action of a leader on the agent $i$ is given by the difference between the velocity of agent $i$ and the velocity of the leader. Since we are considering the case in which the leaders are not moving, the latter flow is zero. The sign convention chosen for the efforts and the power continuity of the interconnection between the agents and the potential elements induces the sign convention for the flows. Thus, the interconnection structure that joins the agents and the leader potential elements is given by

$$\begin{pmatrix} e_a \\ f_{al} \end{pmatrix} = \begin{pmatrix} O_1 & J_2 \\ -J_2^T & O_3 \end{pmatrix} \begin{pmatrix} f_a \\ e_{al} \end{pmatrix} \tag{4.17}$$

where $O_3$ represents a square null matrix of order $3NL$ and $J_2$ is a $3N \times 3NL$ matrix given by

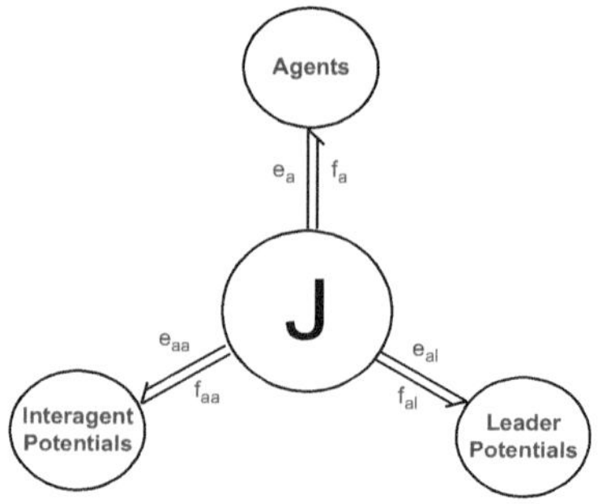


**Fig. 4.1** The controlled system from a port-Hamiltonian perspective

$$J_2 = \left( \begin{array}{cccc|cccc|c|cccc} \multicolumn{4}{c}{\text{N columns}} & \multicolumn{4}{c}{\text{N columns}} & & \multicolumn{4}{c}{\text{N columns}} \\ -\mathbf{1} & -\mathbf{1} & \dots & \mathbf{1} & \mathbf{0} & \mathbf{0} & \dots & \mathbf{0} & & \mathbf{0} & \mathbf{0} & \dots & \mathbf{0} \\ \mathbf{0} & \mathbf{0} & \dots & \mathbf{0} & -\mathbf{1} & -\mathbf{1} & \dots & -\mathbf{1} & & \mathbf{0} & \mathbf{0} & \dots & \mathbf{0} \\ \vdots & \vdots & \vdots & \vdots & \vdots & \vdots & \vdots & \vdots & \cdots & \vdots & \vdots & \vdots & \vdots \\ \mathbf{0} & \mathbf{0} & \dots & \mathbf{0} & \mathbf{0} & \mathbf{0} & \dots & \mathbf{0} & & -\mathbf{1} & -\mathbf{1} & \dots & -\mathbf{1} \end{array} \right) \tag{4.18}$$

where **1** and **0** represent the $3 \times 3$ identity and null matrix respectively.

We are now ready to define the overall interconnection structure.

Using Eq.(4.15) and Eq.(4.18) we have that the interconnection structure along which all the energy processing elements exchange energy is given by

$$\begin{pmatrix} e_a \\ f_{aa} \\ e_{al} \end{pmatrix} = \underbrace{\begin{pmatrix} O_1 & J1 & J2 \\ -J_1^T & O_2 & O_4 \\ -J_2^T & O_4^T & O_3 \end{pmatrix}}_{J} \begin{pmatrix} f_a \\ e_{aa} \\ f_{al} \end{pmatrix} \tag{4.19}$$

where $O_4$ denotes the $3N(N-1)/2 \times NL$ null matrix. Agents and elements associated to the artificial potentials exchange efforts and flows, namely energy, along an interconnection structure that is represented by $J$; the overall controlled system is represented in Fig.4.1 in a bond-graph notation, where the half-arrows indicate a power flow.

Thanks to the skew-symmetry of $J$ it can be immediately seen that the following balance holds:

$$e_a^T f_a + e_{aa}^T f_{aa} + e_{al}^T f_{al} = 0 \tag{4.20}$$

which means that energy is neither produced nor dissipated but simply exchanged along the interconnection. This means that $J$ describes a power preserving interconnection and that it represents a Dirac structure. Using Eq.(4.9) and Eq.(4.11) with Eq.(4.19) we can model a group of agents controlled through the VBAP strategy as a port-Hamiltonian system. In fact we have that

$$\begin{pmatrix} \dot{p}_a \\ \dot{x}_{aa} \\ \dot{x}_{al} \end{pmatrix} = \left[ \begin{pmatrix} O_1 & J1 & J2 \\ -J_1^T & O_2 & O_4 \\ -J_2^T & O_4^T & O_3 \end{pmatrix} - \begin{pmatrix} R & O_5 & O_6 \\ O_5^T & O_2 & O_4 \\ O_6^T & O_4^T & O_3 \end{pmatrix} \right] \begin{pmatrix} \frac{\partial H}{\partial p_a} \\ \frac{\partial H}{\partial x_{aa}} \\ \frac{\partial H}{\partial x_{al}} \end{pmatrix} \tag{4.21}$$

where $O_5$ and $O_6$ are null matrices of proper dimensions and

$$p_a = \begin{pmatrix} p_{a_1} \\ \vdots \\ p_{a_N} \end{pmatrix} \quad x_{aa} = \begin{pmatrix} x_{a_{12}} \\ \vdots \\ x_{a_{N-1N}} \end{pmatrix} \quad x_{al} = \begin{pmatrix} x_{a_1 l_1} \\ \vdots \\ x_{a_N l_L} \end{pmatrix} \tag{4.22}$$

The Hamiltonian function is given by

$$\begin{aligned} H(p_a, x_{aa}, x_{al}) &= H_a(p_a) + H_{aa}(x_{aa}) + H_{la}(x_{la}) = \\ &= \sum_{i=1}^{N} H_{a_i}(p_{a_i}) + \sum_{i=1}^{N} \sum_{j \neq i} V_I(x_{a_{ij}}) + \sum_{i=1}^{N} \sum_{k=1}^{L} V_h(x_{a_i l_k}) \end{aligned} \tag{4.23}$$

and $R$ is the positive definite block diagonal matrices whose diagonal elements are the damping matrices $R_i$ of the agents. The Hamiltonian function represents the total energy of the controlled group of agents, which is the same candidate function used in [6] and, therefore, it has minimum corresponding to the desired configuration of the vehicles. Using the power balance reported in Eq.(4.5) with Eq.(4.21) we obtain that

$$\dot{H} = -\frac{\partial^T H}{\partial p_a} R \frac{\partial H}{\partial p_a} \leq 0 \tag{4.24}$$

which, using LaSalle's invariance principle, proves that the minimum configuration of $H$ is asymptotically stable. This is an alternative way to prove the stability results obtained in [6] using the port-Hamiltonian framework. Nevertheless, the advantage of the port-Hamiltonian representation of the VBAP formation control strategy is the fact that the interconnection structure along which the energy processing elements exchange information is evident and this will be the starting point for extending the VBAP approach over delayed networks.

## 4.4 The Effect of Communication Delay

The interconnection structure reported in Eq.(4.19) tells how to build the inputs of each element by combining the outputs of the others, namely it provides a communication protocol between the energy processing elements. The aim of this section is to develop a communication strategy that allows to obtain the desired formation independently of any delay in the communication. In order to keep the notation simple, in the following we will assume that the delays in the communications among the energy processing elements are all equal and we will indicate the delay with $T$. All the results developed in this section can be easily extended to more general cases. Thus, the exchange of information taking place between the energy processing elements will be described by

$$\begin{pmatrix} e_a(t) \\ f_{aa}(t) \\ e_{al}(t) \end{pmatrix} = \begin{pmatrix} O_1 & J1 & J2 \\ -J_1^T & O_2 & O_4 \\ -J_2^T & O_4^T & O_3 \end{pmatrix} \begin{pmatrix} f_a(t-T) \\ e_{aa}(t-T) \\ f_{al}(t-T) \end{pmatrix} \tag{4.25}$$

The way in which the energy processing elements are interconnected and, in particular, the fact that the interconnection structure is power preserving plays a fundamental role in the formation stabilization process. Because of the communication delay, the balance reported in Eq.(4.20) doesn't hold anymore and thus, the interconnection structure is no more power preserving. This leads to a production of extra energy associated to the delayed exchange of information which invalidates the inequality reported in Eq.(4.24) and which prevents from stabilizing the agents in the desired formation.

A similar problem has been widely studied by the telerobotics community. Loosely speaking, a bilateral telemanipulation system consists of a pair of controlled robots (master and slave) exchanging force and velocity information over a communication channel. Both master and slave sides can be modeled as two port-Hamiltonian systems exchanging power through their power ports. It can be shown that when there is a delay in the communication, the exchange of force and velocity information leads to a non passive and unstable behavior of the telemanipulation system [8]. Modeling the exchange of power through the scattering framework and, therefore, letting the master and the slave sides exchange scattering variables rather than force and velocity variables, leads to a communication channel that is characterized by a lossless dynamics independently of any constant communication delay. This leads to a telemanipulation system characterized by a stable behavior. In other words, the use of scattering variables instead of power variables makes the exchange of energy robust with respect to the communication delay. For further details see [12].

In case of formation control, we have several elements exchanging energy along a power preserving interconnection through their power ports. The main idea for achieving the desired formation control over a delayed network is to represent the power port of each energy processing element in terms of scattering variables. We will find the communication strategy that is equivalent to Eq.(4.19) in terms of scattering variables and we will see that no regenerative effects due to the communication delay take place and that the agents are stabilized in the desired formation independently of the communication delay.

Given a power port $(e(t), f(t))$, the following relations can be easily obtained by Eq.(4.7)

$$\begin{cases} e(t) = \sqrt{\frac{b}{2}}(s^+(t) + s^-(t)) \\ f(t) = \frac{1}{\sqrt{2b}}(s^+(t) - s^-(t)) \end{cases} \tag{4.26}$$

Using Eq.(4.26) with Eq.(4.19) we have that scattering based expression of the interconnection structure joining the energy processing elements is given by

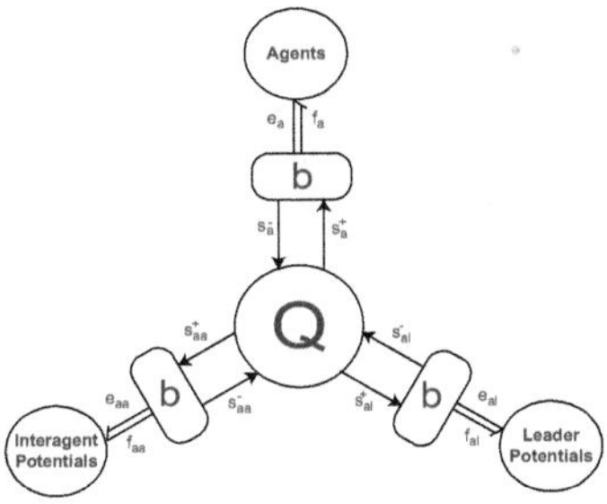


**Fig. 4.2** The overall controlled system

$$\begin{pmatrix} \sqrt{\frac{b}{2}}(s_a^+(t)+s_a^-(t)) \\ \frac{1}{\sqrt{2b}}(s_{aa}^+(t)-s_{aa}^-(t)) \\ \frac{1}{\sqrt{2b}}(s_{al}^+(t)-s_{al}^-(t)) \end{pmatrix} = \begin{pmatrix} O_1 & J1 & J2 \\ -J_1^T & O_2 & O_4 \\ -J_2^T & O_4^T & O_3 \end{pmatrix} \begin{pmatrix} \frac{1}{\sqrt{2b}}(s_a^+(t)-s_a^-(t)) \\ \sqrt{\frac{b}{2}}(s_{aa}^+(t)+s_{aa}^-(t)) \\ \sqrt{\frac{b}{2}}(s_{al}^+(t)+s_{al}^-(t)) \end{pmatrix} \tag{4.27}$$

where the indexes of the scattering variables are the same as those of the power variables they refer to. After some simple computations, we have that Eq.(4.27) can be rewritten as

$$\begin{pmatrix} s_a^+(t) \\ s_{aa}^+(t) \\ s_{al}^+(t) \end{pmatrix} = \underbrace{(I_t - J)^{-1}(I_t + J)\mathscr{I}}_{Q} \begin{pmatrix} s_a^-(t) \\ s_{aa}^-(t) \\ s_{al}^-(t) \end{pmatrix} \tag{4.28}$$

where $I_t$ is the identity matrix of the same size of $J$ and

$$\mathscr{I} = \begin{pmatrix} -I & O & O \\ O & I & O \\ O & O & I \end{pmatrix} \tag{4.29}$$

where $O$ and $I$ are the null and identity matrices of proper dimensions.

It can be easily proven that $Q$ is orthogonal. In fact we have that the matrix $(I_t - J)^{-1}(I_t + J)$ is orthogonal because it is the Cayley transform of the skew symmetric matrix $J$ [14]. Furthermore, it can be immediately seen that $\mathscr{I}^T = \mathscr{I}^{-1}$ and that therefore $\mathscr{I}$ is orthogonal. Since the product of orthogonal matrices is orthogonal, it follows that $Q$ is orthogonal. The overall controlled system is reported in Fig.4.2. Each energy storing element receives a scattering variable from the communication network and it uses it together with its output power variable to compute both the incoming power variable and the outgoing scattering to transmit over the network.

Consider now that a delay is present in the communication between the elements joined through the interconnection. We have that Eq.(4.28) becomes

$$\begin{pmatrix} s_a^+(t) \\ s_{aa}^+(t) \\ s_{al}^+(t) \end{pmatrix} = Q \begin{pmatrix} s_a^-(t-T) \\ s_{aa}^-(t-T) \\ s_{al}^-(t-T) \end{pmatrix} \tag{4.30}$$

The power flowing through the interconnection structure is given by

$$P(t)=\frac{1}{2}\|s_a^-(t)\|^2+\frac{1}{2}\|s_{aa}^-(t)\|^2+\frac{1}{2}\|s_{al}^-(t)\|^2-\frac{1}{2}\|s_a^+(t)\|^2-\frac{1}{2}\|s_{aa}^+(t)\|^2-$$
$$-\frac{1}{2}\|s_{al}^+(t)\|^2=\frac{1}{2}\left(s_a^-(t)\; s_{aa}^-(t)\; s_{al}^-(t)\right)\begin{pmatrix}s_a^-(t)\\ s_{aa}^-(t)\\ s_{al}^-(t)\end{pmatrix}-\frac{1}{2}\left(s_a^+(t)\; s_{aa}^+(t)\; s_{al}^+(t)\right)\begin{pmatrix}s_a^+(t)\\ s_{aa}^+(t)\\ s_{al}^+(t)\end{pmatrix} \tag{4.31}$$

using Eq.(4.30) with Eq.(4.31), we have that

$$P(t)=\frac{1}{2}\left(s_a^-(t)\; s_{aa}^-(t)\; s_{al}^-(t)\right)\begin{pmatrix}s_a^-(t)\\ s_{aa}^-(t)\\ s_{al}^-(t)\end{pmatrix}-\frac{1}{2}\left(s_a^-(t-T)\; s_{aa}^-(t-T)\; s_{al}^-(t-T)\right)Q^TQ\begin{pmatrix}s_a^-(t-T)\\ s_{aa}^-(t-T)\\ s_{al}^-(t-T)\end{pmatrix} \tag{4.32}$$

Since $Q$ is orthogonal, we have that

$$\begin{aligned}P(t)&=\frac{1}{2}\|s_a^-(t)\|^2+\frac{1}{2}\|s_{aa}^-(t)\|^2+\frac{1}{2}\|s_{al}^-(t)\|^2-\\&-\frac{1}{2}\|s_a^-(t-T)\|^2-\frac{1}{2}\|s_{aa}^-(t-T)\|^2-\frac{1}{2}\|s_{al}^-(t-T)\|^2=\\&=\frac{d}{dt}\int_{t-T}^{t}\left(\frac{1}{2}\|s_a^-(\tau)\|^2+\frac{1}{2}\|s_{aa}^-(\tau)\|^2+\frac{1}{2}\|s_{al}^-(\tau)\|^2\right)d\tau=\dot{H}_{ch}(t)\end{aligned} \tag{4.33}$$

All the power injected into the interconnection structure is simply stored and, therefore, using the scattering variable, the delayed network used by the agents for exchanging information is characterized by a lossless behavior, exactly as it happens to the communication channel in bilateral teleoperation. Thus, we have that the power exchanged by the actors of the VBAP approach is stored in the communication channel until it is not delivered and that the delay doesn't introduce any regenerative effects in the interconnection.

We are now ready to prove that the agents are stabilized in the desired formation. Consider as a candidate Lyapunov function the total energy of the controlled system namely

$$\mathscr{H}=H(p_a,x_{aa},x_{la})+H_{ch}(t)=H_a(p_a)+H_{aa}(x_{aa})+H_{la}(x_{la})+H_{ch}(t) \tag{4.34}$$

From the definition of $H_{ch}$, we can see that minimum configuration of $\mathscr{H}$ is the same as that the total energy of the non delayed system defined in Eq.(4.23). Thus, if we prove that the minimum of $\mathscr{H}$ is asymptotically stable, we ensure that agents are controlled in the desired formation independently of the communication delay.

Using Eq.(4.9) and Eq.(4.11) we have that

$$\begin{aligned}\dot{\mathscr{H}}(t)&=\dot{H}_a(t)+\dot{H}_{aa}(t)+\dot{H}_{la}(t)+\dot{H}_{ch}(t)=\\&=e_a^T(t)f_a(t)+e_{aa}^T(t)f_{aa}(t)+e_{al}^T(t)f_{al}^T(t)-\frac{\partial^T H_a}{\partial p_a}R\frac{\partial H_a}{\partial p_a}+\dot{H}_{ch}(t)\end{aligned} \tag{4.35}$$

Using Eq.(4.6) with Eq.(4.33) we obtain that

$$\dot{\mathscr{H}}(t) = -\frac{\partial^T H_a}{\partial p_a} R \frac{\partial H_a}{\partial p_a} \tag{4.36}$$

Thus, recalling the $R$ is a positive definite matrix and using the LaSalle's invariance principle, we have that the minimum configuration of $\mathscr{H}$ are (locally) asymptotically stable. Thus, we have proven the following result

**Proposition 4.1.** *Using the communication strategy proposed in Eq.(4.30) the agents are stabilized in the desired formation independently of any constant transmission delay.*

*Remark 4.1.* Similarly to what happens in bilateral teleoperation, the knowledge of the communication delay is not required for guaranteeing the stabilization of the formation.

The transmission of scattering variables guarantees the stability but it causes a bad transient behavior because of the wave reflection phenomenon. This problem is very well known in telerobotics and the techniques developed in the literature can be used in our case. A simple solution, proposed in [13], is to add a feed-through damping action, with damping coefficient equal to the scattering impedance $b$, to the potential energy storing elements. Thus, Eq.(4.11) becomes

$$\begin{cases} \dot{x}_{a_{ij}} = f_{a_{ij}} \\ e_{a_{ij}} = \frac{\partial V_I(x_{a_{ij}})}{\partial x_{a_{ij}}} + b f_{a_{ij}} \end{cases} \qquad \begin{cases} \dot{x}_{a_i l_k} = f_{a_i l_k} \\ e_{a_i l_k} = \frac{\partial V_h(x_{a_i l_k})}{\partial x_{a_i l_k}} + b f_{a_i l_k} \end{cases} \tag{4.37}$$

The role of this damping effect is to absorb the part of the scattering variable that would be otherwise reflected back into the interconnection and that would give rise to an oscillatory transient behavior. The minimum configuration of the total energy remains the same and, therefore, the agents keep on being stabilized in the desired formation.

## 4.5 Simulations

The aim of this section is to provide some simulations in order to validate the results obtained in the paper. We consider a simple example, treated also in [4], where there are two agents moving on a $xy$ plane. We want to take the agents at a relative distance of 1 m and that the line connecting the two agents corresponds to the $y$ axis of the plane. As reported in [4] this problem can be solved using two leaders disposed as in Fig.4.3, where the black and the grey circles indicate the leaders and the agents respectively. The interagent potential and the leader potential are characterized by the parameters $d_0 = h_0 = 1$ and $d_1 = h_1 = 2$ and the shape of the potentials is the same as in [4]. In order to obtain the desired formation, the leaders are set on the $x$-axis at the configuration $\pm\sqrt{3}/2$ [4]. The initial positions of the agents are $(0.1, 0.1)$ and $(-0.1, -0.1)$. The VBAP control strategy has been implemented within the framework reported in Sec. 4.3.

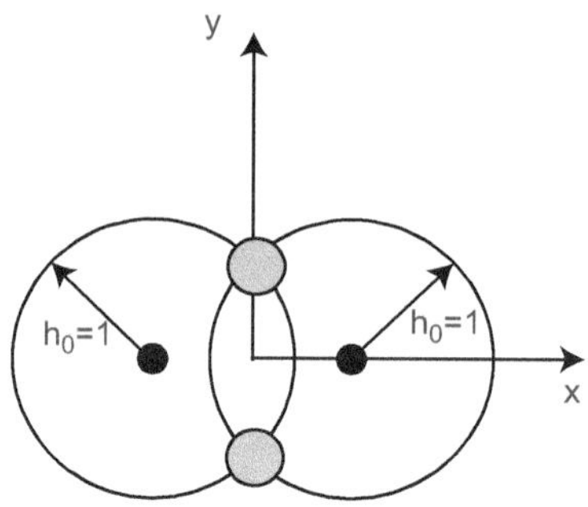


**Fig. 4.3** Agents and leaders on the plane

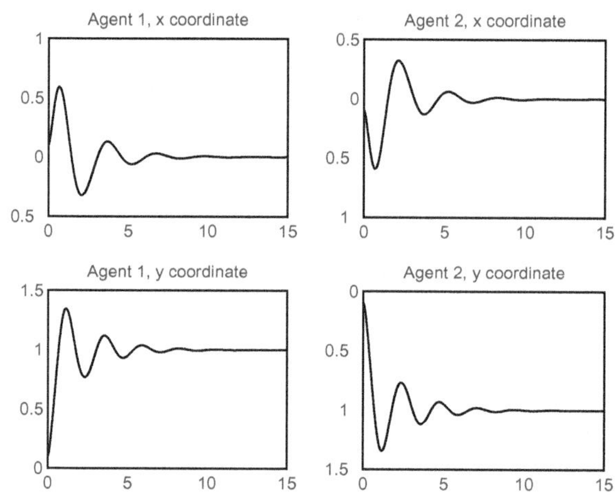


**Fig. 4.4** The positions of the agents when using the VBAP strategy

In Fig.4.4 the position of the agents when using the VBAP control strategy are reported. We can see that the agents are taken in the desired formation, namely in the positions $(0,1)$ and $(0,-1)$.

In the next simulation we have added a delay of 1 s. in the exchange of information between the agents and the artificial potential elements. The results are reported in Fig.4.5. We can see that the delay induces an unstable behavior. In fact, we have that the forces applied to the agents grow very quickly and this implies that the agents are taken at a relative distance and at a distance from the leaders greater than 2. No force but the damping acts on the agents at this point and, consequently, the agents stop when all the energy they have accumulated is dissipated by the damping. Nevertheless, at steady state, the agents are NOT in the desired formation because of the instability induced by the communication delay.

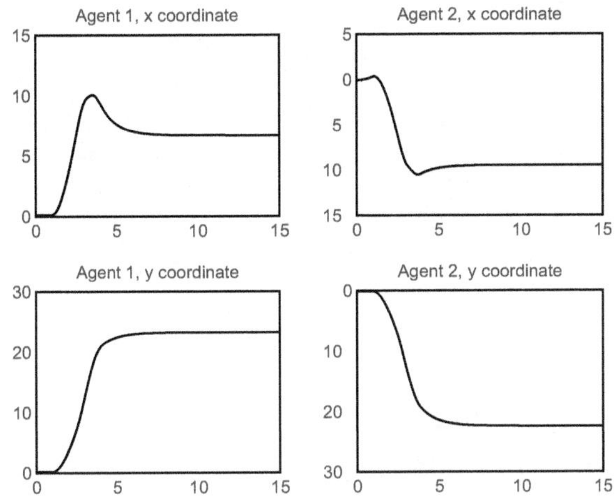


**Fig. 4.5** The positions of the agents when using the VBAP strategy with a communication delay of 1s.

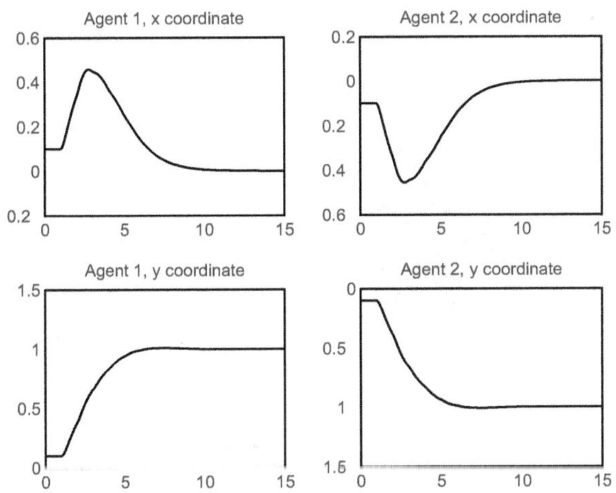


**Fig. 4.6** The positions of the agents when using the VBAP strategy and scattering variables

In the last simulation, we have implemented the communication between the agents and the artificial potential elements using scattering variables (with scattering impedance equal to 1) as described in Sec. 4.4 and we have considered a communication delay of 1 second. The wave reflection problem has been solved by adding a feed-through dissipative action to the artificial potential elements. The results are reported in Fig.4.6. We can see that, despite of the communication delay, the agents are taken to the desired formation.

## 4.6 Conclusions and Future Work

In this paper we have considered the VBAP approach for the stabilization of a formation of a group of point mass agents. We have shown that the controlled group of agents can be modeled as a set of elements exchanging energy over a power preserving interconnection. We have proven that, using the scattering framework, it is possible to keep on stabilizing the agents in the desired formation independently of any communication delay in the exchange of information.

## References

1. Couzin, I.: Collective minds. Nature, 715, 445
2. Dunbar, W., Murray, R.: Distributed receding horizon control for multi-vehicle formation stabilization. Automatica 42(4) (2006)
3. Fax, J., Murray, R.: Information flow and cooperative control of vehicle formations. IEEE Transactions on Automatic Control 49(9) (September 2004)
4. Fiorelli, E.: Cooperative vehicle control, feature tracking and ocean sampling, Ph.D. dissertation, Princeton University (2005)
5. Karnopp, D., Margolis, D., Rosenberg, R.: System Dynamics: A Unified Approach, 2nd edn. John Wiley & Sons Inc., Chichester (1990)
6. Leonard, N., Fiorelli, E.: Virtual leaders, artificial potentials and coordinated control of groups. In: Proceedings of IEEE Conference on Decision and Control, Orlando, Florida (December 2001)
7. Murray, R.: Recent research in cooperative control of multi-vehicle systems. ASME Journal of Dynamic Systems, Measurement and Control (2006)
8. Niemeyer, G., Slotine, J.: Stable adaptive teleoperation. IEEE Journal of Oceanic Engineering 16(1), 152–162 (1991)
9. Ogren, P., Fiorelli, E., Leonard, N.: Cooperative control of mobile sensor networks: adaptive gradient climbing in a distributed environment. IEEE Transactions on Automatic Control 49(8) (2004)
10. Ren, W., Beard, R.: Formation feedback control for multiple spacecraft via virtual structures. IEE Proceedings on Control Theory and Applications 151(3) (May 2004)
11. Ren, W., Beard, R., Atkins, E.: Information consensus in multivehicle cooperative control. IEEE Control Systems Magazine 27(2) (2007)
12. Secchi, C., Stramigioli, S., Fantuzzi, C.: Control of Interactive Robotic Interfaces: a port-Hamiltonian Approach. Springer Tracts in Advanced Robotics. Springer, Heidelberg (2007)
13. Stramigioli, S., van der Schaft, A., Maschke, B., Melchiorri, C.: Geometric scattering in robotic telemanipulation. IEEE Transactions on Robotics and Automation 18(4) (2002)
14. van der Schaft, A.: $L_2$-Gain and Passivity Techniques in Nonlinear Control. Communication and Control Engineering. Springer, Heidelberg (2000)

# Part II
# Social Phenomena

# Chapter 5
# Remarks on Epidemic Spreading in Scale-Free Networks

Carlo Piccardi and Renato Casagrandi

**Abstract.** Classical epidemiological theory has been developed by mainly focusing on the temporal dynamics of the different population compartments, basing on the assumption that individuals are homogeneously mixing. Important results obtained with basic SIS models on scale-free networks have instead given emphasis to the role of heterogeneity in the spreading process. Here, by means of two instrumental examples, we make the point that the interplay between the infection process and the network topology is more important than credited. The first example considers non-vital infectious processes characterized by nonlinear force of infection which causes backward bifurcations. The second example deals with standard infective mechanisms in processes where vital dynamics (births and deaths) cannot be disregarded. In both cases we find evidence that it is too simplistic to claim that scale-free networks are the most efficient media for the spreading of whatever infection.

## 5.1 Introduction

After the works of Pastor-Satorras and Vespignani (PV) [32, 33, 34] concerning the spread of diseases in Scale-Free Networks (SFNs), it is notorious that highly heterogeneous patterns of interaction among individuals in a host population can have interesting epidemiological consequences. The most important of such consequences is the disappearence of the epidemic threshold characterizing the great majority of models of pathogen transmission. Classical models [22, 2, 1] are in fact based on the assumption that each individual has the same number of contacts with others in the population and that the resulting homogeneous network of interactions is infinite in size. PV have shown that the non-zero epidemic threshold of the non-vital SIS model (i.e., a pure infective process) does no longer exist when implemented over

Carlo Piccardi and Renato Casagrandi
Dipartimento di Elettronica e Informazione,
Politecnico di Milano, Piazza Leonardo da Vinci 32, I-20133 Milano, Italy
e-mail: carlo.piccardi@polimi.it, casagran@elet.polimi.it

SFNs whose size tends to infinity. Therefore, the strong heterogeneous nature of SFNs crucially enhances the spreading ability of the epidemic agent in comparison to what happens in Homogeneous Networks (HNs).

The result is so sharp to claim for very general statements like [7]:

> SFNs [are] the ideal media for the propagation of infections, bugs, or unsolicited information. [...] SFNs have the peculiar property of being prone to the spreading of infections.

The reader is therefore left with the impression that [28]

> SFNs are completely prone to epidemic spreading allowing the onset of large epidemics whatever the spreading rate of infection.

More recent studies have shown that, when more realistic aspects of the infection process are accounted for, an epidemic threshold reemerges in SFNs. Besides the finite size of the network [20], such features include, e.g., connectivity-dependent transmission rates [31] or degree-dependent deactivation of links [12]. Here we focus on the influence of two significant characteristics of the epidemic process that may cause either a reduced efficiency of disease spread in SFNs or the reappearance of the epidemic threshold, or both. To this end we discuss two different classes of epidemiological processes that are general enough to cover a wide spectrum of infectious diseases, yet complex enough to avoid the somehow simplistic assumptions underlying the basic SIS model.

The first instrumental example, introduced in [35], concerns mean-field epidemiological models that exhibit saddle-node bifurcations, i.e. infections that can survive and establish at high endemic levels in populations that they will be unable to invade from zero [11]. Saddle-node bifurcations emerge in models of important infectious diseases such as hepatitis B [25], tuberculosis [15] or HIV [19]. The prototypical mechanisms that lead to saddle-node bifurcations are non-constant transmission rates [10], non-constant recovery rates [23], or both of the above [26]. Departures from the basic SIS model are relevant not only to the transmission of human diseases but also when the "infectious material" [17] being exchanged is of intellectual [5], commercial [14, 24], or social [29] nature.

The second instrumental example considers a SIS-like infective process [2] occurring in a population that evolves through time because of demographic reasons. In fact, the epidemiology of many infectious diseases – from respiratory syncytial virus [39] to pertussis [18], and from influenza [8] to gonorrhea [16] – cannot be effectively studied over timescales that disregard the birth-death process of hosts. As we discuss below, demography has the power of profoundly altering the structure of SFNs frequently used to model interactions among individuals.

## 5.2 Spread of Diseases with Nonlinear Transmission Rates

Consider the host population as a network composed of $N$ individuals (nodes). The $i$-th node has degree $k_i$, i.e., it is connected by $0 < k_{\min} \leq k_i \leq k_{\max}$ edges to other nodes. At any time instant $t$, every node $i$ is either susceptible (infectable) or infected (thus infective). During the time interval $\Delta$, an infected node can return susceptible with probability $\gamma\Delta$, while a susceptible node can become infected with probability $\beta G_i \Delta$. The quantity $\beta$ is the disease transmission rate, whereas the function $G_i$ (detailed below) accounts for the number of infected neighbors of node $i$. The state of all nodes is updated synchronously with time step $\Delta$. There are two main generalizations of the basic SIS model in the family of contact processes studied in this section. First, the force of infection [21] is nonlinear, as the transmission rate takes the form

$$\beta = \beta(t) = \beta_0 + \beta_1 I(t), \qquad \beta_0 > 0, \beta_1 \geq 0, \tag{5.1}$$

where $0 \leq I(t) \leq 1$ represents the current disease prevalence, i.e., the fraction of infected individuals in the network at time $t$. By varying the values attributed to the parameters characterizing the state-independent ($\beta_0$) and the state-dependent ($\beta_1 I$) component of the transmission rate $\beta$, a wide spectrum of epidemic processes can be described, from the basic SIS model ($\beta_1 = 0$) to some marketing models [14] for product diffusion ($\beta_0 = 0$). Second, the function $G_i$ is assumed to increase, yet saturate, with the number $n_i$ of infectives among the neighbors of node $i$, i.e., $G_i = \min\{n_i, M\}$, where $M \in \{1, 2, \ldots, k_{\max}\}$ is the saturation level. A value $M = 1$ is adopted in all the network studies implementing the basic SIS model we are aware of. Practically, setting $M$ to unity is equivalent to assuming that the chance for a susceptible node of being infected does not further increases if more than one of its neighbors are carrying the virus. This is quite a typical situation in the word-of-mouth spread of information and rumors or in highly pathogenic diseases.

The spread of every epidemic process depends very much on the network structure. Consistently with [32, 33], we compare the spread of the above process in random HNs and SFNs. We constructed our HN by linking pairs of randomly selected nodes, until all of them have exactly the same number $\bar{k}$ of connections, thus the resulting degree distribution $p_k$ is spiky ($p_{\bar{k}} = 1$, and $p_k = 0$ for all $k \neq \bar{k}$). Other random networks that have been used in the literature can fairly be considered as homogeneous in their connectivity properties. They include the Erdős–Rényi network [13] and the Watts-Strogatz "small-world" network [38], in particular with rewiring probability equal to one [32, 33]. We anticipate that all the results presented in this section are qualitatively identical for all the above mentioned types of random HNs and, moreover, largely independent of the numerical value assigned to the average degree $\bar{k}$. Letting $\Delta \to 0$, the dynamics of the disease prevalence $I$ in HNs can be described by the following mean-field model:

$$\dot{I}(t) = -\gamma I(t) + (\beta_0 + \beta_1 I(t))\,(1 - I(t))\, g_{\bar{k},M}(I(t)), \tag{5.2}$$

where $0 \leq 1 - I(t) = S(t) \leq 1$ is the fraction of susceptible individuals in the network at time $t$, and

$$g_{k,M}(I) = \sum_{n=0}^{k} \min\{n, M\} \binom{k}{n} I^n (1 - I)^{k-n} \tag{5.3}$$

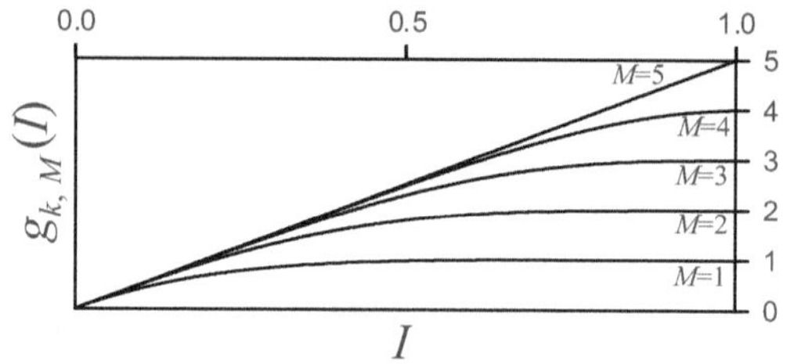


**Fig. 5.1** Examples of the functions $g_{k,M}(I)$ for $k = 5$.

(see Fig. 5.1) is the expected number of infectives (saturated to $M$) among the neighbors of any node, because $\binom{k}{n} I^n (1-I)^{k-n} = P(n)$ is the probability that a degree $k$ node has exactly $n$ infected neighbors. Note that $g_{k,M}(0) = 0$ and $g'_{k,M}(0) = k$, i.e., the functions $g_{k,M}(I)$ are equivalent up to the first derivative for all values of $M$ if $I \to 0$. In absence of saturation ($M = k$), $g_{k,k}(I) = kI$ is the standard contagion term of the basic SIS model. Here we focus on the effects of low saturation values. In the limit case of $M = 1$, equation (5.3) reads as $g_{k,1}(I) = 1 - (1-I)^k$ and represents the probability that a node has at least one infected neighbor.

The analysis of the mean-field model (5.2) reveals that a saddle-node bifurcation is possible in the family of epidemic processes under study. The trivial equilibrium $I = 0$ exists for all parameter values, but it is stable if and only if $\beta_0 < \bar{\beta}_0^{HN} = \gamma/\bar{k}$ (easy to prove via linearization). In contrast, the nontrivial equilibria branch $I = I(\beta_0)$ is implicitly defined by the condition

$$\beta_0 = \frac{\gamma I - \beta_1 I(1-I) g_{\bar{k},M}(I)}{(1-I) g_{\bar{k},M}(I)}, \tag{5.4}$$

and it is such that (a) $I \to 0$ as $\beta_0 \to \bar{\beta}_0^{HN}$ and (b) $I \to 1$ as $\beta_0 \to \infty$. Property (a) shows that the nontrivial equilibria branch intersects the trivial one at the bifurcation point $\beta_0 = \bar{\beta}_0^{HN}$, exactly as in the basic SIS model. However, as shown in Fig. 5.2, the slope of $I(\beta_0)$ at the intersection can be negative for sufficiently large $\beta_1$'s, because

$$\left.\frac{d\beta_0}{dI}\right|_{I \to 0} = \gamma \left( \frac{1}{\bar{k}} - \frac{g''_{\bar{k},M}(0)}{2\bar{k}^2} \right) - \beta_1 \tag{5.5}$$

and $g''_{\bar{k},M}(0) \le 0$ for all $\bar{k}$, $M$ (see again Fig. 5.1). The negative slope at the intersection and property (b) imply the existence of a limit point occurring at another parameter threshold $\beta_0 = \tilde{\beta}_0^{HN} < \bar{\beta}_0^{HN}$, where a saddle-node bifurcation occurs. To summarize the general behavior of the epidemiological process over a HN, we find that if $\beta_0 < \tilde{\beta}_0^{HN}$ the disease cannot persist; if $\beta_0 > \bar{\beta}_0^{HN}$ there is a unique stable high endemic equilibrium, and if $\tilde{\beta}_0^{HN} < \beta_0 < \bar{\beta}_0^{HN}$ the disease can persist or not depending upon initial conditions.

In contrast to HNs, SFNs are characterized by a heterogeneous degree distribution which takes the form $p_k \approx k^{-c}$ when $N$ (thus $k_{\max}$) tends to infinity

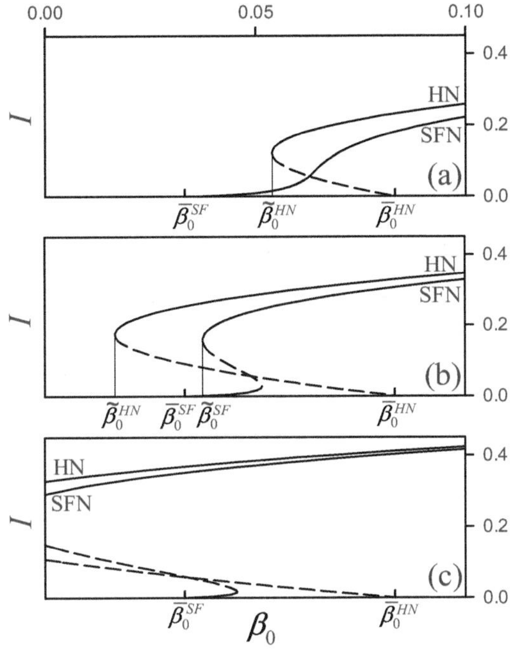


**Fig. 5.2** The nontrivial equilibria $I = I(\beta_0)$ of the HN and SFN mean-field models (5.2) and (5.6), for $M = 1$, $\bar{k} = 12$, $\gamma = 1$, and $\beta_1$ set to 1 (a), 1.25 (b) and 1.5 (c). Solid [dashed] curves denote stable [unstable] equilibria and have been numerically obtained by continuation [9]. For the SFN model we used a number $d = k_{\max} - k_{\min} + 1 = 600$ of node degrees, which is consistent with a finite-size network of $N \approx (d/k_{\min})^2 = 10^4$ nodes [34].

and can be created with the Barabási-Albert algorithm of preferential attachment [3] yielding for large $N$'s to $c = 3$, $\bar{k} = 2k_{\min}$, and a degree distribution $p_k = 2k_{\min}(k_{\min}+1)/[k(k+1)(k+2)]$ (e.g. [30]). Analogously to Refs. [32, 33], we describe the mean-field behavior of the epidemic process in a SFN by the following set of $d = k_{\max} - k_{\min} + 1$ equations:

$$\dot{I}_k(t) = -\gamma I_k(t) + (\beta_0 + \beta_1 I(t))(1 - I_k(t))\, g_{k,M}(\tilde{I}(t)), \tag{5.6}$$

where $0 \le I_k(t) \le 1$ is the fraction of infected nodes at time $t$ among those with degree $k$, $I(t) = \sum_{k=k_{\min}}^{k_{\max}} p_k I_k(t)$ is the current disease prevalence, and

$$\tilde{I}(t) = \sum_{k=k_{\min}}^{k_{\max}} \frac{k p_k I_k(t)}{\bar{k}} \tag{5.7}$$

is the expected proportion of infectives at time $t$ among the neighbors of a node, because $kp_k/\bar{k} = q_k$ is the degree distribution of the neighbors (e.g. [6]). The epidemiological scenario emerging from the analysis of the SFN model (5.6) has some significant differences with respect to that seen in HNs. The major similarity between the two is in that, despite model (5.6) is a high-dimensional system, its qualitative asymptotic behavior cannot be other than stationary (i.e., convergence to equilibria). In fact, the off-diagonal entries of the Jacobian matrix are given by

$$\frac{\partial \dot{I}_k}{\partial I_h} = (1 - I_k)\left(\beta_1 p_h g_{k,M}(\tilde{I}) + \frac{(\beta_0 + \beta_1 I) g'_{k,M}(\tilde{I}) h p_h}{\bar{k}}\right), \tag{5.8}$$

and are strictly positive for all $0 < I_k < 1$. Thus, system (5.6) is monotone and irreducible. Given that the domain $(0,1)^n$ is invariant, this implies that (almost) all trajectories converge to an equilibrium state [36]. The equilibria $I_k$ of model (5.6) are the solutions of the $d$ equations

$$I_k = \frac{(\beta_0 + \beta_1 I)\, g_{k,M}(\tilde{I})}{\gamma + (\beta_0 + \beta_1 I)\, g_{k,M}(\tilde{I})}, \qquad k = k_{\min}, \ldots, k_{\max}. \tag{5.9}$$

Using the above definitions of $I$ and $\tilde{I}$, we obtain the following pair of algebraic equations:

$$\begin{aligned} F_1\left(I, \tilde{I}\right) &= I - \sum_{k=k_{\min}}^{k_{\max}} p_k \frac{(\beta_0+\beta_1 I) g_{k,M}(\tilde{I})}{\gamma+(\beta_0+\beta_1 I) g_{k,M}(\tilde{I})} = 0, \\ F_2\left(I, \tilde{I}\right) &= \tilde{I} - \frac{1}{\bar{k}} \sum_{k=k_{\min}}^{k_{\max}} k p_k \frac{(\beta_0+\beta_1 I) g_{k,M}(\tilde{I})}{\gamma+(\beta_0+\beta_1 I) g_{k,M}(\tilde{I})} = 0. \end{aligned} \tag{5.10}$$

The trivial solution $I = \tilde{I} = 0$ of (5.10) has a branching point at

$$\det \begin{pmatrix} \partial F_1/\partial I & \partial F_1/\partial \tilde{I} \\ \partial F_2/\partial I & \partial F_2/\partial \tilde{I} \end{pmatrix}_{I=\tilde{I}=0} = 1 - \frac{\beta_0}{\gamma} \frac{\langle k^2 \rangle}{\bar{k}} = 0, \tag{5.11}$$

namely at $\beta_0 = \bar{\beta}_0^{SF} = \gamma \bar{k}/\langle k^2 \rangle < \bar{\beta}_0^{HN}$. This threshold is the same lower bound value that marks the existence of a (stable) nontrivial solution in the basic SIS process over a finite SFN [34]. Differently from the outcome of the SIS process, however, we notice that model (5.6) can have nontrivial solutions also for $\beta_0 < \bar{\beta}_0^{SF}$. This can be numerically verified by analyzing how the solutions of (5.10) vary with respect to the transmission coefficients $\beta_0$ and $\beta_1$, as shown in Fig. 5.2. Sufficiently large values of $\beta_0$, no matter the value $\beta_1$, guarantee the disease persistence at high endemic levels in both HNs and SFNs. The most interesting result arises at intermediate values of $\beta_1$. Fig. 5.2b shows the existence of a saddle-node bifurcation in both the HN (at $\beta_0 = \tilde{\beta}_0^{HN}$) and the SFN (at $\beta_0 = \tilde{\beta}_0^{SF}$) models. Decreasing the value attributed to parameter $\beta_0$ causes an abrupt transition from a high endemic state to extremely low or null prevalences. Notably, in all the numerical analyses we have performed, it turns out that $\tilde{\beta}_0^{HN} < \tilde{\beta}_0^{SF}$. In words, this is like saying that HNs can support diseases that would be unable to circulate over SFNs. The fact that for large values of $\beta_1$ (see Fig. 5.2c) the thresholds $\tilde{\beta}_0^{HN}$, $\tilde{\beta}_0^{SF}$ become negative does not change the overall picture.

One could conjecture that the gap between the two thresholds $\tilde{\beta}_0^{HN}$ and $\tilde{\beta}_0^{SF}$ is due to the finite-size approximation of the SFN, and that it can vanish as $k_{\max}$ (thus

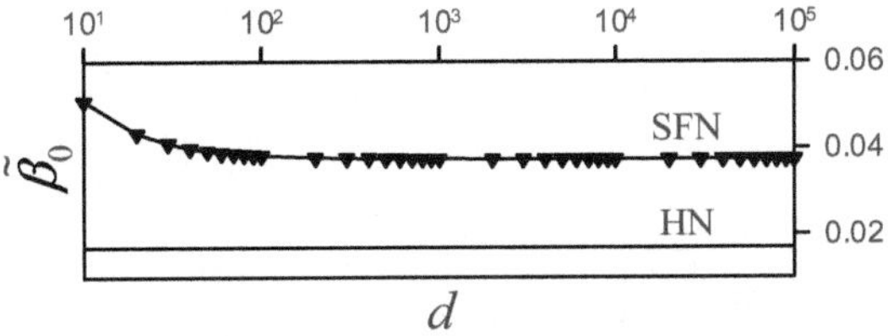


**Fig. 5.3** The value of $\beta_0$ at the saddle-node bifurcation in both HN and SFN models as a function of $d = k_{max} - k_{min} + 1$. The value $d = 10^5$ corresponds to a finite-size SFN of $N \approx (d/k_{min})^2 \approx 2.8 \times 10^8$ nodes [34]. Other parameters set to $\bar{k} = 2k_{min} = 12$, $M = 1$, $\gamma = 1$ and $\beta_1 = 1.25$.

$d$) becomes increasingly large. This is not true, though, as it can be checked by computing $\tilde{\beta}_0^{SF}$ as a function of $d$ over many orders of magnitude. Fig. 5.3 makes evident that $\tilde{\beta}_0^{SF}$ settles to a constant positive limit, which is well above $\tilde{\beta}_0^{HN}$.

This example has started showing that it is necessary to fully account for the characteristics of the considered epidemic process before concluding that SFNs are more prone than other network structures to the propagation of an infection. In any case, it is false to conclude that *whatever* infection is favored by heterogenous rather than homogenous networks or viceversa. In fact, while the basic SIS epidemic is favored by the former network structure, the spread of the above presented process is enhanced by the latter.

## 5.3 Spread of Diseases with Vital Dynamics

The basic SIS model is the most usually studied epidemics in the context of complex networks. Notably, however, it does not account for vital dynamics. The structure of the network is in fact considered to be strictly fixed through time. Such a strong assumption is valid as long as the epidemics occurs over timescales that are much shorter than the typical timescales characterizing the dynamics of the host population. The case study used by Kermack and McKendrick (1927) to discuss the effectiveness of the basic non-vital SIR model (equation (29) in [22]) is a plague occurred in the island of Bombay over the period December 17, 1905, to July 21, 1906. According to demographic data from the 2004 World Population Datasheet of the Population Reference Bureau (available online at `http://www.prb.org`), the life expectancy at birth in India is about 62 years, i.e., two orders of magnitude longer than the spread of the epidemics. If the dynamics of that same plague (*Yersinia pestis*) would have to be studied over longer time horizons in human (see the 56-year time-series in [4]) or animal populations (see the 46-year time-series in [37] concerning great gerbils *Rhombomys opimus*), the mortality and natality of the hosts could hardly be disregarded.

The main departure of the instrumental example presented in this section from the previous scenario is in assuming that each member of the population (node of the network) has come to life from a birth event and will face death. More precisely, as typically done in vital SIR-like processes [1], we assume that the birth and death rates are equal, so that the total number of nodes in the network stochastically fluctuates around an average value ($N$). Technically, we assume that the death probability during the time interval $\Delta$ is equal to $\mu\Delta$, it is independent of the node degree and of its current state. This latter condition is valid for diseases with low virulence, i.e., that cause little or no extra-mortality to infected hosts. As for the birth process, we assume that the *per capita* probability of generating a newborn individual in susceptible state is $\mu\Delta$. Despite the number of individuals in the population can remain rather close to the average value $N$, the birth-death process has the power of strongly modifying the network degree distribution, as we discuss below.

As for the contagious component of the birth-death-infection process considered here, similarly to what has been done in the previous section, we study diseases characterized by two possible states: susceptible or infective. The probability that a node in infective state returns susceptible during time $\Delta$ is $\gamma\Delta$, while the probability that a susceptible node (say node $i$) be infected is $\beta G_i \Delta$ (the symbols $\beta$ and $G_i$ have been defined in Sec. 5.2). Here we restrict our attention to diseases with constant transmission rates $\beta$'s, i.e., with $\beta_1 = 0$ in equation (5.1). Also, we set $G_i = n_i$, namely we assume no limitations to the number of infected neighbors of the susceptible node $i$ that might cause its infection. Although biologically important, we outline that the two simplifications just introduced are uninfluent to the results presented in the sequel. In fact, the main focus of our analysis concerns the disease persistence, i.e., the characteristics of epidemics when $I \to 0$.

Under the above assumptions, the temporal evolution of the fraction $I_k(t)$ of nodes that are infected among those with degree $k$ is governed by

$$\dot{I}_k(t) = -(\mu+\gamma) I_k(t) + \beta\,(1 - I_k(t))\, k\tilde{I}(t). \tag{5.12}$$

By following an analytical procedure similar to what explained in the previous section, we impose the equilibrium to (5.12) and derive

$$I_k = \frac{\beta k \tilde{I}}{(\mu+\gamma) + \beta k \tilde{I}}, \qquad k = k_{\min}, \ldots, k_{\max}. \tag{5.13}$$

Then, relying on the definition of $\tilde{I}$ (see equation (5.7)), we arrive at the following algebraic equation, which must be satisfied at the equilibrium:

$$F\left(\tilde{I}\right) = \tilde{I} - \frac{1}{\bar{k}} \sum_{k=k_{\min}}^{k_{\max}} k p_k \frac{\beta k \tilde{I}}{(\mu+\gamma) + \beta k \tilde{I}} = 0.$$

The branching condition for the trivial solution $\tilde{I} = 0$ is obtained by imposing

$$\det\left(\frac{\partial F}{\partial \tilde{I}}\right)_{\tilde{I}=0} = 1 - \frac{\beta}{(\mu+\gamma)} \frac{\langle k^2 \rangle}{\bar{k}} = 0,$$

thus getting

$$\beta = \beta_{cr} = (\mu + \gamma) \frac{\bar{k}}{\langle k^2 \rangle}. \tag{5.14}$$

This threshold marks the lower bound for the existence of a nontrivial solution. We will discuss its implications below, after considering the impact of the birth/death process on the topology of the network.

Let us first use an Erdős–Rényi network (ERN) [13] as an example of (almost) HN. Its degree distribution can be approximated, for large $N$'s, by a Poisson distribution [30]:

$$p_k = e^{-\bar{k}} \frac{\bar{k}^k}{k!}. \tag{5.15}$$

Under suitable assumptions, this type of network preserves its degree distribution under the birth-death process. Indeed, when a node $i$ dies, all its $k_i$ links are removed so that, on average, $\mu \Delta \bar{k}$ links are removed per unit time. If the following conditions hold: (a) death selects nodes at random, (b) the birth process adds nodes whose degree is Poisson distributed according to (5.15), thus $\mu \Delta \bar{k}$ links are added on average per unit time, and (c) the newly added nodes attach their links at random; then it can be proved that the network retains its degree distribution (5.15) [27].

On the contrary, things are much different when the birth-death process occurs on a SFN. If we assume the following: (a) the birth-death process is begun on a power-law degree distribution $p_k \approx k^{-c}$ with average degree $\bar{k}$; (b) the death process deletes nodes at random; (c) the birth process adds nodes with degree $\bar{k}$ (this keeps the average degree constant); and (d) the newly added nodes attach their links preferentially to the existing nodes with higher degree, exactly as in the Barabási-Albert algorithm [3]; then it can be proved [27] that the degree distribution asymptotically loses its power-law form to become, for large $k$'s, a so-called "stretched exponential":

$$p_k \approx k^{-3/4} e^{-2\sqrt{k}}. \tag{5.16}$$

A network resulting from the above described process will be denoted as *evolved scale-free network* (ESFN). Notably, the second moment $\langle k^2 \rangle$ of the distribution (5.16) remains finite for $k_{\max} \to \infty$, contrarily to what happens for the power-law distribution in a suitable range of values for $c$. This means that, under the birth-death process, the degree distribution loses its characteristic "fat tail" (see Fig. 5.4).

Technically speaking, the degree distributions $p_k$ we used in model (5.12) have been obtained as follows. We first fixed the maximum degree $k_{\max}$, needed because of computational constraints. When considering ERNs, the $p_k$'s were directly computed according to (5.15) – the only parameter to be specified being the average degree $\bar{k}$. As already pointed out, no evolution under the birth-death process is necessary on ERNs, since the degree distribution remains unchanged.

On the contrary, when a SFN was considered, an initial power-law distribution $p_k \approx k^{-c}$ with a pre-specified average degree $\bar{k}$ was first generated with the Barabási-Albert algorithm [3]. Then the evolution of the $p_k$'s under the birth-death process

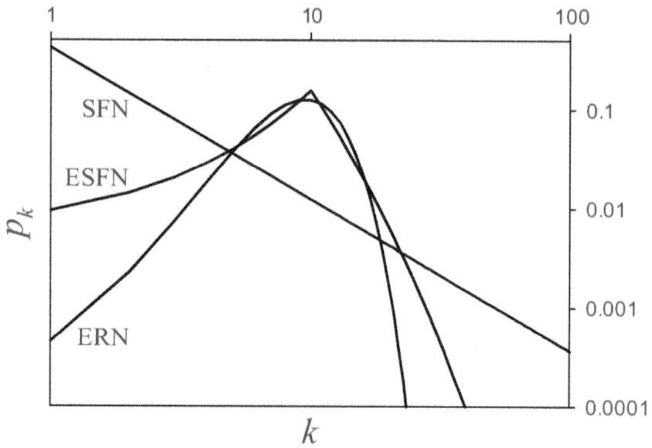


**Fig. 5.4** A log-log plot with examples of degree distributions: (a) Erdős-Rényi network (ERN), (b) scale-free network (SFN), (c) evolved scale-free network (ESFN). All the three distributions have $\bar{k} = 10$.

was computed according to the following discrete-time equation [27], which models the statistical evolution of the degree distribution:

$$\begin{aligned} Np_k(t+\Delta) &= Np_k(t) \\ &+\mu\Delta\left(-p_k(t) - kp_k(t) - \bar{k}\pi_k p_k(t)\right) \\ &+\mu\Delta\left((k+1)p_{k+1}(t)\right) \\ &+\mu\Delta\left(\bar{k}\pi_{k-1}p_{k-1}(t)\right) + \mu\Delta\,\delta_{k\bar{k}}, \end{aligned} \tag{5.17}$$

where $\delta_{k\bar{k}} = 1$ if $k = \bar{k}$ and 0 otherwise, and $\pi_k = k/\bar{k}$ is the preferential attachment kernel. The various terms in the above equation describe the removal due to death $(-p_k)$, the passage of a node from degree $k+1$ to $k$ and from $k$ to $k-1$ when a neighbor is dead ($(k+1)p_{k+1}$ and $-kp_k$), the passage of a node from degree $k-1$ to $k$ and from $k$ to $k+1$ when a newborn node attaches to it ($\bar{k}\pi_{k-1}p_{k-1}$ and $-\bar{k}\pi_k p_k$), and the insertion of new nodes with degree $\bar{k}$ (the term $\delta_{k\bar{k}}$). Starting from a power-law form, we let the network evolve according to (5.17) until we asymptotically obtained an invariant degree distribution, which turned out to be of the type (5.16).

Figure 5.5 shows how the disease prevalence scales with $\beta$ at the equilibrium in the two different network topologies: ERNs and ESFNs. For comparison, the same value has been calculated for SFNs (by simply letting $\mu = 0$ in (5.12) and (5.17)) which must have, in principle, a vanishing epidemic threshold (5.14) for $N \to \infty$, as a consequence of the diverging second moment $\langle k^2 \rangle$ [32, 33, 34]. However, under the pressure of the birth-death process, on the long run the initially SFN changes its topology to an ESFN. The stretched exponential degree distribution has a form which is not far from that of ERNs (see again Fig. 5.4). In particular, $\langle k^2 \rangle$ is finite even with diverging $N$'s. The consequence of incorporating vital dynamics into the SIS process is that of having made reappear a non vanishing epidemic threshold, even if $N \to \infty$ and the node addition is governed by a linear preferential attachment rule, as in the Barabási-Albert SFN.

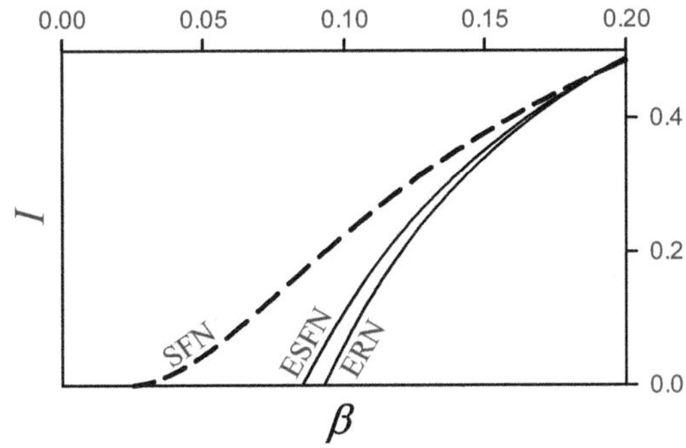


**Fig. 5.5** The nontrivial equilibrium $I = I(\beta) = \Sigma_k p_k I_k(\beta)$ of model (5.12) for an Erdős-Rényi (ERN), a scale-free (SFN) and an evolved scale-free network (ESFN) ($\bar{k} = 10$, $\mu = 0.02$, $\gamma = 1$). The curves have been numerically obtained by continuation [9]. For the SFN and ESFN curves a number $d = 5000$ of node degrees has been used (corresponding to a finite-size network of $N \approx 10^6$ nodes) [34]). For the ERN we used $d = 100$.

## 5.4 Concluding Remarks

The take-home message of the present contribution is to resist the temptation of condensing the elegance of some technical results about epidemic processes over complex networks into far too general and sharp, yet simplistic sentences. In particular, as shown in the first instrumental example presented above, it is wrong to attribute the spread efficiency of specific infectious diseases exclusively to the topology of the networks they propagate through. At the same time, as clearly evidenced by the vital infectious disease discussed in the second example, the dynamical process has the power of profoundly modifying the degree distribution of the network. Therefore, we can conclude that, when analyzing non-elementary spreading processes on complex networks, much more attention must be given to the interplay between the network topology and the dynamical characteristics of the infection process. Such interplay is, to our better understanding, the real responsible of the interesting findings of recent research.

## References

1. Anderson, R., May, R.: Infectious Diseases of Humans. Oxford University Press, Oxford (1992)
2. Bailey, N.: The Mathematical Theory of Infectious Diseases and Its Application. Griffin, London (1975)
3. Barabási, A., Albert, R.: Emergence of scaling in random networks. Science 286(5439), 509–512 (1999)
4. Ben Ari, T., Gershunov, A., Gage, K.L., Snall, T., Ettestad, P., Kausrud, K.L., Stenseth, N.C.: Human plague in the USA: the importance of regional and local climate. Biol. Lett. 4(6), 737–740 (2008)

5. Bettencourt, L., Cintrón-Arias, A., Kaiser, D., Castillo-Chavez, C.: The power of a good idea: Quantitative modeling of the spread of ideas from epidemiological models. Physica A 364, 513–536 (2006)
6. Boccaletti, S., Latora, V., Moreno, Y., Chavez, M., Hwang, D.: Complex networks: Structure and dynamics. Phys. Rep. 424(4-5), 175–308 (2006)
7. Boguñá, M., Pastor-Satorras, R., Vespignani, A.: Absence of epidemic threshold in scale-free networks with degree correlations. Phys. Rev. Lett. 90(2), 28701 (2003)
8. Casagrandi, R., Bolzoni, L., Levin, S., Andreasen, V.: The SIRC model and influenza A. Math. Biosci. 200(2), 152–169 (2006)
9. Dhooge, A., Govaerts, W., Kuznetsov, Y.: MATCONT: A MATLAB package for numerical bifurcation analysis of ODEs. ACM Trans. Math. Softw. 29(2), 141–164 (2003)
10. van den Driessche, P., Watmough, J.: A simple SIS epidemic model with a backward bifurcation. J. Math. Biol. 40(6), 525–540 (2000)
11. Dushoff, J., Huang, W., Castillo-Chavez, C.: Backwards bifurcations and catastrophe in simple models of fatal diseases. J. Math. Biol. 36(3), 227–248 (1998)
12. Eguíluz, V., Klemm, K.: Epidemic threshold in structured scale-free networks. Phys. Rev. Lett. 89(10), 108701 (2002)
13. Erdős, P., Rényi, A.: On random graphs. Publ. Math.-Debr. 6, 290–297 (1959)
14. Feichtinger, G.: Hopf-bifurcation in an advertising diffusion-model. J. Econ. Behav. Organ. 17(3), 401–411 (1992)
15. Feng, Z., Castillo-Chavez, C., Capurro, A.: A model for tuberculosis with exogenous reinfection. Theor. Popul. Biol. 57(3), 235–247 (2000)
16. Garnett, G., Anderson, R.: Contact tracing and the estimation of sexual mixing patterns - The epidemiology of gonococcal infections. Sex. Transm. Dis. 20(4), 181–191 (1993)
17. Goffman, W., Newill, V.: Generalization of epidemic theory - application to transmission of ideas. Nature 204(495), 225–228 (1964)
18. Hethcote, H.: An age-structured model for pertussis transmission. Math. Biosci. 145(2), 89–136 (1997)
19. Huang, W., Cooke, K., Castillo-Chavez, C.: Stability and bifurcation for a multiple-group model for the dynamics of HIV AIDS transmission. SIAM J. Appl. Math. 52(3), 835–854 (1992)
20. Hwang, D., Boccaletti, S., Moreno, Y., Lopez-Ruiz, R.: Thresholds for epidemic outbreaks in finite scale-free networks. Math. Biosci. Eng. 2(2), 317–327 (2005)
21. Keeling, M., Eames, K.: Networks and epidemic models. J. R. Soc. Interface 2(4), 295–307 (2005)
22. Kermack, W., McKendrick, A.: A contribution to the mathematical theory of epidemics. Proc. R. Soc. A 115(772), 700–721 (1927)
23. Kribs-Zaleta, C., Velasco-Hernández, J.: A simple vaccination model with multiple endemic states. Math. Biosci. 164(2), 183–201 (2000)
24. Meade, N., Islam, T.: Modelling and forecasting the diffusion of innovation - A 25-year review. Int. J. Forecast. 22(3), 519–545 (2006)
25. Medley, G., Lindop, N., Edmunds, W., Nokes, D.: Hepatitis-B virus endemicity: heterogeneity, catastrophic dynamics and control. Nat. Med. 7(5), 619–624 (2001)
26. Moghadas, S., Alexander, M.: Bifurcations of an epidemic model with non-linear incidence and infection-dependent removal rate. Math. Med. Biol. 23(3), 231–254 (2006)
27. Moore, C., Ghoshal, G., Newman, M.: Exact solutions for models of evolving networks with addition and deletion of nodes. Phys. Rev. E 74(3), 036121 (2006)
28. Moreno, Y., Vázquez, A.: Disease spreading in structured scale-free networks. Eur. Phys. J. B 31(2), 265–271 (2003)

29. Nakamaru, M., Levin, S.: Spread of two linked social norms on complex interaction networks. J. Theor. Biol. 230(1), 57–64 (2004)
30. Newman, M.: The structure and function of complex networks. SIAM Rev. 45(2), 167–256 (2003)
31. Olinky, R., Stone, L.: Unexpected epidemic thresholds in heterogeneous networks: The role of disease transmission. Phys. Rev. E 70(3), 030902 (2004)
32. Pastor-Satorras, R., Vespignani, A.: Epidemic dynamics and endemic states in complex networks. Phys. Rev. E 63(6), 066117 (2001)
33. Pastor-Satorras, R., Vespignani, A.: Epidemic spreading in scale-free networks. Phys. Rev. Lett. 86(14), 3200–3203 (2001)
34. Pastor-Satorras, R., Vespignani, A.: Epidemic dynamics in finite size scale-free networks. Phys. Rev. E 65(3), 035108 (2002)
35. Piccardi, C., Casagrandi, R.: Inefficient epidemic spreading in scale-free networks. Phys. Rev. E 77(2), 026113 (2008)
36. Smith, H.: Systems of ordinary differential-equations which generate an order preserving flow - A survey of results. SIAM Rev. 30(1), 87–113 (1988)
37. Stenseth, N.C., Samia, N.I., Viljugrein, H., Kausrud, K.L., Begon, M., Davis, S., Leirs, H., Dubyanskiy, V.M., Esper, J., Ageyev, V.S., Klassovskiy, N.L., Pole, S.B., Chan, K.S.: Plague dynamics are driven by climate variation. Proc. Natl. Acad. Sci. U.S.A. 103(35), 13110–13115 (2006)
38. Watts, D.J., Strogatz, S.H.: Collective dynamics of 'small-world' networks. Nature 393(6684), 440–442 (1998)
39. Weber, A., Weber, M., Milligan, P.: Modeling epidemics caused by respiratory syncytial virus (RSV). Math. Biosci. 172(2), 95–113 (2001)

# Chapter 6
# Complex Networks and Critical Infrastructures

Roberto Setola and Stefano De Porcellinis

**Abstract.** The term "Critical Infrastructures" indicates all those technological infrastructures such as: electric grids, telecommunication networks, railways, healthcare systems, financial circuits, etc. that are more and more relevant for the welfare of our countries. Each one of these infrastructures is a complex, highly non-linear, geographically dispersed cluster of systems, that interact with their human owners, operators, users and with the other infrastructures. Their augmented relevance and the actual political and technological scenarios, which have increased their exposition to accidental failure and deliberate attacks, demand for different and innovative protection strategies (generally indicate as CIP - Critical Infrastructure Protection). To this end it is mandatory to understand the mechanisms that regulate the dynamic of these infrastructures. In this framework, an interesting approach is those provided by the complex networks. In this paper we illustrate some results achieved considering structural and functional properties of the corresponding topological networks both when each infrastructure is assumed as an autonomous system and when we take into account also the dependencies existing among the different infrastructures.

## 6.1 Introduction

The developed countries largely depend on several infrastructures generally indicated as *Critical Infrastructures*. These include, among the others: energy production, transportation and distribution, telecommunications networks, water management and supply networks, transportation (air, rail, marine, surface), banking and financial services [1]. The term "Critical Infrastructure" stresses the fact that they are so vital for the country's welfare that "if disrupted or destroyed, would have a serious impact on the health, safety, security or economic well-being of Citizens or the effective functioning of governments" [1].

Roberto Setola and Stefano De Porcellinis
Complex Systems & Security Lab., Universitá CAMPUS Bio-Medico, Roma (Italy)
e-mail: r.setola@unicampus.it, s.deporcellinis@unicampus.it

Moreover, the use of the single term "Critical Infrastructures" to indicate the so huge set of heterogeneous entities stresses that these infrastructures cannot be longer assumed as isolated systems, because they are strongly and deeply coupled by means of numerous and complex networks of mutual dependency links. Indeed, although originally designed as logically separated systems, in order to improve their efficiency and optimise their management, modern infrastructures rely (directly or indirectly) on many external services, becoming highly dependent from each other. The presence of such dependency networks contributed to create a global system of systems, whose complexity makes it prones to large cascade failures, as dramatically emphasised by several recent accidents.

A famous example of the consequences of such a condition of pervasive interdependency, was the wide critical scenario which originated, in 1998, from the failure of the telecommunication satellite Galaxy IV. Indeed, the failure of this satellite in geo-stationary orbit over the U.S. west cost [2], besides to create the predictable blackouts in telecommunications (almost 90% of U.S. pagers were affected), significantly affected the operativeness of many transport systems: numerous flights were delayed due to the absence of high-quote weather information, while refueling on highways became difficult as gas stations could not process credit card transactions (for more information on this and other interdependency related incidents see [3]).

A second citation about the risks deriving from the "unpredictable" (or "still–not–predicted") dependencies, regards the events happened, on January 2004, when the air conditioning system of an important telecom node near to Rome got out of order [4] . This apparently trivial and isolated fault caused, instead, large blackouts into land and wireless telecommunications (affecting almost all the service providers), the quitting of the financial transaction into 5.000 banks and in 3.000 postal offices and, also, difficulties in the international airport of Fiumicino, where about 70% of check-in desks were forced to use manual procedures.

To make the situation even worse, a lot of critical infrastructures are more and more exposed to deliberate sabotages. Indeed, due to their increased relevance, they frequently became targets for terroristic attacks or criminal actions [5]. Indeed, attacks carried on these infrastructures could easily create widespread ill services or increased the effects of acts against more visible targets, e.g. slowing down emergency services and delaying rescue operations [7]. Moreover, limited or paltry attacks on public utility infrastructures may easily (and cheaply) induce panic and mistrust in the government, also with long-term decrement effects on the national economies, as illustrated by the 9/11 and its consequences on the air transportation.

Such considerations imposed to governments and international organizations to improve the security, robustness and resilience of these infrastructures [6]. To this end, and especially due to the increased presence of the dependency links among the infrastructures, it took place the need for considering, besides sectorial specific security strategies, coordinated inter-sectorial strategies, with the aim to integrate, into a single framework, security requirements and constraints about critical infrastructures as a whole. These strategies are usually referred as Critical Infrastructure Protection (CIP).

In this paper we illustrate some results obtained in the field of the complex network analysis with a specific focus on the robustness properties of the critical infrastructures. Specifically, Section 7.2 is devoted to illustrate the structural and functional properties of a critical infrastructure (i.e. the Italian High Voltage network) when it can be modelled as isolated system. Section 6.3 introduces the interdependency phenomena and Section 6.5 illustrates a methodology to analyse the consequences of the presence of these dependency links on the robustness properties of the infrastructures. Section 6.5 approaches the multi-dependencies model, i.e. how to represent inside a single framework a multitude of dependency mechanisms. Finally some conclusive comments are collected in Section 7.5.

## 6.2 Critical Infrastructure as Complex Network

Modelling Critical Infrastructures is a hot research topic. In the literature we can find different approaches that spans from simplified holistic models to very complex simulation frameworks (see [8, 9, 10] and references there in).

In this framework an interesting area of research is that of the complex networks of which Critical Infrastructures represent an excellent metaphor. Indeed, Critical Infrastructures show a number of structural and "behavioural" features which have been widely investigated in the last years [11, 12, 13, 14, 15].

The promise of these efforts is to unveil relevant insights on growth mechanisms, causes of vulnerability, dynamic behaviour under perturbation, onset of emerging phenomena, etc. even neglecting some peculiar characteristics.

According to the recent developments there are two aspects which, if properly analysed, may allow to gain relevant insights on Critical Infrastructures:

- the study of the topology properties of the graph representing the infrastructure.
- the study of their "behaviour", as it can be deduced from the analysis of some functional model able to reproduce the dynamic process (mainly transport of some entity, like electricity, data, vehicles, etc.) taking place on them.

The first one is generally referred to as *structural* properties of the network, while the latter one represents its *functional* properties.

### 6.2.1 *Structural Analysis*

Topological analysis of Critical Infrastructures received a renewed interest after the pioneering works of Strogartz and Barabasi [16, 17]. In these studies they emphasized that many technological, social and biological networks may evolve without any central authority, nevertheless showing peculiar structural patterns like *small world* and *scale-free* with immediate consequences on many properties and characteristics of the corresponding infrastructure.

Specifically, it has been shown that scale-free models, with few hubs (i.e. nodes with a high degree) and a poorly connected peripheries, show good resilience against

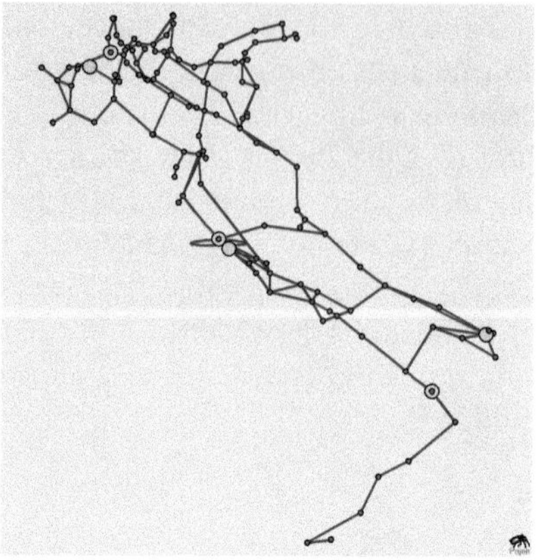

Fig. 6.1 Italian high-voltage (380 kV) transmission grid graph. The nodes emphasised are those with the highest degree.

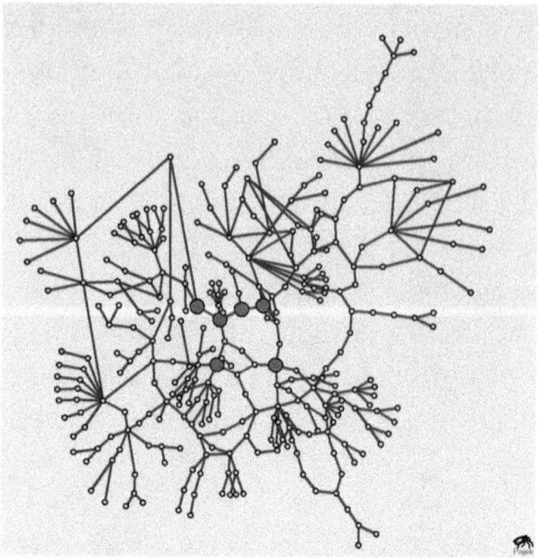

Fig. 6.2 Italian high-voltage transmission grid graph. The nodes emphasised are those with highest betweeness coefficient.

accidental (or random) failures but are prone to deliberate attacks [18]. At the same time, the small-world property, which implies low average path lengths with respect to the number of nodes in the network, can lead to a very fast propagation of pandemic events, carried out by viruses, failures or ill-services.

In spite of these considerations, it is worth to notice that, quite often, large technological infrastructures do not show a clear "scale–free" structure, especially due to technological constraints which limit the growth of node's degrees [19]. As a consequence, results about robustness and resilience, based on the assumption of pure scale–free, cannot be directly applied in many technological contexts, leaving the issues related to structural vulnerability to be differently evaluated on each case. A recent review of the issue of topological analysis of networks is presented in [12].

In the following, we will analyse the structural properties of the Italian high–voltage (380 kV) electrical transmission network. The grapgh of which consists of $N = 310$ nodes and $E = 359$ arcs (also referred to as "lines") reported in Figure 6.1.

In order to classify the network, in [20] it was analysed the distribution $P(k)$ of the node's *degree* $k$ (the *degree* is the number of links connecting each node to its nearest neighbours). As shown in Figure 6.3 this network has a limited number of hubs, whose maximum degree is $k_{max} = 11$. $P(k)$ and the cumulative degree distribution $P(k > K)$ are both likely to be fit by an exponential (single–scale network [11, 12, 19]). The latter distribution can be fitted $P(k > K) \sim e^{-0.55K}$ in agreement with the findings for the North-american power grid [11].

Other useful parameters that can be evaluated on this graph are the average *closeness* coefficient that is equal to 0.1, the *betweenness* is 0.03, the *diameter* measures 27, and the *averange path lentght* is 10.22 A further property measured on the network is the average *clustering* coefficient $C$ [12] which results to be as small as $C = 2.06 \cdot 10^{-2}$.

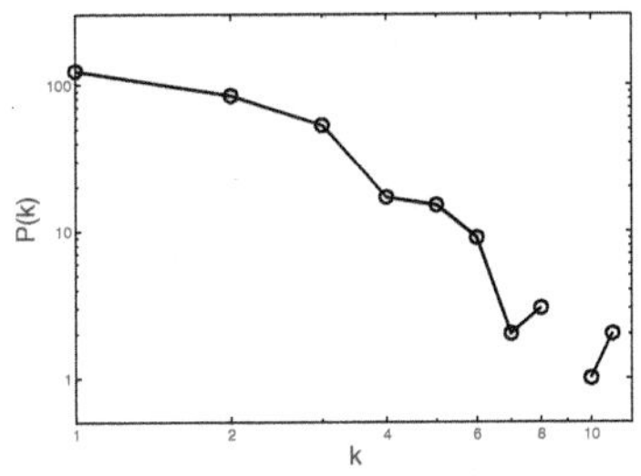


**Fig. 6.3** The distribution of the node's degree for the Italian high voltage (380kV) electric network (log–log scale).

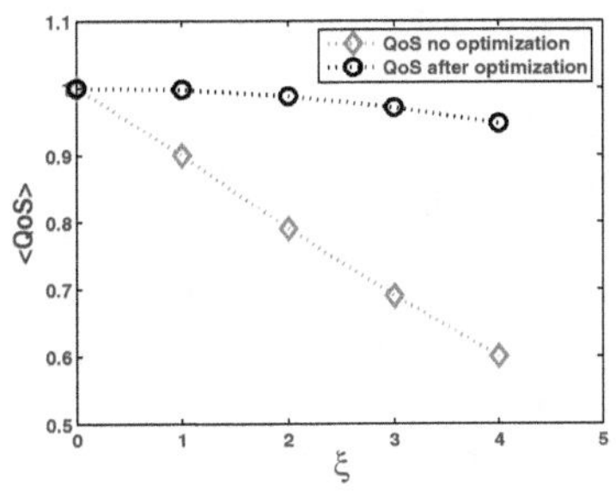


**Fig. 6.4** *QoS* values as a function of the perturbation $\xi$ (number of removed lines). Circles represent *QoS* when no optimisation is performed; diamonds when re–dispatching is performed.

Such topological analyses may contribute to identify the most critical (from the structural point of view) elements of the network. For example, in Figure 6.1 the six hubs with the highest degree are highlighted, while in Figure 6.2 it is possible to identify the node with high betweenness value (i.e. the nodes which belong to the highest number of minimal paths).

## 6.2.2 Functional Analysis

Unlike the structural analysis, functional analysis of Critical Infrastructure is an hard task. This is due, besides the most intrinsic complexity of the topic, to the lack of accurate and complete functional data of the infrastructures. Indeed, in order to formulate consistent functional models, a deeper knowledge of the infrastructures is required. Such a necessity clashes with the common unavailability of these information, often treated as confidential and classified by the stakeholders.

In order to overcome such limitations some simplifying hypothesis are generally assumed, enabling the development "simpler" functional models, still able to capture the basic features of the networks but disregarding the most complex effects related to the exact technological implementations.

Then, the main aim of functional analysis is to evaluate the effects on the flows induced by simple topological perturbations.

To be more concrete we consider an other time the Italian high-voltage electrical network introduced previously and assume that the power flow on it can be modelled via the *DC Power Flow Model* described in Appendix 6.7.

As illustrated in [21], the cut of one or more links induces a change of the power flow distributions over the network, which might be evaluated, assuming the simplified *DC Power Flow Model*. However, exploiting this simple model, it results that, for several different cuts, a solution is inhibited by the physical constraints imposed by model itself. In other terms, due to overload conditions or unbalancing situations, the electric grid is no longer able to supply energy, implying the presence possible blackouts if any corrective action is taken.

However, these such a kind of dramatic consequences do not comply with the real data (and the common sense). Indeed, in order to prevent so catastrophic conditions, the operators of the electric networks continuously perform corrective and adjustment actions to limit the insurgence of black-outs. Hence, in order to take into account such mechanisms of self-tuning, we need to consider also some *re-dispatching* strategies that, miming the typical policies adopted by electrical operators, enable to modulate produced and dispatched powers.

To this end, we introduced, as better described in [21], a re-dispatching procedure which, in the presence of unfeasible power flow configurations, modulates the injected electric power in order to minimize the distance of the new configuration with respect to the "normal" load conditions $\mathbf{P}^{(0)}$ preserving, at the same time, the system compliance to the model constraints which were assumed.

In [20] it is assumed that the "distance" of this solution from the *normal* distribution, represents a measure of the quality of service under perturbed conditions,

$$QoS = 1 - \frac{\sum_{i \in loads}[P_i - P_i^0]}{\sum_{i \in loads} P_i^0} \tag{6.1}$$

Figure 6.4 reports the $QoS$ as a function of number of lines ($\xi$) removed either in the case in which "re-dispatching" is performed (circles) and when it is not performed (diamonds). In the latter case, $QoS$ is set to zero when the system cannot be solved and set to one if the system has a solution within the physical constraints. Data reported in Figure 6.4 are averaged over a large number of runs.

## 6.3 Interdependency

All the studies mentioned in the previous section dealt with a single infrastructure, analysed on both the structural and, in some cases, the functional viewpoints. However, far from being "stand–alone" systems, Critical Infrastructures are mutually, and sometimes strongly, dependent. The improper functioning of one infrastructure might, indeed, have dramatic effects on many other infrastructures somehow related (like, e.g., the effects of an electrical outage which rapidly spreads over railways and communication networks, through a *domino* cascade phenomenon).

In [22] it was emphasised that in order to correctly understand the complex behaviour of critical infrastructures, it is mandatory to adopt a three-layers model:

- *Physical Layer*: The physical component of the infrastructure, e.g., the grid for the electrical network;
- *Cyber Layer*: Hardware and software components of the system devoted to control and manage the infrastructure, e.g., SCADA and DCS;
- *Organizational Layer*: Procedures and functions used to define activities of human operators and to support cooperation among infrastructures.

Indeed, the authors emphasized that each component of an infrastructure largely interacts by means of *inter-domains* dependency links, besides with the elements belonging to the same infrastructure (i.e. *intra-domain* dependency), also with elements belonging, in the same layers, to other infrastructures. It is also worth to notice that the importance of such a decomposition has been recognised also by the US and Canada government commission charged to analyse the 2003 black-out in US and Canada [23]. As a matter of fact, to explain the multitude of causes that produced that episode, they describe the events in terms of grid (physical), computer and human layers. Only considering all the three layers together it was possible to identify and correctly understand what really led to the black-out.

Another episode that we can cite is the electric black-out that affected Italy on September 2003. It induced large degradation in railway network, in healthcare systems, in financial services and also in different types of communication networks. On the other side, the partial failure of communication systems affected the capability of the SCADA network (i.e. the system used to manage the electric grid) to perform its function, thus producing a negative feedback on the restore phase [4].

In this episode, electrical and communication networks showed a bi–directional functional dependency. This phenomenon, generally indicated as *interdependency*, should be carefully considered in risk assessment strategies due to the positive feedbacks it may imply, which could represent the major causes of the amplification of the negative consequences, up to catastrophic levels.

In [24] the authors emphasized how such interdependencies should be analysed with respect to different dimensions. In particular, they catalogue the mechanisms that underlie the interdependencies phenomena in:

- *Physical Interdependency*. Two infrastructures are physically interdependent if the operations of one infrastructure depend on the physical output(s) of the other.
- *Cyber Interdependency*. An infrastructure presents a Cyber dependency if its state depends on information transmitted through the information infrastructure.
- *Geographical Interdependency*. A geographic interdependency occurs when elements of multiple infrastructures are in close spatial proximity. In this case, events such as an explosion or a fire may create a failure into the nearby infrastructures.
- *Logical Interdependency*. Two infrastructures are logically interdependent if the state of one of them depends on the state of the other via control, regulatory or other mechanisms that cannot be considered physical, geographical or cyber.

The different inter–dependencies among Critical Infrastructures are, sometimes, much more subtle and difficult to be described, due to the presence of several relations and complex feedback paths. The problem is further complicated by the poor knowledge of dependencies, and by the (sometimes) inaccurate knowledge of the Critical Infrastructure's structures themselves. Indeed, they often grow up through many subsequent (and largely undocumented nor unplanned) upgrades, modifications and integration of existing apparatuses.

Moreover, as stressed in [25], negative consequences often propagate along *indirect* links. These last are connections that have not been specifically constructed nor planned and are very often originated by proximity which do not exist under "normal" conditions, but emerge during crisis scenarios.

## 6.4 Interdependency Modelling

Studies in the field of Complex Network related to inter-dependent Critical Infrastructures are more rare. Recent works have reported on layered networks [26] where several *homegenous* networks (i.e. of similar nature) interact by exchanging loads.

In [27] the authors deal with heterogeneous networks (i.e. formed by infrastructures of different nature). Specifically they analysed a system composed by two connected networks ($L$ and $M$), assuming that in the presence of a failure in one component of the system (say $L$) such , e.g., an over-load condition, this induces the reconfiguration of the loads in the system $M$, while increasing the loads in the model $L$ itself. The authors showed that this interdependent load growth may induce the shifting in the critical point or, in other terms, the coupling makes the system more susceptible to large failure.

The same result is obtained in [28] where the infrastructures are modelled via the probabilistic Demon model. This model has a percolation threshold above which cascading failures of all size are possible. They show that the coupling of two infrastructure decrease this threshold, hence coupled infrastructures are more susceptible to large-scale failures.

In [29] the effects of geographical interdependencies are analysed considering a simplified version of real water and electric network. They assume that a tunable parameter captures the conditional probability that a failure in the electric network is propagated into the water one. In this way they show that the interdependent effects are more significant at low removal fractions for aspects related with the functionality of the network, and at high removal fractions for aspects related with the topology.

In the following, we will detail the results reported in [21], about the possible coupling effects between an electric and a telecommunication network. Specifically, we will describe how a given perturbation affecting the Italian high voltage electrical transmission network (according to the model introduced before), may be propagated on the GARR Internet network (the GARR model is illustrated in the Appendix 6.8).

In order to describe the first order dependencies among these two networks, we assumed that nodes "geographically" close are also functionally related. Then, each GARR node has been assumed to be functionally connected to the closest electric node. Moreover, it is assumed that each GARR node is connected only to one electric node, whereas any electric node can be linked to more than one GARR node.

Under normal conditions, the energy supplied by the electric grid is supposed to be sufficient to feed all the GARR nodes (i.e. all GARR nodes are in the on state). Then, if any kind of fault begins to affect the electric power grid, the current power flow configuration is perturbed and, consequently, it varies the capability of the grid to properly dispatch electricity to its loads. After such a kind of perturbation, if the actual power results no longer sufficient to supply a given GARR node such a node is put in an `off` state, according to:

$$\text{GARR node } k = \begin{cases} \texttt{on} & \text{if } \tilde{P}_i \geq \alpha P_i^{(0)} \\ \texttt{off} & \text{otherwise} \end{cases} \tag{6.2}$$

where $0 \leq \alpha \leq 1$ is a suitable parameter which determines the strength of the "coupling" between the two networks. In other words, if the electrical node is not able to dispatch a sufficient power (i.e. at least the $\alpha$ fraction of the normal power) then the communication node is switched–off and the GARR network results to be perturbed. In [21] it is assumed for the coupling parameter $\alpha$ the value 0.75 which represents a trade-off between a strong interdependence coupling ($\alpha$ close to one) and a substantial absence of interdependence (with $\alpha$ close to zero).

Hence, to evaluate the functional dependency of the GARR network with respect to the high voltage network, this latter has been perturbed by randomly removing a line ($\xi = 1$). For each instance, the actual power distribution is evaluated using (6.6) and, eventually, adopting the re-dispatching strategy.

The procedure has been iterated considering the concurrent removal, from the electric power grid, of two ($\xi = 2$) and three ($\xi = 3$) lines at once.

Figure 6.5 reports the result for different perturbation severity ($\xi = 1,2,3$), it shows that the effect of the perturbation on the GARR network consists in the increase of the average delivery time in the "normal" regime, where the congested phase onsets, remains essentially unchanged.

To evaluate the global impact that a failure in the electric grid induces on the GARR network, one can compare the degradation in the electric QoS defined as

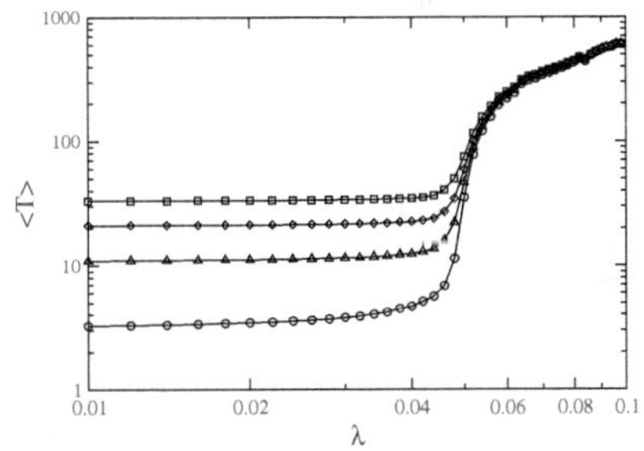


**Fig. 6.5** Log–log scale representation of the behaviour of the average delivery time $< T >$ as a function of the traffic level $\lambda$ in the unperturbed GARR network (circles), and in presence of a perturbation on the electric grid of strength $\xi = 1$ (triangles), $\xi = 2$ (diamonds), $\xi = 3$ (squares).

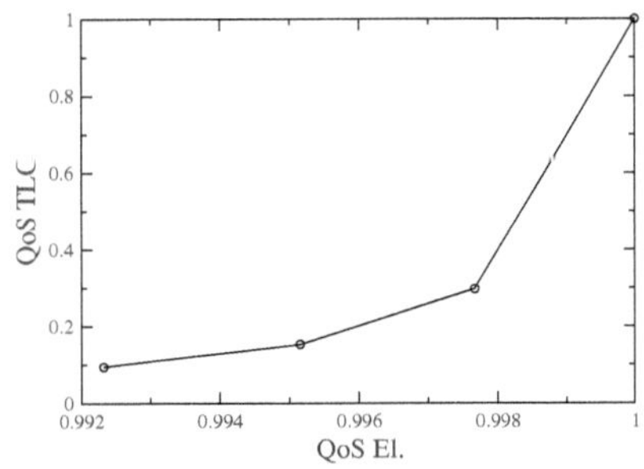


**Fig. 6.6** Variation of the *QoS* value for the communication infrastructure (*QoS TLC*) with respect to the variation of the *QoS* of the electrical network (*QoS EL*).

in (6.1) with the correspondent degradation experiensed by the QoS of the GARR. To this end in [20] the *QoS* of a communication network is defined as

$$QoS_{TLC} = \frac{\frac{m}{M}}{\frac{<T>}{<T_0>}} \tag{6.3}$$

where $m$ and $M$ are the count of generated and dispatched packets, respectively, while $< T >$ and $< T_0 >$ are, respectively, the average delivery times in the normal phase for the perturbed and the unperturbed network.

Figure 6.6 shows the relation that exist between the two QoSs. It is evident that, under the undertaken hypotheses, there is a significant amplification of the effect, on the communication network of a fault occurring in the electrical infrastructure. This emerges because, when the electric network is affected by a fault, it is anyhow able (through the re–dispatching procedure) to partially recover its operability level but the same does not hold for communication network which, despite the adoption of an alternative routing strategy (after the introduction of the faults in the GARR network, new RTs are evaluated for eliminating the `off` state nodes from the data paths) it is not able to fully recover its operability and to grant an acceptable *QoS* (due to the small size of the GARR network).

## 6.5 Multi Interdependency Model

The models described in the previous sections assume, implicitly, that dependency phenomena are induced by a single type of mechanism, generally the geographical proximity. However, as emphasised before, we need to consider a multitude of not exclusive mechanism of dependencies. Each of them represents a peculiar concept of proximity that defines, for each element, a specific set of the first order neighbours. This is schematically illustrated in Figure 6.7. Here each layer represents a specific mechanism of dependency, i.e. a dimension along which the different elements interact. For example, the presence of a link in the *geographical* layer, among the $A_1$ and $B_1$, means that they are geographically close. On the other side, the presence of a link among $A_2$ and $C_2$ in the *cyber* layer means that this two element are neighbours from a cyber point of view, and so on.

Some layers represent the exchange paths for physical quantities (e.g. raw material, electricity), others the exchange paths for immaterial elements (e.g. services), others codify the propagation of some type of failures, etc. Due to the different underlying mechanisms, in each layer the links are characterised by specific set of properties. Indeed, in some cases the links can be assumed bi-directional (as in the case of geographical proximity) while in other case they have a specific direction; some are characterised by a time delay other by an attenuation/amplification gain coefficient. Notice that in Figure 6.7 the "real" elements are represented by the vertical oval, while the circles present in each layer represent the projection of the element with respect to the specific concept of proximity.

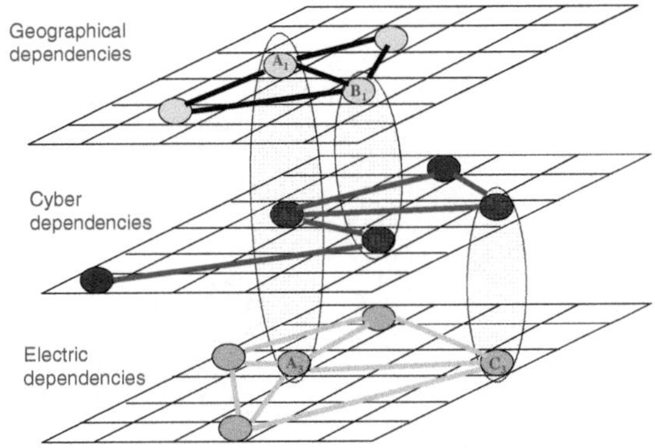


**Fig. 6.7** Multi-layer vision of interdependent infrastructure. Each layer represent a mechanism of dependency that defines a specific set of the element's neighbours.

Having to deal with a scenario as the one presented in Figure 6.7 results in a very hard task, especially because it is required to model both the dynamics of each network (i.e. in each layer) and the propagation mechanisms among the networks (i.e. the transversal dynamics). To this end in [30] a specific modelling framework, labelled "CISIA", is proposed. Specifically, in CISIA each entity is modelled as a black-box able to acquire inputs from the different networks, merge them in accordance with the input-output behaviour of the element and then spread several outputs each one on its specific network. Hence, a CISIA model is composed by a $n$ elements that interact via $m$ different mechanisms each one codified via a specific incidence matrix (i.e. a layer in Figure 6.7).

In Figure 6.8 we report the networks discovered analysing with CISIA a portion of the cost near to Rome. Here co-exist 7 infrastructures: sea ports (2), high voltage electric grid (35); TLC (141); railway (27); high-way (23); urban areas (6); airports (2) (the number in the branches indicates how many entities have been considered inside the model). Each one of these infrastructures is a layer in Figure 6.8 where the horizontal links model the *intra-domain* relationships. The *inter-domains* relationships, i.e. the link existing among entities belonging to different infrastructures, are represented by the transversal connections. To model this scenario, 13 different concept of proximity (each one modelled by its incidence matrix) were introduced for a total of more than 350 dependency links.

It is interesting to note that, if one neglects the peculiarities that characterise each mechanism of dependency, or in other words collapsing the different layer of Figure 6.7 into a single one, it emerges a scale-free like structure (see Figure 6.9). This figure shows the presence of a huge number of peripheral nodes with a limited number of connections and with low clustering coefficient. On the other side, the nodes into the centre show an high number of links and they form strongly connected clusters. However, the studies about the properties of a network of networks are still very preliminarily and there not effective results about such a topic.

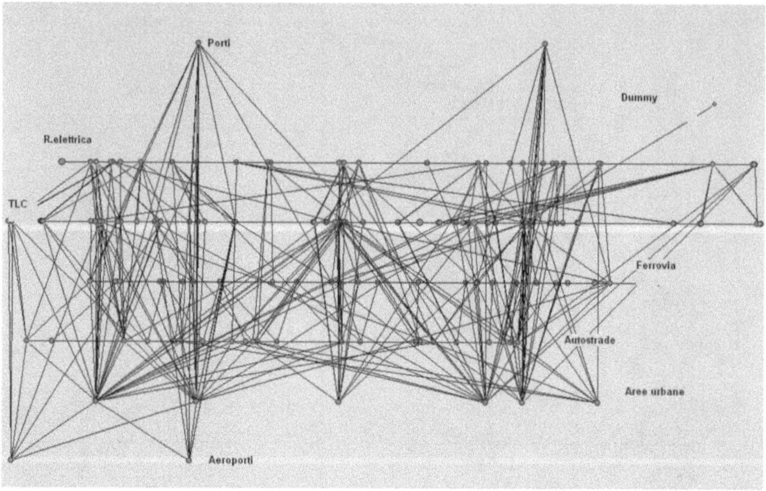


**Fig. 6.8** Topological graph of seven infrastructures that are coupled via 13 dependency's mechanisms.

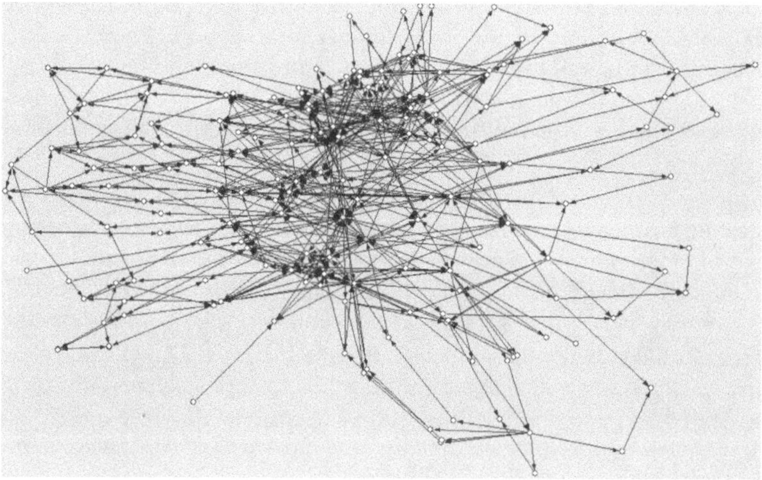

**Fig. 6.9** Planar graph of all dependencies existing among the analysed infrastructures.

## 6.6 Conclusions

In this paper we illustrate some of the results achieved using the Complex Networks approach in the framework of the Critical Infrastructures. Such a framework represents an interesting and important field of application for the studies on complex networks, due to its relevance and the its complexity. Indeed, even if the models based only on the topological structure represent a very crude approximation of any real infrastructure, they have been able to emphasise some peculiarities previously neglected (e.g. the scale-free nature of many of them, with the consequent implications for their robustness). Actually, the studies are proceding also to include some sort of dynamics, in order to analyse the functional properties, and to consider the phenomena induced by the coupling of two or more infrastructures. However, these goals appear more challenging due to the intrinsic complexity and the difficulties to achieve valuable data. The preliminarily results emphasise both that the infrastructures appear more fragile when they are interdependent and that the functional and structural properties change differently under topological perturbations.

## 6.7 Appendix - DC Power Flow Model

A widely adopted method to estimate the electrical power flow in a grid is the *DC power flow model* [31]. Such a model assumes a linear relationship between the active power flow through the lines and the power input into the nodes. They can be formulated as follows:

$$F_{km} = \frac{\theta_k - \theta_m}{x_{km}} \tag{6.4}$$

where $x_{km}$ is the reactance of the line connecting nodes $k$ and $m$, $F_{km}$ is the active power flow on this line and $\theta_k, \theta_m$ are the voltage phase of the $k$-th and $m$-th node. Summing on all branches connected to node $i$, the power flow of that node $P_i$ is

$$P_i = \sum_j F_{ij} = \theta_i \sum_j x_{ij}^{-1} - \sum_j \frac{\theta_j}{x_{ij}} \tag{6.5}$$

The power flow on the network can be written in a matrix form as:

$$\mathbf{P} = \mathbf{B}\theta \tag{6.6}$$

where $\theta$ and $\mathbf{P}$ are, respectively, the vector composed by voltage phase and the electrical power at each of the $N$ nodes of the grid, and $\mathbf{B}$ is a $N \times N$ matrix ($B_{km} = -1/x_{km}$ and $B_{kk} = \sum_l 1/x_{kl}$). The rank of $\mathbf{B}$ is $N-1$ since the network must comply the conservation condition $\sum_{i=1}^{N} P_i = 0$ (notice that source nodes are characterized by positive $P_i$ values, junction by vanishing values, loads by negative values).

To solve the system, an equation is removed and the associated link is chosen in a way to introduce a reference node whose phase angle is arbitrarily set to $\theta = 0$.

Two constraints must be imposed to ensure the physical correctness of the solution. These comes from the fact that the DC power flow results from the elimination of the imaginary part of the current equations, under the hypothesis that power phase angles differences are small. For this reason, it should be, $\forall(k,m)$

1. $\theta_k < 30$ degrees
2. $\mid F_{km} \mid < F_{km}^{max}$ (where $F_{km}^{max}$ is some specified limiting power flux on the link between nodes $k$ and $m$).

If constraint 1.) is not fulfilled, the inductive part of the electrical flux cannot be disregarded and eq. (6.6) does not hold. Constraint 2.) is a technological limit, relating to the specific line's impedance. A too large flux produces an unendurable heat, normally prevented by *ad hoc* elements which disconnect the line.

Then, after having selected a $\mathbf{P}^{(0)}$ one can use (6.6) and (6.4) to evaluate the resulting power flux $F_{ij}$ along the lines and the phase angles $\theta_i$.

## 6.8 Appendix - The GARR Model

As a real telecommunication network, we investigate the traffic dynamics on the Italian high–bandwidth backbone of the Internet network dedicated to link universities and research institutions (GARR) which topological network consists of $N = 39$ nodes and $E = 58$ arcs.

To model the dynamic functioning of the GARR one can consider a simplified model of traffic of data on a network [32]. Here data packets are generated by an Origin Node (ON) and are directed towards a Destination Node (DN). The size of data packets is supposed to be infinitesimal and no hypothesis on the arcs' capacity (in terms of bandwidth) is done.

Hence at each time–step, each node might perform two basic actions:

- send a packet to a nearest neighbour node (i.e. a node directly connected to it);
- receive one or more packets from its nearest neighbour nodes.

In order to define the route that the packets follow from the ON to the DN, each node contains two basic elements: (a) a buffer (unlimited in size) allowing the received packets to form a queue and (b) a routing table (RT) which associates, for each DN, two different nodes $j_1$ and $j_2$, both belonging to its nearest neighbours and each of them being part of a different minimum–path for reaching the DN. The transit packet will be directed toward one of these two nodes according to a probabilistic rule [33]. Packet dispatching takes place using a FIFO strategy.

Notice that the RT is evaluated on the current network's topology. In the presence of a failure (arc or node removal), the RT is re–evaluated to eliminate the removed element from the different paths.

The amount of traffic present in the network is measured by the variable $\lambda$ which measures the frequency with which nodes emit packets ($0 \leq \lambda \leq 1$). According to this definition $\lambda = 0.1$ represents a level of traffic where, at each time step, 10% of the $N$ nodes of the network generate a packet directed toward a randomly chosen set of DNs. This dynamic is iterated for a large number of time–steps. Packets are received by their DN after a certain delivery time $\tau$ which averaged value can be assumed as an indicator of the efficiency of the network. Specifically the "average delivery time" $< T >$ is

$$< T >= \frac{1}{m} \sum_{i=1}^{m} \tau_i \tag{6.7}$$

where $\tau_i$ is the delivery time of the packet $i$ and $m$ are the packets correctly delivered within the simulation time.

The behaviour of the network, produced by the action of the basic rules of the model, produces a packet's dynamics which, as a function of the traffic level $\lambda$, can be ascribed to two different phases: a *normal* phase, at $\lambda < \lambda_c$, where the $< T >$ behaviour is a (slowly) linearly increasing function of $\lambda$. When $\lambda > \lambda_c$ a *congested* phase takes place, producing a rapid, non linear increase of $< T >$. A typical behaviour of the quantity $< T >$ as a function of the traffic $\lambda$ for our model of the GARR network is reported in Figure 6.5. The *congested* phase originates by the presence of buffers which, for large enough traffic values (depending on network size and topology) start filling at a rate larger than their discharge rate.

## References


1. EU Commission, Communication from the Commission on a European Programme for Critical Infrastructure Protection, COM(2006)786 (2006)
2. Rosenbush, S.: Satellite's death puts millions out of touch. USA Today, May 21 (1998)
3. Bologna, S., Setola, R.: The Need to Improve Local Self-Awareness in CIP/CIIP. In: Proc. First IEEE International Workshop on Critical Infrastructure Protection (IWCIP 2005), November 3-4, pp. 84–89 (2005)
4. Italian Government Working Group on Critical Information Infrastructure Protection (PIC), La Protezione delle Infrastrutture Critiche Informatizzate - La Realtá Italiana (March 2004)
5. Government of Canada, Office of Critical Infrastructure Protection and Emergency Preparedness (OCIPEP), Threats to Canada's Critical Infrastructure, TA03-001 (March 2003)
6. Suter, M., Brunner, E.: International CIIP Handbook 2008/2009. Center for Security Studies, ETH Zurich (2008)
7. U.S. General Accounting Office, Critical Infrastructure Protection: Challenges for Selected Agencies and Industry Sectors, GAO-03-233 (February 2003), `http://www.gao.gov`
8. Casalicchio, E., Donzelli, P., Setola, R., Tucci, S.: Modelling and Simulation of Interdependent Critical Infrastructure: The Road Ahead. In: Modelling to Computer Systems and Networks, pp. 143–157. Imperial College Press, London (2006)
9. Pederson, P., Dudenhoeffer, D., Hartley, S., Permann, M.: Critical Infrastructure Interdependency Modelling: A Survey of US and International Research. Idaho National Lab, INL (2006)
10. Tools and Techniques for Interdependency Analysys, Delivery D2.2.1 of IRRIIS project, `http://www.irriis.org`
11. Albert, R., Albert, I., Nakarado, G.I.: Structural vulnerability of the North American power grid. Phys. Rev. E 69, 025103(R) (2004)
12. Crucitti, P., Latora, V., Marchiori, M.: A topological analysis of the Italian electric power grid. Physica A 338, 92 (2004)
13. Pastor–Satorras, R., Vazquez, A., Vespignani, A.: Dynamical and correlation properties of the Internet. Phys. Rev. Lett. 87, 258701 (2001)

14. Issacharoff, L., Bologna, S., Rosato, V., Dipoppa, G., Setola, R., Tronci, F.: A dynamical model for the study of complex system's interdependence. In: Proc. of the International Workshop on Complex Network and Infrastructure Protection (CNIP 2006), Rome, March 28-29, p. 276 (2006)
15. Tiriticco, F., Bologna, S., Rosato, V.: Topological properties of high-voltage electrical transmission networks. Electr. Power Syst. Res. 77, 99 (2006)
16. Watts, D.J., Strogartz, S.H.: Collective dynamics in small–world networks. Nature 393, 440 (1998)
17. Jeong, H., Mason, S.P., Barabasi, A.L., Oltvai, Z.N.: Lethality and centrality in protein networks. Nature 411, 41 (2001)
18. Albert, R., Barabasi, A.L.: Statistical mechanics of complex networks. Rev. Mod. Phys. 74, 47 (2002)
19. Amaral, L.A.N., Scala, A., Barthelemy, M., Stanley, H.E.: Classes of small-world networks. Proc. Nat. Acad. Sci. USA 97, 11149 (2000)
20. Rosato, V., Issacharoff, L., Bologna, S.: Influence of the topology on the power flux of the Italian high-voltage electrical network (submitted to Europhysic Letter)
21. De Porcellinis, S., Issacharoff, L., Meloni, S., Rosato, V., Setola, R., Tiriticco, F.: Modelling Interdependent Infrastructures using Interacting Dynamical Models. Int. J. Critical Infrastructures (IJCIS) 4(1/2), 63–79 (2008)
22. Macdonald, R., Bologna, S.: Advanced Modelling and Simulation Methods and Tools for Critical infrastructure Protection. Tech. Rep., ACIP project report (June 2001)
23. U.S. and Canada Power System Outage Task Force, Final Report on the August 14, 2003 Blackout in the United States and Canada: Causes and Recommendations (April 2004), `https://reports.energy.gov/`
24. Rinaldi, S., Peerenboom, J., Kelly, T.: Identifying Understanding and Analyzing Critical Infrastructure Interdependencies. IEEE Control System Magazine, 11–25 (2001)
25. Benoit, R.: A Method For The Study Of Cascading Effects Within Lifeline Networks. Int. Journal of Critical Infrastructure 1, 86–99 (2004)
26. Kurant, M., Thiran, P.: Layered Complex Networks. Phys. Rev. Lett. 96, 138701 (2006)
27. Newman, D.E., Nkei, B., Carreras, B.A., Dobson, I., Lynch, V.E., Gradney, P.: Risk Assessment in Complex Interacting Infrastructure Systems. In: Proc. of IEEE 38th Annual Hawaii International Conference on System Sciences, January 03-06, p. 63 (2005)
28. Carreras, B.A., Newman, D.E., Gradney, P., Lynch, V.E., Dobson, I.: Interdependent Risk in Interacting Infrastructure Systems. In: Proc. 40th Annual IEEE Hawaii International Conference on System Sciences (2007)
29. Duenas-Osorio, L., Craig, J.I., Goodno, B.J., Bostrom, A.: Interdependent Response of Networked Systems. Journal of Infrastructure Sytems, 185–194 (2007)
30. De Porcellinis, S., Panzieri, S., Setola, R., Ulivi, G.: Simulation of Heterogeneous and Interdependent Critical Infrastructures. Int. J. Critical Infrastructures (IJCIS) 4(1/2), 110–128 (2008)
31. Wood, P.A.J., Wollenberg, B.F.: Power Generation, Operation and Control. John Wiley, New York (1984)
32. Rosato, V., Issacharoff, L., Caligiore, D., Tiriticco, F., Meloni, S.: Is the topology of the Internet network really fit to its function? Physica A 387, 1689–1704 (2008)
33. Echenique, P., Gomez-Gardenes, J., Moreno, Y.: Improved routing strategies for Internet traffic delivery. Phys. Rev. E 70, 056105 (2004)

# Part III
# Sensor Networks

# Chapter 7
# Distributed Maximum Likelihood Estimation over Unreliable Sensor Networks

Giuseppe C. Calafiore and Fabrizio Abrate

**Abstract.** In this chapter we consider a network of sensing nodes where each sensor may take at each time iteration a noisy linear measurement of some unknown parameter that we wish to estimate. We study a distributed consensus diffusion scheme, which allows every node to compute an estimate of the unknown parameter that asymptotically converges to the "true" parameter value. The diffusion scheme relies only on bidirectional communication among neighboring nodes. A measurement update and a network diffusion phase are performed across the network at each iteration, and then each node computes a local least-squares estimate of the unknown parameter. We prove that the local estimates converge to the true parameter value, under suitable hypotheses. The proposed scheme works on networks with dynamically changing communication topology, thus being robust to unreliable communication links and failures in measuring nodes.

## 7.1 Introduction

Small low-cost and low-power devices with sensing abilities and which can perform local data processing and can communicate with other sensors in a network are nowadays commonly available. Even if each sensor node has limited storage capacity and computation power, the network as a whole can perform complex tasks. Sensor networks are progressively substituting wired sensors, mainly for applications in commercial and industrial endeavors, in order to manage data that would be difficult or expensive to deal with using wired sensors. Many applications are now available in various areas, including environmental monitoring, surveillance, object

Giuseppe C. Calafiore
Politecnico di Torino, Corso Duca degli Abruzzi 24, 10129 Torino, Italy
e-mail: giuseppe.calafiore@polito.it

Fabrizio Abrate
Politecnico di Torino, Corso Duca degli Abruzzi 24, 10129 Torino, Italy
e-mail: fabrizio.abrate@polito.it

tracking, collaborative information processing, traffic monitoring and mobile agents control, see for instance [1, 5, 10].

Estimation and fusion of data coming from sensors is one of the most challenging tasks in each of these application fields, and various schemes for sensor data fusion exist, both centralized or distributed. In centralized schemes, data is sent to a fusion center, that computes the best possible estimate of the parameter (e.g., the Maximum Likelihood (ML) estimate), but the drawback is that high communication overhead is imposed on the network. Moreover, persistent communication induces high energy consumption over the network, and the energy budget is often a critical parameter for smart sensors. In distributed processing schemes, instead, data is only exchanged among neighbors, and sensors carry out local computation in order to obtain a good estimate of the unknown parameter of interest. Distributed processing has several advantages with respect to centralized processing: there is no central data fusion center, each sensor can compute the estimates on its own without having any knowledge of the whole network, and communication occurs only when needed and among neighbors. Many algorithms for distributed estimation and tracking exist, see for instance [2, 6, 9, 17].

In [6], an iterative distributed algorithm for linear minimum mean-squared-error (LMMSE) estimation in sensor networks is proposed, while in [2] consensus among distributed noisy sensors observing an event is addressed. In [12, 15, 16], a distributed version of the Kalman filter (DKF) is analyzed for distributed estimation of time-varying parameters.

In this chapter we start from the setup of [18], and analyze a completely distributed consensus diffusion scheme for linear parameter estimation on networks with unreliable links. Each node in the network may take at each time $t$ a noisy linear measurement of the unknown parameter. The nodes measurement noise covariances are allowed to be time-varying, thus permitting to model, for instance, sensor failures or measurement precision degradation. The network topology may also change with time. We prove that if the *frequency of connectedness* of the superposition of sequences of communication graphs is lower-bounded by a quantity proportional to the logarithm of time then, as $t \to \infty$, the estimates at each local node converge to the true parameter value in the mean square sense. This result may be considered as an extension of the results obtained in [18], where the convergence of estimation is only proved when sensors take a finite number of measurements; further details in this respect are discussed in Remark 7.3.

The rest of this chapter is organized as follows. The proposed distributed scheme for parameter estimation is introduced in Section 7.2. Section 7.3 contains our main convergence results. Numerical examples are presented in Section 7.4, and conclusions are drawn in Section 7.5.

### 7.1.1 Notation

For a matrix $X$, $X_{ij}$ denotes the element of $X$ in row $i$ and column $j$, and $X^\top$ denotes the transpose of $X$. $X > 0$ (resp. $X \geq 0$) denotes a positive (resp. non-negative) matrix, that is a matrix with all positive (resp. non-negative) entries. $\|X\|$ denotes the spectral (maximum singular value) norm of $X$, or the standard Euclidean norm, in case of vectors. For a square matrix $X \in \mathbb{R}^{n,n}$, we denote with $\sigma(X) = \{\lambda_1(X), \ldots, \lambda_n(X)\}$ the set of eigenvalues, or *spectrum*, of $X$, and with $\rho(X)$ the spectral radius: $\rho(X) \doteq \max_{i=1,\ldots,n} |\lambda_i(X)|$, where $\lambda_i(X)$, $i = 1, \ldots, n$, are the eigenvalues of $X$. $I_n$ denotes the $n \times n$ identity matrix, and $\mathbf{1}_n$ denotes a $n$-vector of ones; subscripts with dimensions are omitted whenever they can be inferred from context.

## 7.2 The Consensus-Based Estimation Scheme

### *7.2.1 Preliminaries*

Consider $n$ distributed sensors (nodes), each of which may take at time $t$ a measurement of an unknown parameter $\theta \in \mathbb{R}^m$ according to the linear measurement equation

$$y_i(t) = A_i(t)\theta + v_i(t), \quad i = 1, \ldots, n; \; t = 0, 1, \ldots$$

where $y_i(t) \in \mathbb{R}^{m_i}$ is the noisy measurement from the $i$-th sensor at time $t$, $v_i(t) \in \mathbb{R}^{m_i}$ is measurement noise, and $A_i(t) \in \mathbb{R}^{m_i,m}$ is the time-varying regression matrix.

We assume $v_i(t)$ to be independent zero mean Gaussian random vectors, with possibly time-varying covariances $\Sigma_i(t)$. Allowing the covariance matrices to be time-varying helps modeling realistic circumstances. If a sensor has a correct measurement at time $t$, we set its covariance matrix to $\Sigma_i(t) = \Sigma_i$, where $\Sigma_i$ is fixed and determined by the technical characteristics of the $i$-th sensor. If instead the sensor does not have a valid measurement at time $t$ (for any reason, including sensor failures), then we set $\Sigma_i(t)^{-1} = 0$, thus neglecting the measurement.

Notice that if full centralized information were available, the optimal Maximum Likelihood (ML) estimate $\hat{\theta}_{\mathrm{ml}}$ of the parameter $\theta$ could be obtained. Defining the quantities

$$P_{\mathrm{ml}}(t) \doteq \sum_{k=0}^{t-1} \sum_{j=1}^{n} A_j^\top(k)\Sigma_j^{-1}(k)A_j(k), \quad q_{\mathrm{ml}}(t) \doteq \sum_{k=0}^{t-1} \sum_{j=1}^{n} A_j^\top(k)\Sigma_j^{-1}(k)y_j(k), \tag{7.1}$$

the ML estimate of $\theta$ is

$$\hat{\theta}_{\mathrm{ml}}(t) \doteq P_{\mathrm{ml}}{}^{-1}(t)q_{\mathrm{ml}}(t),$$

and the ML error covariance matrix is

$$Q_{\mathrm{ml}}(t) \doteq P_{\mathrm{ml}}{}^{-1}(t). \tag{7.2}$$

However, we assume it is not possible (due to communication constraints, etc.) to construct the optimal centralized estimate. Instead, our objective is to exploit peer-to-peer information exchange among communicating nodes in order to build "good" *local* estimates of $\theta$. We shall prove in Section 7.2.3 that under suitable hypotheses, all local estimates converge asymptotically to the true parameter $\theta$, in mean square sense.

We describe the communication structure among nodes using graph formalism, see Sections A.1 and A.2 of [3] for notation and preliminary results on graphs. Let $\mathscr{V} = \{1,2,\ldots,n\}$ denote the set of nodes of the sensor network, and let $\mathscr{E}(t)$ denote the set of active links at time $t$; i.e., nodes $(i,j)$ can communicate at time $t$ if and only if $(i,j) \in \mathscr{E}(t)$. The time-varying communication network is represented by the graph $\mathscr{G}(t) = (\mathscr{V}, \mathscr{E}(t))$.

We denote with $\mathscr{N}_i(t)$ the set of nodes that are linked to node $i$ at time $t$ (note that it is assumed that $i \notin \mathscr{N}_i(t)$), and with $|\mathscr{N}_i(t)|$ the cardinality of $\mathscr{N}_i(t)$; $|\mathscr{N}_i(t)|$ is called the *spatial degree* of node $i$ in graph $\mathscr{G}(t)$. Following the notation in [18], we define the *time degree* of node $i$ as the number of measurements that node $i$ has collected up to time $t$, that is $d_i(t) = t+1$, and the space-time degree as

$$d_i^{ST}(t) = d_i(t) + \sum_{j \in \mathscr{N}_i(t)} d_j(t) = (t+1) + (t+1)|\mathscr{N}_i(t)| = (1 + |\mathscr{N}_i(t)|)(t+1).$$

With this position, we introduce the weights that shall be employed for information averaging among neighboring nodes. To this end, we use the Metropolis weights (see [18]), defined as

$$\begin{aligned} \tilde{W}_{ij}(t) &= \min\left(1/d_i^{ST}(t), 1/d_j^{ST}(t)\right) = \frac{1}{\max\left(d_i^{ST}(t), d_j^{ST}(t)\right)} \\ &= \frac{1}{1+\max\left(|\mathscr{N}_i(t)|, |\mathscr{N}_j(t)|\right)} \cdot \frac{1}{t+1}, \quad \text{for } (i,j) \in \mathscr{E}(t), i \neq j. \end{aligned} \tag{7.3}$$

The distributed space-time diffusion scheme is described in the next section.

### 7.2.2 Distributed Space-Time Diffusion Scheme

The proposed distributed iterative scheme performs a temporal update phase and a (network) spatial update phase. Using the same notations of [18], we assume that each node keeps as local information a *composite information matrix* $P_i(t)$ and a *composite information state* $q_i(t)$.

At time $t$ a measurement is collected at each node, and a temporal (measurement) update phase is performed locally at the nodes. This phase amounts to computing

$$P_i(t_+) = \frac{t}{t+1} P_i(t) + \frac{1}{t+1} A_i^\top(t) \Sigma_i^{-1}(t) A_i(t) \tag{7.4}$$

$$q_i(t_+) = \frac{t}{t+1} q_i(t) + \frac{1}{t+1} A_i^\top(t) \Sigma_i^{-1}(t) y_i(t), \tag{7.5}$$

where each node only has to know its local information $P_i(t)$, $q_i(t)$, and the current time degree $d_i(t)$, which is actually constant for all nodes and equal to $d_i(t) = t+1$.

Note that the temporal updates are finished instantaneously at each node, thus $t+$ and $t$ are essentially the same integer. Notice also that the initial value $P_i(0)$, $q_i(0)$ are irrelevant.

After the temporal update, each node has to broadcast its space degree and its current values of $P_i(t_+)$ and $q_i(t_+)$ to its neighbors. At this point, a spatial update phase is performed. Considering (7.3) and defining

$$W_{ij}(t) \doteq \begin{cases} (t+1)\tilde{W}_{ij}(t) & \text{if } (i,j) \in \mathscr{E}(t) \\ 1 - \sum_{j \in \mathscr{N}_i(t)} W_{ij}(t) & \text{if } i = j \\ 0 & \text{otherwise,} \end{cases} \tag{7.6}$$

the $i$-th node updates the composite information matrix and composite information state at time $t+1$ as follows:

$$P_i(t+1) = P_i(t_+) + \sum_{j \in \mathscr{N}_i(t)} W_{ij}(t) \big( P_j(t_+) - P_i(t_+) \big) \tag{7.7}$$

$$q_i(t+1) = q_i(t_+) + \sum_{j \in \mathscr{N}_i(t)} W_{ij}(t) \big( q_j(t_+) - q_i(t_+) \big). \tag{7.8}$$

Merging the temporal update phase and the spatial update phase leads to the following proposition (see Section A.5 of [3] for a proof).

**Proposition 7.1.** *For $t = 1, 2, \ldots$ the composite information matrix and composite information state at each node $i = 1, \ldots, n$ are given by the expressions*

$$P_i(t) = \frac{1}{t} \sum_{k=0}^{t-1} \sum_{j=1}^{n} \Phi_{ij}(t-1;k) A_j^\top(k) \Sigma_j^{-1}(k) A_j(k)$$

$$q_i(t) = \frac{1}{t} \sum_{k=0}^{t-1} \sum_{j=1}^{n} \Phi_{ij}(t-1;k) A_j^\top(k) \Sigma_j^{-1}(k) y_j(k),$$

*where*

$$\Phi(t-1;k) \doteq W(t-1) \cdots W(k). \tag{7.9}$$

*Remark 7.1.* [ Notice that the recursions in (7.7), (7.8) are well suited for distributed implementation, since at each step each node only needs to know the current time instant, and the space-time degrees and local informations of its neighbors. In particular, the nodes do not need global knowledge of the communication graph, or even of the number of nodes composing the network. Also, no matrix inversion need be performed in this recursion. Notice further that the expressions in Proposition 7.1, which are useful for a-posteriori analysis, do *not* describe the actual computations performed by the nodes, which use instead the recursions (7.7), (7.8).]

*Remark 7.2.* [ (Measurement and consensus time scales) In some practical cases it may happen that communication occurs at more frequent intervals than observations, or vice versa, and this would lead to a distinction between the measurement time scale and the distributed averaging one. For notational simplicity in this work we use a single time scale for both the averaging and the observation processes. However, this is done without loss of generality, due to the flexibility introduced by the time-varying nature of the process parameters. More precisely, we assume that $t = 0, 1, \ldots$ represent the time indices at which either a measurement occurs at some node, and/or a consensus averaging step should be performed through the network. If some sensor $i$ does not have a valid measurement at time $t$, we just set $\Sigma_i^{-1}(t) = 0$, thus covering the situation when communication occurs more frequently than observations. Vice versa, if some sensor takes a measurement at $t$ but no actual consensus iteration should be performed at that time, we simply set the consensus weight matrix equal to the identity, i.e., $W(t) = I$ (this means that nodes are only connected with themselves at these specific time instants), thus covering the situation when measurements occur more frequently than consensus steps. Note further that in any time interval $[T, T+\tau]$ in which measurements persistently occur without averaging, the algorithm evolves according to (7.4), (7.5), with (7.7), (7.8) simply reduced to $P_i(t+1) = P_i(t_+)$, $q_i(t+1) = q_i(t_+)$. It can be readily checked that in this case the algorithm yields optimal Maximum Likelihood estimates at each observation step. ]

The properties of the local estimates are discussed in the next section.

### 7.2.3 *Properties of Local Estimates*

At each time when the composite information matrix $P_i^{-1}(t)$ is invertible, each node $i$ in the network is able to compute its local estimate at time $t$ as

$$\hat{\theta}_i(t) \doteq P_i^{-1}(t) q_i(t), \quad i = 1, \ldots, n.$$

The following fact holds.

**Proposition 7.2.** *The local estimate $\hat{\theta}_i(t)$ is an unbiased estimator of $\theta$, that is*

$$\mathrm{E}\left\{\hat{\theta}_i(t)\right\} = \theta.$$

*Moreover, the covariance of the local estimate is given by the expression*

$$\begin{aligned} Q_i(t) &\doteq \mathrm{var}\left\{\hat{\theta}_i(t)\right\} = \mathrm{E}\left\{(\hat{\theta}_i(t) - \theta)(\hat{\theta}_i(t) - \theta)^\top\right\} \\ &= \frac{1}{t^2} P_i^{-1}(t) \left( \sum_{k=0}^{t-1} \sum_{j=1}^{n} \Phi_{ij}^2(t-1;k) A_j^\top(k) \Sigma_j^{-1}(k) A_j(k) \right) P_i^{-1}(t). \quad (7.10) \end{aligned}$$

Note from (7.10) that if we actually want to compute numerically the local covariance $Q_i(t)$, we need to know the occurred sequence of graphs, since this is needed for constructing the entries of $\Phi(t-1;k)$. Notice also that we can upper bound the local covariance as detailed in Lemma 7.1 below. To this end, define

$$\bar{P}_i(t) \doteq tP_i(t) = \sum_{k=0}^{t-1}\sum_{j=1}^{n} \Phi_{ij}(t-1;k)H_j(k) \tag{7.11}$$
$$H_j(k) \doteq A_j^\top(k)\Sigma_j^{-1}(k)A_j(k).$$

The following result holds.

**Lemma 7.1.** *Whenever $\bar{P}_i^{-1}(t)$ is invertible, the covariance matrix of the i-th local estimate satisfies*

$$Q_i(t) \preceq \bar{P}_i^{-1}(t). \tag{7.12}$$

## 7.3 Mean Square Convergence Results

We show in this section that, under suitable hypotheses, as the number of measurements goes to infinity, *all* the local estimates $\hat{\theta}_i(t)$ converge to the true parameter value $\theta$, in the mean square sense. That is, $\lim_{t\to\infty} \|Q_i(t)\| = 0$, for $i = 1,\ldots,n$. Notice that this is not obvious since we allow for the existence in the network of nodes that do not collect an infinite number of measurement, or that even do not collect measurements at all, hence the law of large numbers cannot be trivially applied. The convergence result holds for time-varying network topology, and it is derived under two assumptions. The first condition is a very natural one, and requires that the centralized ML estimate mean square error goes to zero as $t \to \infty$. This condition is actually necessary, since one cannot hope to make the local estimates converge when even the centralized optimal estimate (who ideally has all the available information) does not converge. The second condition is a technical condition needed for proving convergence of the distributed scheme, and it is detailed in the next section. Loosely speaking, this condition requires that the time-varying communication graphs form, at least "rarely" in time, subsequences whose union graph is connected, see Section A.1 of [3] for a definition of graph union. Subsequences such that the union of the graphs in the subsequence forms a connected graph are here called *jointly connected*.

We first state some technical preliminaries. The main theorem is stated in Section 7.3.1.

We start by looking more closely at the structure of the $W(t)$ matrices in (7.6) and of the transition matrices $\Phi(t-1;k)$ in (7.9). First notice that $W(t)$ is non-negative, symmetric, and $W_{ij}(t) > 0$ for $(i,j) \in \mathscr{E}(t)$. Moreover, the diagonal entries of $W(t)$ are strictly positive,[1] and the sum over each row or column of $W(t)$ is equal to one ($W(t)\mathbf{1} = \mathbf{1}$, $\mathbf{1}^\top W(t) = \mathbf{1}^\top$). This means that $W(t)$ is a symmetric and doubly stochastic matrix belonging to the set $\mathscr{M}_{\mathrm{ss}}$, defined as

$$\mathscr{M}_{\mathrm{ss}} \doteq \{R \in \mathscr{M}_{\mathrm{s}} : R = R^T\}. \tag{7.13}$$

[1] This follows from the definition (7.6), since $\sum_{j\in\mathscr{N}_i(t)} W_{ij}(t) = \sum_{j\in\mathscr{N}_i(t)}(1+\max\{|\mathscr{N}_i(t)|,|\mathscr{N}_j(t)|\})^{-1} \leq \sum_{j\in\mathscr{N}_i(t)}(1+|\mathscr{N}_i(t)|)^{-1} = |\mathscr{N}_i(t)|/(1+|\mathscr{N}_i(t)|) < 1$.

where the set $\mathcal{M}_s$ is defined as the set of (row) stochastic matrices. See Section A.2.1 in [3] for more details. Moreover $W(t)$ is *compatible* with graph $\mathcal{G}(t)$, in the sense of the following definition

**Definition 7.1.** For $R \in \mathcal{M}$, we say that the matrix/graph pair $(R, \mathcal{G}(\mathcal{V}, \mathcal{E}))$ is *compatible* if $R_{ij} > 0 \Leftrightarrow (i, j) \in \mathcal{E}$.

$\mathcal{M}$ is the set of non-negative matrices with positive diagonal entries.
Note further that the set of all possible $W(t)$ generated by the time-varying graphs is *finite*, since the set of Metropolis weights one can obtain from a fixed number of nodes is of finite cardinality.

Using the notation defined in (29)–(31) of [3], we write $W(t)$ in the form

$$W(t) = \frac{1}{n}\mathbf{1}\mathbf{1}^\top + Z(t), \quad Z(t) = V(t)D(t)V^\top(t) \tag{7.14}$$

where $V(t) \in \mathbb{R}^{n,n-1}$ is such that $V^\top(t)V(t) = I_{n-1}$, $V(t)V^\top(t) = I_n - \frac{1}{n}\mathbf{1}\mathbf{1}^\top$, $\mathbf{1}^\top V(t) = 0$, and $D(t) = \mathrm{diag}(\lambda_2(t), \ldots, \lambda_n(t)) \in \mathbb{R}^{n-1,n-1}$ is a diagonal matrix containing the last $n-1$ eigenvalues of $W(t)$ arranged in order of non-increasing modulus. Define

$$\Phi(t-1;k) \doteq W(t-1)W(t-2)\cdots W(k), \tag{7.15}$$

$$\Upsilon(t-1;k) \doteq Z(t-1)Z(t-2)\cdots Z(k), \tag{7.16}$$

where it holds that

$$\Phi(t-1;k) = \frac{1}{n}\mathbf{1}\mathbf{1}^\top + \Upsilon(t-1;k). \tag{7.17}$$

### 7.3.1 Main Result

In order to prove our main result on convergence of $\|Q_i(t)\|$ we need to impose an assumption on the connectivity properties of the graph sequence. In its essence, this assumption just requires that any sequence of consecutive graphs of length $k > \bar{k}$ (for some $\bar{k} > 0$) contains a suitable number $N(k)$ of jointly connected subsequences, that is subsequences that contain graphs whose union is connected. Formally, we state the following property.

**Definition 7.2 (RJC property).** An ordered set of graphs $S = \{\mathcal{G}_k(\mathcal{V}, \mathcal{E}_k), k = 0, 1, \ldots\}$ is said to be repeatedly jointly connected (RJC) if there exist finite integers $\bar{k} \geq 0$, $M \geq 0$ such that any ordered sequence from $S$ of length $k > \bar{k}$ contains $N(k) > 0$ subsequences of length no larger than $M$ that are jointly connected, where $N(k) : \mathbb{N} \to \mathbb{N}$ is a non-decreasing function such that $N(k) \to \infty$ for $k \to \infty$.

Notice that the RJC property does not require connectivity at any given time. It only requires that, for sufficiently large $k$, any ordered sequence of $k$ graphs contains $N(k)$ subsequences such that the *union* of graphs in each of these subsequences is connected. The number $N(k)$ of such subsequences is left unspecified for the time being. However, we shall prove shortly that this number need only increase slowly with the logarithm of $k$. This means in turn that as $k$ grows we shall require the existence of only few graph subsequences whose union is connected, an assumption that appears to be very mild in practice. We preliminarily state the following technical lemma.

**Lemma 7.2.** *Let $S = \{\mathscr{G}_k(\mathscr{V}, \mathscr{E}_k), k = 0, 1, \ldots\}$ be an ordered set of graphs having the RJC property, and let $\mathscr{R} = \{R_k, k = 0, 1, \ldots\}$ be a corresponding set of compatible matrices, such that for all k, $R_k \in \mathscr{M}_{ss} \bigcap \bar{\mathscr{R}}$, where $\bar{\mathscr{R}}$ is a set of finite cardinality.*

*For $k > \bar{k}$, $t \geq 1$, let $S(t-k, t-1) = \{\mathscr{G}_{t-k}, \ldots, \mathscr{G}_{t-1}\}$ be an ordered sequence of $k$ graphs from S, and let $\{R_{t-k}, \ldots, R_{t-1}\}$ be a corresponding sequence of compatible matrices from $\bar{\mathscr{R}}$. Define $Z_{t-\tau} = R_{t-\tau} - \frac{1}{n}\mathbf{1}\mathbf{1}^\top$. Let further $[s_1, e_1], [s_2, e_2], \cdots, [s_{N(k)}, e_{N(k)}]$, $s_1 \geq t-k$, $e_{N(k)} \leq t-1$, denote the indices delimiting the $N(k)$ subsequences of $S(t-k, t-1)$ that are jointly connected. Then there exist a $\lambda < 1$ such that*

$$\|Z_{e_i} Z_{e_i - 1} \cdots Z_{s_i}\| \leq \lambda, \quad i = 1, \ldots, N(k); \quad \text{for all } t \geq 1, k > \bar{k}. \tag{7.18}$$

See Section A.3 of [3] for a proof of Lemma 7.2. Notice that the sequence of Metropolis weight matrices $R_t = W(t)$ are compatible with the corresponding communication graphs $\mathscr{G}_t$ and belong to a set of finite cardinality, therefore Lemma 7.2 applies in particular to the sequence $\{Z(t-k), \ldots, Z(t-1)\}$ of matrices of the form (7.14).

**Definition 7.3 (Joint connectivity index).** Let $S$ and $\bar{\mathscr{R}}$ be defined as in Lemma 7.2. The *joint connectivity index* of S with respect to $\bar{\mathscr{R}}$ is defined as

$$\bar{\lambda} \doteq \min \lambda \text{ such that (7.18) holds.} \tag{7.19}$$

Let us briefly discuss the meaning of the result in Lemma 7.2 and of Definition 7.3. We know that if the graph set $S$ is RJC then for sufficiently large $k$ every sequence of length $k$ contains $N(k)$ finite subsequences that are jointly connected. We are interested in the products of $R_k$ matrices (which are actually the Metropolis weight matrices $W(k)$ in our specific application) corresponding to the graphs in each of these jointly connected subsequences. More specifically we are interested in the products of the related $Z_k$ matrices, and Lemma 7.2 states that the norm of any such product is upper bounded by a quantity that is strictly less than one. The joint connectivity index in Definition 7.3 is simply the smallest of these upper bounds.

We now define a specific class of lower-bound functions for $N(k)$.

**Definition 7.4 (Log-RJC property).** Let $S = \{\mathscr{G}_k(\mathscr{V}, \mathscr{E}_k), k = 0, 1, \ldots\}$ be an ordered set of graphs having the RJC property. Let $\mathscr{R} = \{R_k, k = 0, 1, \ldots\}$ be a corresponding set of compatible matrices, such that for all $k$, $R_k \in \mathscr{M}_{ss} \bigcap \bar{\mathscr{R}}$, where $\bar{\mathscr{R}}$ is a set of finite cardinality, and let $\bar{\lambda} < 1$ be the joint connectivity index of $S$ with respect to $\bar{\mathscr{R}}$.

Then, $S$ is said to be logarithmical repeatedly jointly connected (log-RJC) with index $\bar{\lambda}$ with respect to $\bar{\mathscr{R}}$, if there exist a constant $\alpha > 0$ and a finite integer

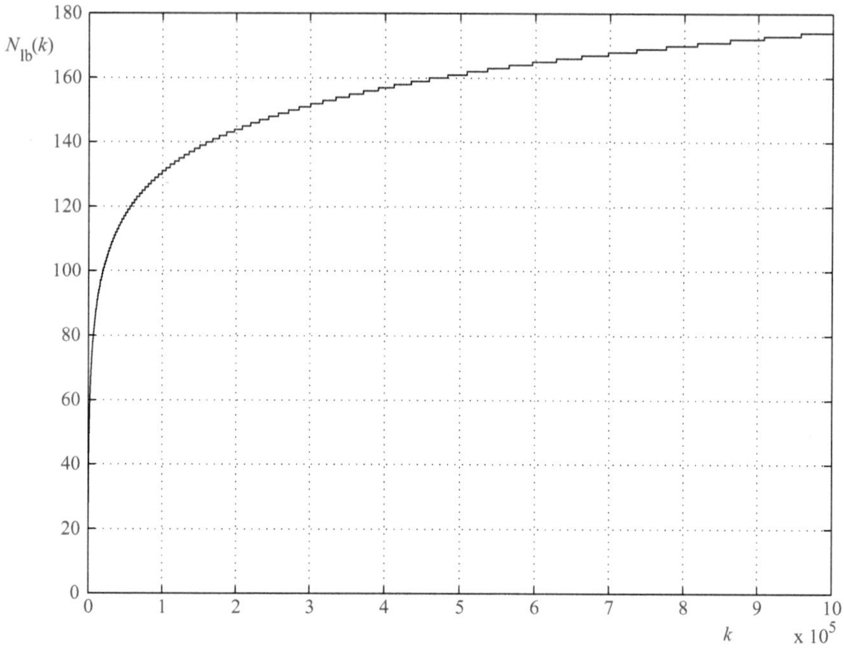


**Fig. 7.1** Plot of $N_{\text{lb}}(k)$, for $\bar{\lambda} = 0.9$, $\alpha = 100$

$\bar{k} \geq \alpha$ such that any ordered sequence from $S$ of length $k > \bar{k}$ contains $N(k) \geq N_{\text{lb}}(k)$ subsequences that are jointly connected, where

$$N_{\text{lb}}(k) \doteq \lceil \max\left(c \log(k/\alpha), 0\right) \rceil, \quad k = 1, 2, \ldots; \quad c \doteq \frac{2}{\log(1/\bar{\lambda})} \tag{7.20}$$

and where $\lceil x \rceil$ denotes the smallest integer larger than or equal to $x$.

Notice that $N_{\text{lb}}(k) = 0$ for $k \leq \alpha$, whereas $N_{\text{lb}}(k)$ grows at sub-linear (specifically logarithmic) rate for $k > \alpha$. Figure 7.1 shows an example of plot of $N_{\text{lb}}(k)$.

The following key technical result holds, see Section A.4 of [3] for a proof.

**Lemma 7.3.** *Let* $S = \{\mathscr{G}_k(\mathscr{V}, \mathscr{E}_k), k = 0, 1, \ldots\}$ *be an ordered set of graphs having the log-RJC property with respect to a compatible sequence of matrices* $\{R_k \in \mathscr{M}_{\text{ss}} \bigcap \mathscr{R}, k = 0, 1, \ldots\}$, *where* $\mathscr{R}$ *is a set of finite cardinality. Define* $Z_k = R_k - \frac{1}{n}\mathbf{1}\mathbf{1}^\top$, $k = 0, 1, \ldots$ *Then,*

$$\lim_{t \to \infty} \sum_{k=1}^{t} \|Z_{t-1} Z_{t-2} \cdots Z_{t-k}\| = \text{constant} < \infty.$$

We can now state the main result of this section in the following theorem.

**Theorem 7.1.** *Let the occurring communication graph sequence* $S = \{\mathscr{G}_k(\mathscr{V}, \mathscr{E}_k), k = 0, 1, \ldots\}$ *have the log-RJC property with respect to the compatible sequence of Metropolis weight matrices* $\{W(k), k = 0, 1, \ldots\}$, *and let* $\|H_j(k)\| \leq C$, *for all* $j = 1, \ldots, n$, $k = 0, 1, \ldots$

*If* $\lim_{t\to\infty} \|P_{\text{ml}}(t)\| = \infty$ *(or, equivalently, if the centralized maximum likelihood error covariance goes to zero), then*

$$\lim_{t\to\infty} \|Q_i(t)\| = 0, \quad i = 1, \ldots, n.$$

**Proof.** Consider the expression of $\bar{P}_i(t)$ in (7.11), and substitute (7.17) to obtain

$$\begin{aligned} \bar{P}_i(t) &= \sum_{k=0}^{t-1}\sum_{j=1}^{n} \frac{1}{n} H_j(k) + \sum_{k=0}^{t-1}\sum_{j=1}^{n} \Upsilon_{ij}(t-1;k) H_j(k) \\ &= \frac{1}{n} P_{\text{ml}}(t) + \sum_{k=0}^{t-1}\sum_{j=1}^{n} \Upsilon_{ij}(t-1;k) H_j(k). \end{aligned} \tag{7.21}$$

Recall now that for any two matrices $A$, $B$ and any norm, applying the triangle inequality to the identity $A = (-B) + (B + A)$, it results that $\|A + B\| \geq \|A\| - \|B\|$. Applying this inequality to (7.21), and taking the spectral norm, we have

$$\|\bar{P}_i(t)\| \geq \frac{1}{n} \|P_{\text{ml}}(t)\| - \|\sum_{k=0}^{t-1}\sum_{j=1}^{n} \Upsilon_{ij}(t-1;k) H_j(k)\|. \tag{7.22}$$

Notice further that

$$\begin{aligned} \|\sum_{k=0}^{t-1}\sum_{j=1}^{n} \Upsilon_{ij}(t-1;k) H_j(k)\| &\leq \sum_{k=0}^{t-1}\sum_{j=1}^{n} |\Upsilon_{ij}(t-1;k)| \cdot \|H_j(k)\| \\ &\leq C \sum_{k=0}^{t-1}\sum_{j=1}^{n} |\Upsilon_{ij}(t-1;k)| \\ [\text{from } \|\Upsilon\|_\infty \leq \sqrt{n}\|\Upsilon\|, \text{ see Section 5.6 of [7]}] &\leq \sqrt{n} C \sum_{k=0}^{t-1} \|\Upsilon(t-1;k)\| \\ [\text{reversing the summation}] &= \sqrt{n} C \sum_{k=1}^{t} \|\Upsilon(t-1;t-k)\|. \end{aligned}$$

Going back to (7.22), we hence obtain that

$$\begin{aligned} \|\bar{P}_i(t)\| &\geq \frac{1}{n} \|P_{\text{ml}}(t)\| - \|\sum_{k=0}^{t-1}\sum_{j=1}^{n} \Upsilon_{ij}(t-1;k) H_j(k)\| \\ &\geq \frac{1}{n} \|P_{\text{ml}}(t)\| - \sqrt{n} C \sum_{k=1}^{t} \|\Upsilon(t-1;t-k)\|. \end{aligned}$$

Recalling (7.16) and applying Lemma 7.3 we now have that there exist $K < \infty$ such that

$$\lim_{t \to \infty} \sum_{k=1}^{t} \|\Upsilon(t-1;t-k)\| = K,$$

therefore

$$\lim_{t \to \infty} \|\bar{P}_i(t)\| \geq \frac{1}{n} \lim_{t \to \infty} \|P_{\mathrm{ml}}(t)\| - \sqrt{n}CK.$$

Since by hypothesis $\lim_{t \to \infty} \|P_{\mathrm{ml}}(t)\| = \infty$, we obtain that

$$\lim_{t \to \infty} \|\bar{P}_i(t)\| = \lim_{t \to \infty} \|P_{\mathrm{ml}}(t)\| = \infty.$$

Finally, from (7.12) it follows that whenever $\bar{P}_i(t)$ is invertible

$$\|Q_i(t)\| \leq \|\bar{P}_i^{-1}(t)\| = \frac{1}{\|\bar{P}_i(t)\|}$$

hence

$$\lim_{t \to \infty} \|Q_i(t)\| \leq \lim_{t \to \infty} \frac{1}{\|\bar{P}_i(t)\|} = 0,$$

which concludes the proof. □

*Remark 7.3.* [ A few remarks are in order with respect to the result in Theorem 7.1. First, we notice that it is not required by the theorem that each individual node collects an infinite number of measurements as $t \to \infty$. Indeed, the local estimate at a node may converge even if this node *never* takes a measurement, as long as the other hypotheses are satisfied. As an extreme situation, even if only one node in the network takes measurements, then local estimates at *all* nodes converge, if the hypotheses are satisfied. These hypotheses basically require that the graph process forms at least seldom in time subsequences whose union is connected, and that the total information collectively gathered by all nodes is sufficient to make an hypothetical centralized maximum likelihood estimate converge to the "true" parameter value as $t \to \infty$. Our hypothesis of joint graph connectivity is consistent with similar hypotheses that appeared in the literature on consensus, formation and agreement problems, see, e.g., [8, 13, 14].

Further, it is worth to underline that the convergence result presented here is quite different from related results given in [18]. The main situation considered in [18] assumes that the total number of measurements collected by the whole set of sensors remains finite as $t \to \infty$; contrary, we allow this number to grow as time grows, which seems a more natural requirement. Besides technicalities, considering the number of measurements to remain finite essentially amounts to assuming that, from a certain time instant on, the network evolves with "spatial" (consensus) iterations only. This in turn permits the authors of [18] to apply standard tools for convergence of products of stochastic matrices, see [18] and the references therein. These results cannot be directly applied to our setup, due to the persistent presence of new measurements, which acts as a forcing term in the local iterations (7.4), (7.5). In some sense, [18] deals with convergence of a particular type of homogeneous (unforced) system, whereas we here deal with convergence of a system persistently "forced" by external measurements. As a matter of fact, Section 6 of [18] also contains an extension to the case of infinite measurements, which is stated in Theorem 4 of this reference for a special scalar case. On the one hand this result does not require assumptions on graph connectivity, but on the other hand it guarantees convergence only for some of the nodes that indeed collect an infinite number of measurements as $t \to \infty$. Since such nodes, left alone, would be able to build estimates that converge to the true parameter value, the mentioned result shows that the diffusion scheme does not worsen the situation with respect to a case when no communication exchange is made among nodes, but do not actually prove that some benefit is taken from communication. Contrary, the convergence result stated in Theorem 3.1 of the present manuscript do require an hypothesis on graph connectivity (the log-RJC condition), but then guarantees convergence for *all* the nodes of the network, both those that collect an infinite as well as a finite number of measurements.] ◊

## 7.4 Numerical Examples

In this section, we illustrate the distributed estimation algorithm on some numerical examples. We considered two different situations. A first example shows the estimation performance in a middle-sized network with fixed topology, for three different scenarios with increasing sensor measurement rate. The second example shows the convergence of the proposed distributed scheme in a time-varying topology setting with a ring network structure.

### *7.4.1 Example 1*

We considered a network with fixed topology constructed by drawing $n = 50$ nodes at random on the unit square $[0,1] \times [0,1]$, and assuming that any two nodes can communicate whenever their distance is less than 0.25.

We consider a vector $\theta$ of unknown parameters with dimension $m = 5$. Each sensor takes a scalar measurement $y_i = a_i^\top \theta + v_i$, where the vectors $a_i$ have been chosen from an uniform distribution on the unit sphere in $\mathbb{R}^5$. When the sensor takes a measurement its measurement noise $v_i$ is i.i.d. Gaussian with unit variance. We model the possibility of intermittent measurements by assuming that each sensor takes a valid measurement every $p \geq 1$ iterations. As discussed in Remark 7.2, this situation is captured by assuming that the $i$-th sensor covariance is such that

$$\Sigma_i^{-1}(t) = \begin{cases} I \text{ if } (t \bmod p) = 0 \\ 0 \text{ otherwise.} \end{cases}$$

To quantify the estimation performances, we define an average index of the local mean square estimation errors:

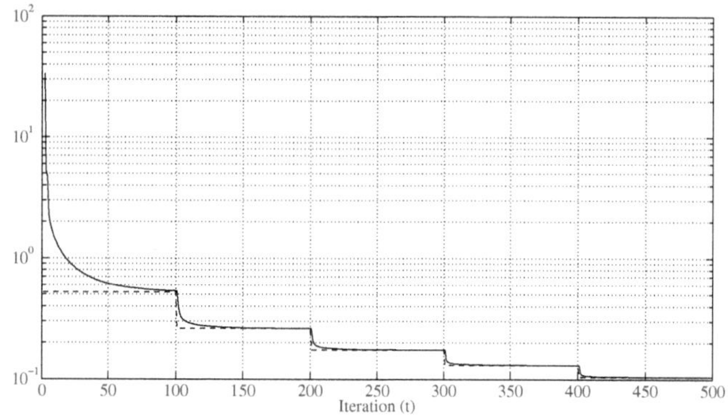


**Fig. 7.2** First experiment $p = 100$: $\mathbf{MLE}(t)$ (dashed), $\mathbf{MSE}(t)$ (solid)

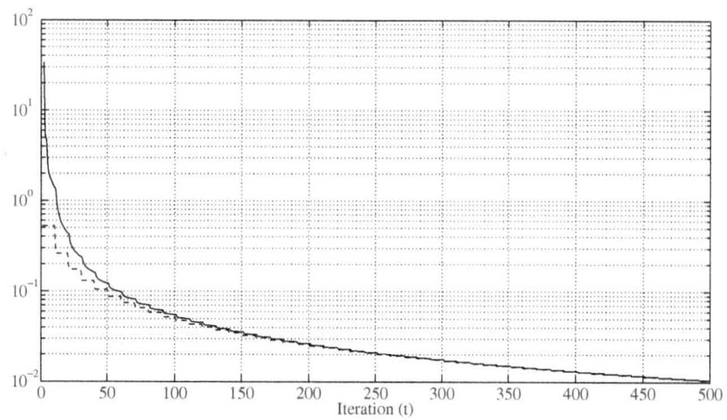


**Fig. 7.3** Second experiment $p = 10$: $\mathbf{MLE}(t)$ (dashed), $\mathbf{MSE}(t)$ (solid)

$$\mathbf{MSE}(t) = \frac{1}{n} \sum_{i=1}^{n} \mathrm{Tr}\ (Q_i(t)),$$

and, for the purpose of comparison, we also compute the *Maximum Likelihood Error* (MLE) as

$$\mathbf{MLE}(t) = \mathrm{Tr}\ (Q_{\mathrm{ml}}(t)).$$

Three experiments have been carried out, with measurement rates $1/p = 0.01$, $1/p = 0.1$, $1/p = 1$, see Figures 7.2, 7.3, 7.4. In case $p = 1$, measurements and consensus iterations happen at the same rate, whereas in cases $p = 10$, $p = 100$, consensus iterations are more frequent than measurement iterations, hence in the instants among measurements the algorithm performs "consensus-only" steps, leading the local estimates to approach the current Maximum Likelihood estimate. This effect

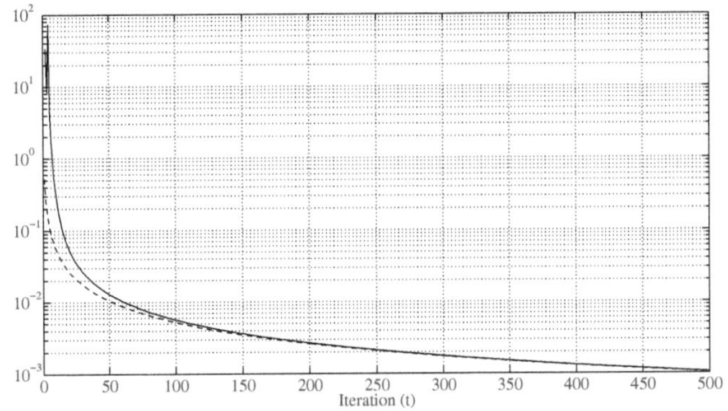


**Fig. 7.4** Third experiment $p = 1$: $\mathbf{MLE}(t)$ (dashed), $\mathbf{MSE}(t)$ (solid)

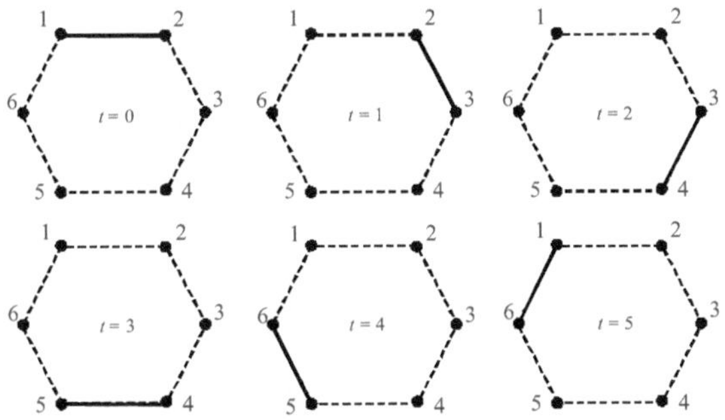


**Fig. 7.5** Time-varying ring network with six nodes

is more evident as $p$ increases, see Figure 7.2 and Figure 7.3. Since more information is globally gathered by the network as $p$ decreases, we observe as expected that the final local estimation error decreases with $p$.

### *7.4.2 Example 2*

In a second example, we considered a network with ring structure and time-varying topology, in two cases with $n = 3$ and $n = 6$ nodes. Specifically, we assumed that at each time $t = 0, 1, \ldots$, only two sensors are able to communicate and collect measurements. We denote the two sensors that are active at $t$ with $s_1(t)$ and $s_2(t)$, respectively, where these indices are defined as

$$s_1(t) = [t \bmod n] + 1$$
$$s_2(t) = [(t+1) \bmod n] + 1.$$

Figure 7.5 shows the time varying graph topology for $n = 6$.

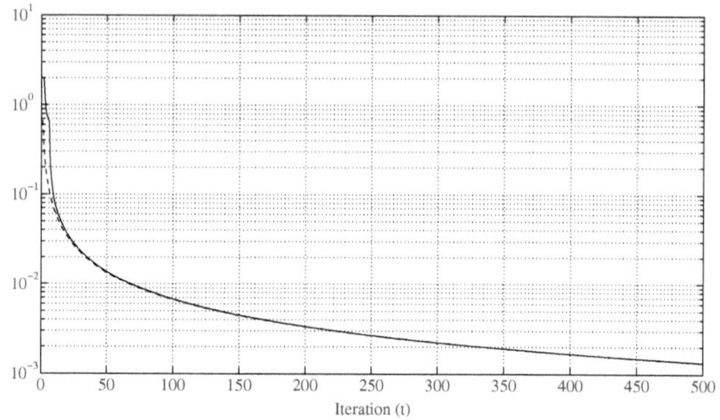


**Fig. 7.6** Ring with 6 nodes. **MLE**($t$) (dashed), and **MSE**($t$) (solid)

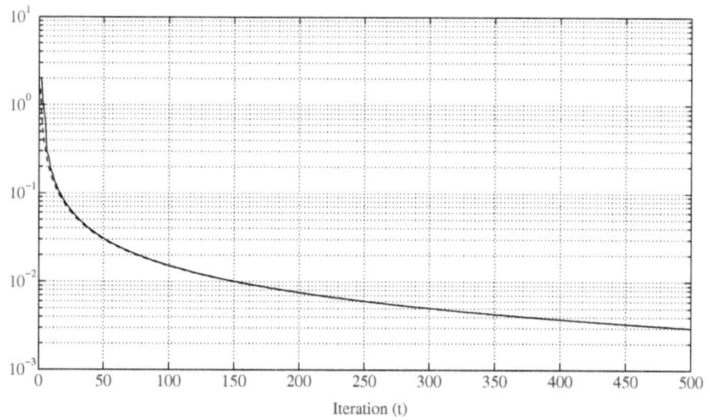


**Fig. 7.7** Ring with 3 nodes. **MLE**($t$) (dashed), and **MSE**($t$) (solid)

In this situation, the Metropolis weight matrix is written as

$$W(t) = I + \frac{1}{2}e_{s_1(t)}e_{s_2(t)}^\top + \frac{1}{2}e_{s_2(t)}e_{s_1(t)}^\top - \frac{1}{2}e_{s_1(t)}e_{s_1(t)}^\top - \frac{1}{2}e_{s_2(t)}e_{s_2(t)}^\top,$$

where $e_i \in \mathbb{R}^n$ is a vector having all entries equal to zero except for the $i$-th position, which is equal to one. The vector of unknown parameters has dimension $m = 2$, and the scalar measurement equation for node $i$ is

$$y_i = a_i^\top \theta + v_i$$

with $a_i = [1\ 0]$ if $i$ is odd, and $a_i = [0\ 1]$ otherwise. The inverse measurement noise covariance is

$$\Sigma_i^{-1}(t) = \begin{cases} 1 \text{ if } i \in \{s_1(t), s_2(t)\} \\ 0 \text{ otherwise.} \end{cases}$$

Notice that the communication graph is not connected at any time instant. However, each sequence of graphs of length $k$ has $\lfloor k/n \rfloor$ jointly connected subsequences. Therefore, the frequency of joint connectedness grows linearly with the sequence length, hence our convergence assumptions that require logarithmic growth are largely satisfied. Theorem 7.1 thus guarantees that all local estimates converge asymptotically to the true parameter value. This is indeed confirmed by the plots resulting from numerical simulations, shown in Figures 7.6 and 7.7, where it can also be observed as expected that the rate of convergence of the local estimates decreases as the number of nodes increases.

## 7.5 Conclusions and Future Work

In this chapter we discussed a distributed estimation scheme for sensor networks. The nodes maintain a common data structure and can communicate with their instantaneous neighbors. At each time iteration, a node may collect a new measurement, compute a local estimate of the unknown parameter and then average its local information with the neighbors' information. We showed in Theorem 7.1 that *all* local estimates converge asymptotically to the true parameter, even for nodes that collect only a finite number of measurements. Convergence is proved under a necessary condition of convergence of a virtual centralized estimate and under a rather mild hypothesis on the frequency of connectivity for the superposition (union) of subsequences of occurring communication graphs. It is worth underlining that our results on the convergence of "consensus under persistent excitation" (the "persistent excitation" is in our case due to the possible continued presence of new measurements) required an analysis somewhat different from the one usually found in the consensus literature, [4, 8, 11, 18]. More precisely, the key technical tool required in the classical consensus approach is the convergence of certain infinite matrix products, whereas under persistent excitation we also need that the infinite sum of such products remains finite (see Lemma 7.3). We expect in turn that the approach developed here in the context of estimation problems could be exported to other agreement-type problems such as those arising in decentralized coordination and control, [8, 11, 13, 14].

## References

1. Akyildiz, I., Su, W., Sankarasubramniam, Y., Cayirci, E.: A survey on sensor networks. IEEE Communication Magazine 40(8), 102–114 (2002)
2. Alanyali, M., Venkatesh, S., Savas, O., Aeron, S.: Distributed Bayesian hypothesis testing in sensor networks. In: Proceedings of the American Control Conference, pp. 5369–5374 (2004)

3. Calafiore, G., Abrate, F.: Distributed linear estimation over sensor networks. International Journal of Control (in press, 2009) doi: 10.1080/00207170802350662
4. Cao, M., Morse, A., Anderson, B.: Reaching a consensus in a dynamically changing environment: A graphical approach. SIAM Journal on Control and Optimization 47(2), 575–600 (2008)
5. Chu, M., Haussecker, H., Zhao, F.: Scalable information-driven sensor querying and routing for ad hoc heterogeneous sensor networks. International Journal of High-Performance Computing Applications 16(3), 293–313 (2002)
6. Delouille, V., Neelamani, R., Baraniuk, R.: Robust distributed estimation in sensor networks using the embedded polygons algorithm. In: Proceedings of The Third International Conference on Information Processing in Sensor Networks, pp. 405–413 (2004)
7. Horn, R.A., Johnson, C.R.: Matrix Analysis. Cambridge (1985)
8. Jadbabaie, A., Lin, J., Morse, A.: Coordination of groups of mobile autonomous agents using nearest neighbor rules. IEEE Transactions on Automatic Control 48(6), 988–1001 (2003)
9. Luo, Z.: An isotropic universal decentralized estimation scheme for a bandwidth constrained ad hoc sensor network. IEEE Journal on Selected Areas in Communication 23(4), 735–744 (2005)
10. Martinez, S., Bullo, F.: Optimal sensor placement and motion coordination for target tracking. Automatica 42(4), 661–668 (2006)
11. Moreau, L.: Stability of multiagent systems with time-dependent communication links. IEEE Transactions on Automatic Control 50(2), 169–182 (2005)
12. Olfati-Saber, R.: Consensus problems in networks of agents with switching topology. IEEE Transactions on Automatic Control 49(9), 1520–1533 (2004)
13. Roy, S., Saberi, A.: Static decentralized control of a single-integrator network with Markovian sensing topology. Automatica 41(11), 1867–1877 (2005)
14. Roy, S., Saberi, A., Herlugson, K.: A control-theoretic perspective on the design of distributed agreement protocols. International Journal of Robust and Nonlinear Control 17(10-11), 1034–1066 (2006)
15. Spanos, D.P., Olfati-Saber, R., Murray, R.: Approximate distributed Kalman filtering in sensor networks with quantifiable performance. In: Fourth International Symposium on Information Processing in Sensor Networks (IPSN 2005), pp. 133–139 (2005)
16. Spanos, D.P., Olfati-Saber, R., Murray, R.: Distributed sensor fusion using dynamic consensus. In: Proceedings of the 16th IFAC World Congress (2005)
17. Tsitsiklis, J.: Decentralized Detection. In: Poor, H.V., Thomas, J.B. (eds.) Advances in Signal Processing, vol. 2, pp. 297–344. JAI Press (1993)
18. Xiao, L., Boyd, S., Lall, S.: A space-time diffusion scheme for peer-to-peer least-squares estimation. In: Proceedings of Fifth International Conference on Information Processing in Sensor Networks (IPSN 2006), pp. 168–176 (2006)

# Chapter 8
# Optimal Sensor Scheduling for Remote Estimation over Wireless Sensor Networks

Roberto Ambrosino, Bruno Sinopoli, and Kameshwar Poolla

**Abstract.** Wireless Sensor Networks (WSNs) enable a wealth of new applications where remote estimation is essential. Individual sensors simultaneously sense a dynamic process and transmit measured information over a *shared* lossy channel to a central base station. The base station computes an estimate of the process state. We consider this remote estimation problem for wireless single-hop communication with the widely used IEEE 802.15.4 Media Access Control (MAC) protocol. Using a Markov chain model for the MAC protocol, we derive an expression for the probability of successful packet transmission $\lambda(N)$. We show that $\lambda(N)$ is a monotone decreasing function of the number of sensors $N$ attempting to access the channel. State estimation is better served by having data from more sensor nodes, but this results in decreased probability of successful packet transmission. As a consequence, we are faced with a design trade-off in determining how many sensors should attempt to communicate their observations to the base station and which sensors are most informative for the purpose of state estimation. We show that this problem of *optimal sensor selection* can be cast as an optimization problem which can be solved approximately using convex programming. The optimal selection of sensors is dynamic and leads, in turn, to the problem of optimal sensor *scheduling*. We offer a synthetic example to illuminate our ideas.

---

Roberto Ambrosino
Department of Technology, University of Naples Parthenope, Naples, Italy
e-mail: `ambrosino@uniparthenope.it`

Bruno Sinopoli
Department of Electrical and Computer Engineering,
Carnegie Mellon University, Pittsburgh, PA 15213, USA
e-mail: `brunos@ece.cmu.edu`

Kameshwar Poolla
Department of Electrical Engineering and Computer Science,
University of California at Berkeley, Berkeley, CA 94720, USA
e-mail: `poolla@eecs.berkeley.edu`

## 8.1 Introduction

Wireless sensor networks (WSNs) offer fertile ground for research in several interconnected research domains within computer science, communication systems, control theory. A WSN is composed of low power devices (variously called sensors, motes, and nodes) that integrate computing with heterogeneous sensing and wireless communication. In particular, the devices interact with the physical world, collect the information of interest and transmit these collected data to a base station. The medium access control (MAC) protocol sits directly on top of the physical layer and regulates access to and the communication on the network.

WNSs enable observation of the physical world at an unprecedented level of granularity, with a potentially enormous impact on applications such as industrial and home automation, consumer electronic, military security and health care [5]. In the monitoring and control of moving machinery, for example, wireless sensor networks have compelling economic and engineering advantages over their wired counterparts. They may also deliver crucial information in real-time from environments and processes where data collection is impossible or impractical with wired sensors. WSNs can be installed in the Heating Ventilation Air Conditioning (HVAC) systems of commercial buildings at a fraction of the cost of wired systems [14]. They effectively enable increases energy efficiency by providing the control systems with more complete information about the building environment.

In each of these WSN application examples, remote estimation is a central problem. In computing optimal estimates of the system state, having access to data from more sensors in clearly beneficial. However, using too many sensors can generate bottlenecks in the communication infrastructure when they compete for bandwidth and will reduce the probability that the sensor data is successfully transmitted to the base station. As a consequence, we are faced with a design trade-off in determining how many sensors should attempt to communicate their observations to the base station and which sensors are most informative for the purpose of state estimation.

In this paper we consider the remote estimation problem over WSNs operating with the widely used IEEE 802.15.4 MAC protocol. Under this protocol, we study the problems of optimal sensor selection and scheduling.

Another important class of problems in remote estimation over WSNs is that of optimal sensor placement/density. Here, we have to determine optimal locations and numbers of sensors for a desired performance threshold (or state estimation accuracy). This problem is further confounded in the event that the sensing agents are mobile or if the underlying process dynamics is non-stationary. Our results in this paper can be viewed as determining the optimal *local* density of sensors in the context of state estimation.

### 8.1.1 Related Work

In recent years, state estimation over lossy networks has received considerable attention. The work by Sinopoli et al. [15] shows the existence of a critical value of the packet loss rate for bounded estimation error covariance. For multi-sensor scenario, Liu and Goldsmith [8] extend the analytical work [15] to a two-sensor case. Matveev and Savkin [9] consider the effect of packet latencies on state estimation. Gupta et al. [6] address the multi-sensor joint state-estimation of a plant, allowing only one sensor to take measurement and access communication channel at each time step.

The tradeoff between communication and estimation performance is explored as *controlled communication* in the works of Yook et al. [21], Xu and Hespanha [19] [20], to actively reduce network traffic.

Zhu et al. [22] first attempted to explicitly determine the optimal density of homogeneous sensor nodes in a large wireless sensor network. Ambrosino et al. [1] extended their result to the case of identical sensors but with different measurement noise covariances and provide a method to compute not only the optimal number of sensors, but they also indicate which sensors need to be used at any given time.

In this paper we consider the more general case of remote estimation problem over a WSN composed by nonhomogeneous sensors and we provide the optimal scheduling of sensors to maximize the estimation performance as measured by the error covariance of the state estimate.

The remainder of this paper is organized as follows. A summary of major MAC protocols used in wireless sensor network, along with characterization of packet reception probability based on a Markov model of node transmission is described in Section 8.2. In Section 8.3 the remote estimation problem is treated using a centralized information filter. Section 8.4 presents the optimization algorithm to find the optimal set of sensors that maximize the estimation performance. An illustrative example is offered in Section 8.5. Finally, in Section 8.6, we summarize our results and draw conclusions.

## 8.2 Modeling of Media Access Control

The Media Access Control (MAC) layer arbitrates which sensor node is allowed to access the radio channel at any given time. In general, MAC protocols can be categorized into Time Division Multiple Access (TDMA) and contention based Carrier Sense Multiple Access (CSMA) protocol. TDMA requires global synchronization and each node to maintain a list of its neighbors' schedules, which may be difficult to achieve in resource constrained wireless sensor networks. In the case of power deprivation, sensor nodes may stop functioning, thus this class of protocols is not easily scalable.

The channel access schemes for wireless sensor networks are generally contention based and are scalable because of their simplicity and flexibility. Among them, the two most widely used MAC protocols are B-MAC [13] and IEEE 802.15.4 MAC [16], [10]. B-MAC is still the de-facto standard networking stack while IEEE 802.15.4 is emerging as the new global standard in low rate wireless personal area networks (WPANs). Zigbee, a global wireless industry open standard that is built on top of IEEE 802.15.4, has become the preferred technology for home automation and control, among other applications.

Following these consideration we focus on CSMA-type MAC protocols, in particular the IEEE 802.15.4, to study the the correlation between packet reception probability $\lambda(N)$ and the number of sensors $N$ attempting to access the channel.

### 8.2.1 IEEE 802.15.4 MAC

The IEEE 802.15.4 standard for low rate wireless personal area networks (WPANs) is emerging as the premier candidate for WSNs [5], since it supports small, inexpensive, energy-efficient devices operating on battery power that require little operating infrastructure.

In an IEEE 802.15.4-compliant WPAN, a central controller device (commonly referred to as the PAN coordinator) builds a WPAN with other devices within a small physical space known as the personal operating space. This protocol supports two network topologies: in the star topology network, all communications, even those between the devices themselves, must go through the PAN coordinator; in the peer-to-peer topology, the PAN coordinator must still be present, but the devices can communicate with one another directly.

The standard also defines two channel access mechanisms, depending on whether a beacon frame (which is sent periodically by the PAN coordinator) is used to synchronize communications or not. Beacon enabled networks use slotted carrier sense multiple access mechanism with collision avoidance (CSMA-CA), while the non-beacon enabled networks use the simpler, unslotted CSMA-CA.

In this work, we consider a star-topology 802.15.4 protocol operating in unslotted CSMA-CA. We suppose there are $N$ sensors observing a dynamical system and reporting to a central location over the wireless sensor network over a single shared radio channel. The aim of this section is to determine the probability that a node successfully transmits its sensed information within a specified sample period.

### 8.2.2 The Unslotted CSMA-CA Algorithm of IEEE 802.15.4

In this subsection we illustrate the basic idea of the unslotted CSMA-CA algorithm of IEEE 802.15.4, the standard for the low rate wireless personal area network. Fig. 8.1 provides its flow chart.

In the unslotted CSMA-CA, all operations must be synchronized to backoff time units. A node with a packet ready for transmission, first backs off for a random waiting time $k$ in the range 0 and $2^{BE}-1$, where $BE$ is the *backoff exponent* which is initially set to *macMinBE*; $2^{BE}-1$ is called *backoffWindow*. This random backoff time, which is an expedient to reduce the probability of collision among contending nodes, is decremented at the boundary of each backoff period. Once the backoff time counter decrements to zero, the node starts to sense whether the channel is busy through a procedure called Clear Channel Assessment (CCA). If the channel

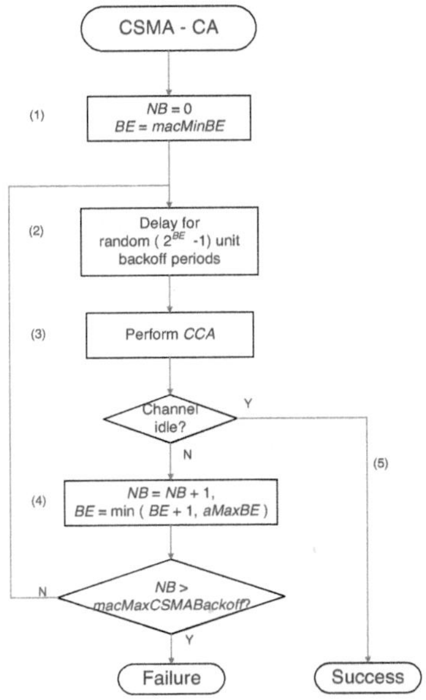


**Fig. 8.1** Flow chart of the contention access period of unslotted IEEE 802.15.4, where *NB* is the number of backoff stage

is found to be busy, the backoff exponent is incremented by one and a new random number of backoff units is drawn for the node to wait, until the channel can be sensed again. This process is repeated until BE equals the parameter *aMaxBE*. At this point, it is frozen at *aMaxBE* until a certain maximum number of permitted random backoff stages (determined by *macMaxCSMABackoffs*) is reached. Then an access failure is declared to the upper layer.

### 8.2.3 Transmission Model of a Device

The behavior of a single device in the un-slotted IEEE 802.15.4 protocol can be modeled by means of a Markov chain, as introduced in [3] for modeling IEEE 802.11 and also used in [10] for modeling the slotted IEEE 802.15.4.

We assume that the busy channel probability $c$ at each access attempt can be evaluated in stationary condition, i. e. the transmission queue at each station is assumed to be always nonempty. The consequences of this approximation are evaluated in Section 8.5.

Let $s(t)$ be the stochastic process representing the backoff stage $(0,...,m)$ of a node at time $t$ with $m$ being determined by the sample period of the dynamical system to estimate. We also suppose that $macMaxCSMABackoff = m$. Let $b(t)$ be

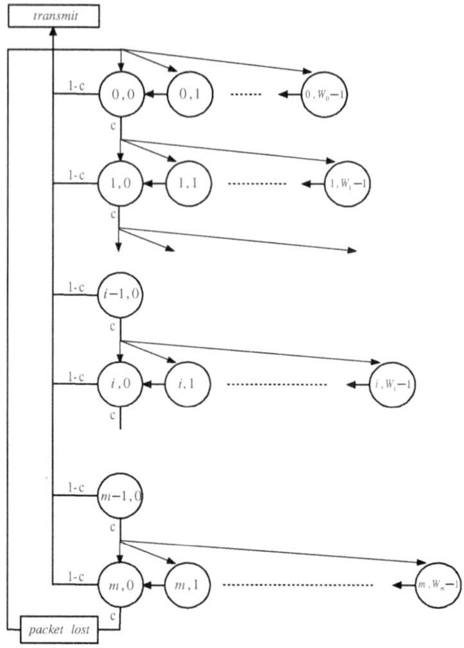


**Fig. 8.2** Markov chain model of the node state including backoff stages and channel sensing. $c$ refers to the probability that the channel is busy. The backoff timer decrements till zero and commences the *Clear Channel Assessment*(CCA).

the stochastic process representing the time counter at each backoff stage. Denote the backoff window size at backoff stage $i$ with $W_i = 2^i W$, where $W = 2^{macMinBE} - 1$.

The key approximation for the stationary model (see [3]) is that we assume that the busy channel probability $c$ at each transmission attempt is constant and independent of the number of backoff stages. It is intuitive that this assumption becomes more accurate as $N$ gets larger.

With these assumptions in place, it is possible to model the bi-dimensional process $\{s(t), b(t)\}$ with the discrete time Markov chain depicted in Fig 8.2.

In this Markov chain the only non zero transition probabilities are

$$\begin{cases} P\{s(t+1) = i, b(t+1) = k | s(t) = i, b(t) = k+1\} = 1 \\ i \in [0, m] \qquad k \in [0, W_i - 2] \\ P\{s(t+1) = 0, b(t+1) = k | s(t) = i, b(t) = 0\} = (1-c)/W_0 \\ i \in [0, m-1] \qquad k \in [0, W_i - 1] \\ P\{s(t+1) = 0, b(t+1) = k | s(t) = m, b(t) = 0\} = 1/W_0 \\ k \in [0, W_0 - 1] \\ P\{s(t+1) = i, b(t+1) = k | s(t) = i-1, b(t) = 0\} = c/W_i \\ i \in [1, m] \qquad k \in [0, W_i - 1]. \end{cases} \tag{8.1}$$

The first equation in (8.1) describes the fact that the time counter always decrements. The second equation says that, after a successful access attempt, a node starts to transmit a new packet and the initial random backoff is uniformly chosen in the interval $(0, W_0 - 1)$. The third one reflects the fact that, after the maximum number of backoff stages, regardless of the status of the access attempt, a node will transmit a new packet. The last equation explains that any unsuccessful access attempt before the maximum backoff stage leads to a new round of backoff with the new backoff value uniformly chosen from a longer backoff window, i.e. $(0, W_i)$.

Let $b_{i,k} = \lim_{t->\infty} P\{s(t) = i, b(t) = k\}, i \in (0,m), k \in (0, W_i - 1)$ be the stationary distribution of the chain. We have

$$b_{i-1,0}c = b_{i,0} \qquad \Rightarrow b_{i,0} = c^i b_{0,0} \qquad i \in [1,m]$$
$$b_{0,0} = (1-c)\sum_{i=0}^{m} b_{i,0} + cb_{m,0}.$$

By the chain regularity, we have $b_{i,k} = \frac{W_i - k}{W_i} b_{i,0}$.

We can express all the values of $b_{i,k}$ as functions of the value $b_{0,0}$ and the conditional channel busy probability $c$. By balancing the Markov Chain, we have

$$1 = \sum_{i=0}^{m}\sum_{k=0}^{W_i-1} b_{i,k} = \frac{b_{0,0}}{2}\left[\frac{1-c^{m+1}2^{m+1}}{1-2c}W + \frac{1-c^{m+1}}{1-c}\right]. \tag{8.2}$$

Thus

$$b_{0,0} = \frac{2(1-2c)(1-c)}{(1-c)(1-c^{m+1}2^{m+1})W + (1-2c)(1-c^{m+1})}. \tag{8.3}$$

We are now in the position to evaluate the probability $p_{tr}$ that a node transmits in a randomly chosen time period. Any transmission occurs when a node finds the channel to be idle after its backoff time counter decrements to zero, regardless of the backoff stage, namely

$$\begin{aligned} p_{tr} = (1-c)\sum_{i=0}^{m} b_{i,0} &= (1-c^{m+1})b_{0,0} \\ &= \frac{2(1-2c)(1-c)(1-c^{m+1})}{(1-c)(1-c^{m+1}2^{m+1})W + (1-2c)(1-c^{m+1})} \end{aligned} \tag{8.4}$$

The fundamental independence assumption implies that each access attempt "sees" the system in the same state, i.e. in steady state. At steady state, each node transmits with probability $p_{tr}$. Thus for a given node, at each channel access attempt, the probability of collision, i.e. the probability that, in a backoff period, at least one of the other $N-1$ remaining nodes transmits, is

$$p_c = 1 - (1-p_{tr})^{N-1} \tag{8.5}$$

Moreover, the probability of channel busy, i.e. the probability that, in a backoff period, at least one of the remaining $N-1$ sensors has already started to transmit, is

$$c = \left(1-(1-p_{tr})^{N-1}\right)(d-1) \tag{8.6}$$

where $d$ is the dimension of the transmitted packet in terms of backoff periods, that we suppose to be fixed. It represents the probability that at least one of the $N-1$ remaining sensors started the transmission in one of the previous $d-1$ backoff periods. We can derive the value of $p_{tr}$ and $c$ by solving the set of non-linear fixed point equations (8.4, 8.6). It is straightforward to show that, for a fixed value of $W$ and $m$, there is only one fixed point for $p_{tr}$ and $c$ for each $N$, so we refer to these variable as $p_{tr}(N)$ and $c(N)$.

### 8.2.4 Successful Packet Reception Probability

The ideal channel condition assumes that a device has a successful transmission during a sample period if it transmits during one of the $m+1$ backoff stages and that there are no collision during the transmission. The probability to have a successful transmission is therefore

$$\begin{aligned}\lambda(N) &= (1-c(N))(1+c(N)+\cdots+c(N)^m)(1-p_c(N))\\ &= (1-c(N)^{m+1})(1-p_c(N)).\end{aligned} \tag{8.7}$$

## 8.3 Remote Estimation Problem

Consider the following discrete time linear dynamical system and measurement model:

$$x_{t+1} = Ax_t + w_t \tag{8.8}$$

$$y_{i,t} = C_i x_t + v_{i,t}\,, \tag{8.9}$$

where $i = 1,2\ldots N$ is the sensor index, $x_t \in \mathrm{R}^{\mathrm{n}}$ is the state vector, $y_{i,t} \in \mathrm{R}^{\mathrm{m}}$ is the measurement from the sensor $i$, $w_t \in \mathrm{R}^{\mathrm{n}}$ models process noise, assumed to be white Gaussian noise with zero mean and covariance $Q > 0$ and $v_{i,t}$'s $\in \mathrm{R}^m$ model the measurement noise and they are white Gaussian noises with zero mean and covariance $R_i > 0$. The initial system state $x_0$ is Gaussian with zero mean and covariance $\Sigma_0$. We assume that $x_0$, $w_t$ and $v_{i,t}$ are independent.

The objective of this section is to determine how many and which sensors under a WPAN with an un-slotted IEEE 802.15.4 protocol should communicate their observations to the base station in order to optimize state estimation performance.

### 8.3.1 Multi-sensor Lossy Estimation

As discussed above, use of the un-slotted 802.15.4 protocol does not ensure that each sensor successfully transmits the sensed information during the sample period.

The impact of packet loss on estimation performance for the case of a single sensor observing a dynamic process is treated in Sinopoli et al. [15]. Liu et al [8] discuss the related case of 2 reporting nodes, in which measurements come from two sensor nodes with different loss rates. In both of these studies, there is no explicit MAC model and no sensor selection in performed. Here, we will explicitly use our MAC model to determine sensor selection for optimal state estimation using the information form of the Kalman filter [12].

Let $N$ denote the number of available sensors. Assume that for any sensor $i$ at time $t$ the arrival at the base station of the corresponding observation $y_{i,t}$ during the sample period $(t, t+1]$ is a Bernoulli random variable, defined as $\gamma_{i,t}$. The successful packet reception probability is $p_{\gamma_{i,t}}(1) = \lambda(N)$, assumed to be equal for all sensors.

Kalman filter with lossy measurement (see [15]) is defined as

$$\begin{aligned}
\hat{x}_{t|t} &= \mathrm{E}[x_t|\mathbf{y_t}, \gamma_\mathbf{t}] \\
P_{t|t} &= \mathrm{E}[(x_t - \hat{x}_{t|t})(x_t - \hat{x}_{t|t})'|\mathbf{y_t}, \gamma_\mathbf{t}] \\
\hat{x}_{t+1|t} &= \mathrm{E}[x_{t+1}|\mathbf{y_t}, \gamma_\mathbf{t+1}] \\
P_{t+1|t} &= \mathrm{E}[(x_{t+1} - \hat{x}_{t+1|t})(x_{t+1} - \hat{x}_{t+1|t})'|\mathbf{y_t}, \gamma_\mathbf{t+1}] \\
\hat{y}_{t+1|t} &= \mathrm{E}[y_{t+1}|\mathbf{y_t}, \gamma_\mathbf{t+1}].
\end{aligned}$$

where $\mathbf{y_t} = [y_0, ..., y_t]$, and $\gamma_\mathbf{t} = [\gamma_0, ..., \gamma_t]$ is a matrix whose columns are $\gamma_i = [\gamma_{1,i}, \dots \gamma_{N,i}]'$ for $i = 1 \dots t$.

In the information filter, the estimated state and the covariance matrix are replaced by the information vector and information matrix respectively. These are defined as:

$$\begin{aligned}
\hat{z}_{t|t} &= P_{t|t}^{-1}\hat{x}_{t|t} \quad , \quad \hat{z}_{t+1|t} = P_{t+1|t}^{-1}\hat{x}_{t+1|t} \\
Z_{t|t} &= P_{t|t}^{-1} \quad , \quad Z_{t+1|t} = P_{t+1|t}^{-1}.
\end{aligned}$$

Let us define the information filter for multi-sensor lossy estimation.

The prediction phase for $\hat{z}_{t+1|t}$ and $Z_{t+1|t}$ of the information filter is independent of the observation process with:

$$\hat{z}_{t+1|t} = Z_{t+1|t}AZ_{t+1|t}^{-1}\hat{x}_{t|t} \tag{8.10}$$

$$Z_{t+1|t} = (AZ_{t|t}^{-1}A' + Q)^{-1}. \tag{8.11}$$

The measurement update is stochastic as the received measurements are functions of $\gamma_t$s, which are random. Let us define the information vectors and matrices for each sensor as

$$i_{i,t} = \gamma_{i,t}C_i'R_i^{-1}y_{i,t} \tag{8.12}$$

$$I_{i,t} = \gamma_{i,t}C_i'R_i^{-1}C_i. \tag{8.13}$$

The principal advantage of the information form of the Kalman filter is that $N$ measurements can be filtered at each time step simply by summing their information matrices and vectors. In particular the filter step of the information filter is given by

$$\hat{z}_{t+1|t+1} = \hat{z}_{t+1|t} + \sum_{i=1}^{N} i_{i,t} \tag{8.14}$$

$$Z_{t+1|t+1} = Z_{t+1|t} + \sum_{i=1}^{N} I_{i,t}. \tag{8.15}$$

By combining (8.11), (8.13) and (8.15), the information filter equation can be written as follows

$$Z_{t+1|t+1} = (AZ_{t|t}^{-1}A' + Q)^{-1} + \sum_{i=1}^{N} \gamma_{i,t} C_i' R_i^{-1} C_i. \tag{8.16}$$

Given the initial condition $Z_{0|-1} = \Sigma_0^{-1}$, the sequence $\{Z_{t|t}\}_{t=0}^{\infty}$ is a random process dependent on the realization $\{\gamma_{i,t}\}_{t=0}^{\infty}$. We only focus on the statistical properties of the error covariance matrix.

We define the modified algebraic Riccati equation (MARE) for the inverse Kalman filter with multiple reporting sensors as

$$g_{\lambda(N)}(X) = (AX^{-1}A' + Q)^{-1} + \lambda(N) \Sigma_{i=1}^{N} C_i' R_i^{-1} C_i .$$

It follows that

$$g_{\lambda}(Z_{t|t}) = \mathrm{E}[Z_{t+1|t+1} | Z_{t|t}], \quad \text{and} \tag{8.17}$$

$$\overline{Z}_{t+1|t+1} = \mathrm{E}[Z_{t+1|t+1}] \quad = \mathrm{E}[g_{\lambda}(Z_{t|t})] . \tag{8.18}$$

## 8.4 Sensor Selection and Scheduling

We choose $E[Z_{t+1|t+1}|Z_{t|t}]$ as estimation performance index. In particular, in the following we select at each step time the set of sensors that maximize a scalar objective function related to $g_{\lambda}(Z_{t|t})$.

For a fixed value $p$ of sensors, the optimization problem can be expressed as:

$$\max_{x} \quad h(f(x)) \tag{8.19}$$

$$s.t. \quad \Sigma_{i=1}^{N} x_i = p \tag{8.20}$$

$$x_i \in \{0, 1\}, \tag{8.21}$$

where $x = [x_1\, x_2 \ldots x_N]$ is the sensor selection vector of binary variables, the function $f(x)$ is defined as

$$f(x) = (AZ_{t|t}^{-1}A' + Q)^{-1} + \lambda(p) \Sigma_{i=1}^{N} x_i C_i' R_i^{-1} C_i ,$$

and the function $h : \mathrm{R}^{n \times n} \to \mathrm{R}$ converts the matrix function $f(x)$ into a scalar index.

With an appropriate choice of the function $h(\cdot)$, this problem can be solved in the framework of convex optimization [4]. In particular, taking into account that we consider a maximization problem and $f(x)$ is linear with respect to $x$, possible functions $h(\cdot)$ are, for example, the trace and logdeterminant [7], [4]. While $trace(f)$ is related to the sum of the squared semi-axes length of the ellipsoid generated by the matrix $f$, $logdet(f)$ is related to its volume. Observe that if we consider the objective function $trace(f)$, the optimization problem will have a static solution, i. e. the selected sensors will be identical at every time step. In contrast, the $logdet(f)$ objective function can result in dynamically varying optimal sensor selections. A critical issue in sensor selection or scheduling is that of power. A static choice of sensors will rapidly deplete available power at the chosen sensor nodes. An interesting and important research direction is to incorporate available power in our problem formulation and optimize estimation performance subject to network lifetime constraints.

All the considered optimization problems are Boolean-convex problems because we have a maximization of a linear or concave objective function, a set of linear constraints, and the last $N$ constraints restrict $x$ to be Boolean.

By replacing the non-convex constraints $x_i \in \{0,1\}$ with the convex constraints $\tilde{x}_i \in [0,1]$ it is possible to obtain a convex relaxation of the sensor selection problem (8.19):

$$\max \quad h(f(\tilde{x})) \tag{8.22}$$

$$s.t. \quad \Sigma_{i=1}^{N} \tilde{x}_i = p \tag{8.23}$$

$$0 \leq \tilde{x}_i \leq 1 \quad i = 1, \ldots N. \tag{8.24}$$

The relaxed problem is not equivalent to the original one because its optimal solution $\tilde{x}^*$ can contain fractional values. Moreover the optimal value of its objective function represents an upper bound for the objective function in problem (8.19) [7]. However, the optimal solution $\tilde{x}^*$ can be used to generate a suboptimal solution $x^*$ for the original problem simply selecting the $p$ sensors associated with the $p$ biggest values of the vector $\tilde{x}^*$. To underline the dependence of $x^*$ from the number of sensors, let us indicate this vector as $x(p)^*$.

Once the considered optimization problem is solved for all possible $p \leq N$, the optimal number of sensors is given by

$$p^* = arg \max_{p} \quad h(f(x(p)^*)) \,. \tag{8.25}$$

Hence the optimal subset of sensors in each time step is $x(p^*)^*$.

## 8.5 Illustrative Example: Monitoring

In this section, a specific estimation example, monitoring over wireless sensor network, is presented to further illustrate the impact of a sensor scheduling strategy on the estimation performance.

### 8.5.1 Model Definition

Let's consider the problem of carbon dioxide ($CO_2$) sequestration to reduce emissions of this gas produced from burning fossil fuels in power generation plants [2]. WSNs can be employed to monitor sequestration sites for leaks [18]. Due to the nature of the problem, acquiring data from every sensors at each sampling period is impractical and expensive for the energy viewpoint. Thus, near-optimal sensor scheduling strategies become crucial in order to accurately estimate the $CO_2$ concentration level with limited sensing.

Starting with a continuous-time PDE model describing a convection-dispersion process, we derive the following discrete-time finite-dimensional model for surface $CO_2$ leakage, as in [17]

$$x_{t+1} = A_t x_t + B_t u_t + w_t \,, \tag{8.26}$$

where we divide the bi-dimensional observed region in $N \times M$ points placed $\Delta x$ apart and we discretize in time with a step $\Delta t$. Hence

- $x_t = [c_t(1,1)\, c_t(1,2) \ldots c_t(1,M) \ldots c_t(i,j) \ldots c_t(N,M)]' \in \mathrm{R}^{NM}$is the set of concentrations of $CO_2$ at time $t\Delta t$ at points $(i\Delta x, j\Delta x)$.
- $u_t = [\lambda_t(1,1)\, \lambda_t(1,2) \ldots \lambda_t(1,M) \ldots \lambda_t(i,j) \ldots \lambda_t(N,M)]' \in \mathrm{R}^{NM}$ is the set of source inputs of $CO_2$ at time $t\Delta t$ at $(i\Delta x, j\Delta x)$.
- $w(t)$ is the process noise.
- the matrices $A, B \in \mathrm{R}^{NM \times NM}$ represent the lumped parameter state dynamics governing x(k) according to the (PDE) model. They depend on the advection coefficient $\phi(i,j,t) = [\phi_x(i,j,t), \phi_y(i,j,t), \phi_z(i,j,t)]$ and the dispersion coefficient $\alpha(i,j,t) = [\alpha_x(i,j,t), \alpha_y(i,j,t), \alpha_z(i,j,t)]$ at time $t\Delta t$ at $(i\Delta x, j\Delta x)$.

In this application, we do not have *a priori* knowledge of location and magnitude of the sources, but we assume constant leaks. Taking into account that the sources represent the input in the discrete-time finite-dimensional model, in order to apply our sensor scheduling methods, we augment the state of the previous model as

$$\tilde{x}_{t+1} = \tilde{A}_t \tilde{x}_t + \tilde{w}_t$$

where $\tilde{x}_t = [x_t \; u_t]'$ and

$$\tilde{A}_t = \begin{pmatrix} A_t & B_t \\ 0 & I \end{pmatrix} . \tag{8.27}$$

### 8.5.2 Sensor Scheduling

Let us consider a network of $\hat{N}$ sensors uniformly distributed over the monitored region. Let's suppose that the sensors measure the concentration of $CO_2$ at the points where they are placed. According to the definition of $x(t)$, let us model the measurement of each sensor as

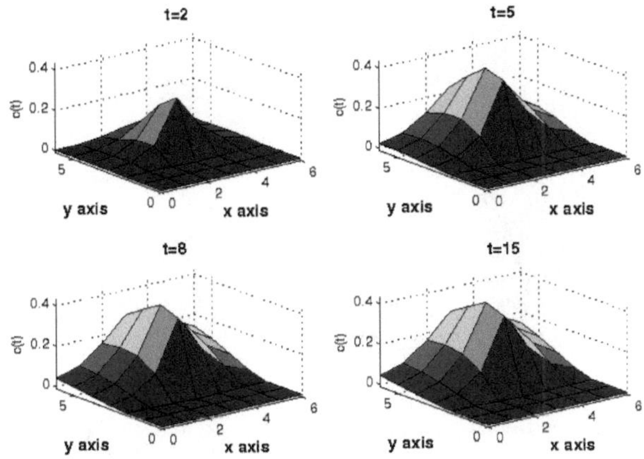


**Fig. 8.3** Concentration of $CO_2$ for different values of time.

$$y_{(i,j),t} = C_{(i,j)}x_t + v_{i,t}\,, \tag{8.28}$$

with

$$C_{(i,j)} = \begin{bmatrix} 0 \dots 0 \; 1_{(i,j)} \; 0 \dots 0_{NM} \; 0 \dots 0_{2NM} \end{bmatrix} \in \mathrm{R}^{1\times 2NM} \tag{8.29}$$

For the simulations, we impose the following values of the parameters:

- $\Delta x = 6m$
- $\Delta t = 3sec$
- $N = M = 7$
- $\phi_x = 0.5\frac{m}{sec}$ ; $\phi_y = 2\frac{m}{sec}$
- $\alpha_x = \alpha_y = \alpha_z = 2.5\frac{m^2}{sec}$
- $\lambda_t(4,4) = 0.0015\frac{mol}{m^3 \cdot sec}$ $\quad \forall t$ else $\lambda_t(\cdot,\cdot) = 0$
- $\hat{N} = 20$
- $R_i = 0.3$ .

The resulting system will have 98 states and 20 sensors. Figure 3 shows the concentration of $CO_2$ for different values of $t$. Under the hypothesis of constant wind, the system reaches the steady state after approximately 30 seconds.

We assume that the wireless sensors constitute a WPAN with star-topology 802.15.4 protocol operating in un-slotted CSMA-CA. In particular, we suppose that

- the communication rate of the channel is r=250 Kbps
- the dimension of the communication packet is D=60 byte.

The backoff period is set to $b_p = 0.002\,s \simeq \frac{1}{2}\frac{D}{r}$, so that we suppose that a packet will take $2\,b_p$ to be transmitted.

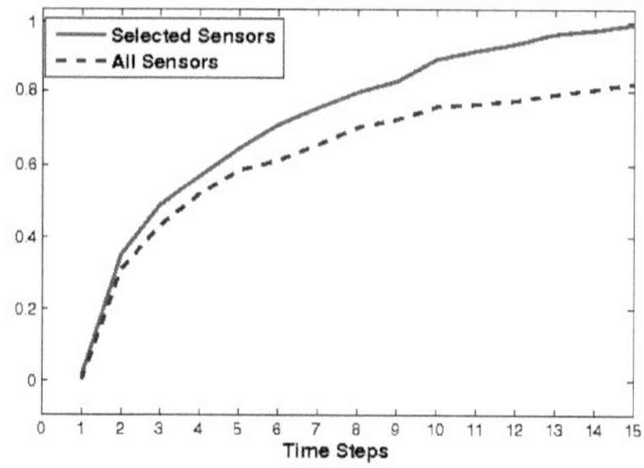


**Fig. 8.4** Normalized estimation performance index vs Time steps

The number of backoff stages can be chosen as:

$$m^* = max\{m : (\sum_{i=0}^{m} 2^i W)b_p < \beta \Delta t\} \tag{8.30}$$

where $W = 2^{macMinBE} - 1$ and $\beta \in [0\ 1]$ is chosen to take into account the time to run the sensor selection algorithm. If we set *macMinBE*= 4, and $\beta = 2/3$ the maximum number of backoff stages is $m^* = 6$. We also choose $d = 2$.

We applied the procedure proposed in Section 8.4 with $h(\cdot) = logdet(\cdot)$ to find, for each time interval, the optimal set of sensors. To underline the importance of a sensor selection strategy, Figure 4 compares the performance of the proposed sensor selection scheme with the case where all sensors attempt to transmit their measurements over a WSN with IEEE 802.15.4 protocol. In particular, Figure 4 shows the mean value of the normalized estimation performance index evaluated over 50 realizations of the stochastic process. After just few seconds, the proposed scheme achieves better performance using less sensors.

### 8.5.3 Protocol Model Validation

In section 8.2, we have assumed that the probability parameters for transmission, collision and channel busy at each channel access attempt can be evaluated in stationary condition.

In this subsection we validate the successful transmission probability derived by the Markov model by means of a Matlab simulator of the 802.15.4 protocol operating in unslotted CSMA-CA.

For fixed values of *macMinBE*, $m$, $N$ and $d$ the simulator provides a realization of the stochastic variable representing the number of sensors that successfully transmits during a sample period. For *macMinBE*= 2, $m = 3$ and $d = 2$ Figure 5 compares the probability of successfully transmission given by the proposed Markov model with the one given by the the simulator with 50000 realizations for each value of $N$. Figure 5 shows that the proposed probability of successfully transmission represents

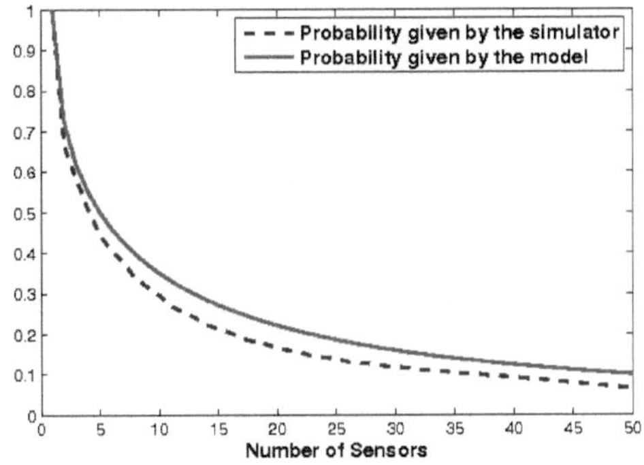


**Fig. 8.5** Probability of successfully transmission vs Number of reporting sensors

an upper bound of the real one and this behavior can be also noticed for different values of *macMinBE* and $m$. This is due to the fact that, at the beginning of each sample period, all the $N$ sensors start from the first backoff stage $W = 2^{macMinBE} - 1$, that is the smallest. So the probability of collision is higher than the one proposed by the steady state analysis. We can conclude that the proposed optimal number of sensors is an upper bound of real one.

## 8.6 Conclusion

In order to provide the optimal scheduling of sensors, i.e. how many and which ones should communicate their readings, the remote estimation problem over a wireless sensor network operating with one of the major MAC protocol is considered.

The MAC protocol IEEE 802.15.4 for wireless sensor network is analyzed using a Markov model. A lossy multi-sensor information filter is proposed to evaluate the estimation performance. An optimization algorithm based on the analysis of a convex optimization problem is proposed. A target monitoring example with wireless sensor network is provided to illustrate the analysis. Simulation results show that sensor selection achieves better performance while saving energy, extending network lifetime. The convex optimization approach ensures that the optimal solution can be efficiently computed even for fairly large scale systems. We are currently extending our approach to optimize sensor selection over a longer horizon, by selecting the sets of sensors for the future $N$ steps. We are also planning to validate sensor selection via experimentation and to integrate our algorithm in the Sensor Andrew project [11] at Carnegie Mellon.

## References

1. Ambrosino, R., Sinopoli, B., Poolla, K., Sastry, S.S.: Optimal Sensor Density for Remote Estimation Over Wireless Sensor Networks. In: Forty-Sixth Annual Allerton Conference (2008)

2. Apt, J., Gresham, L., Morgan, M., Newcomer, A.: Incentives for near-term carbon dioxide geological sequestration. Technical report, EPP Dept., Carnegie Mellon University (2007)
3. Bianchi, G.: Performance Analysis of the IEEE 802. 11 Distributed Coordination Function. IEEE J. Special Areas in Comm. -Wireless Series 18(3), 535–547 (2000)
4. Boyd, S., Vandenberghe, L.: Convex Optimization. Cambridge University Press, Cambridge (2004)
5. Callaway, E.: Wireless Sensor Networks: Architectures and PRotocols. CRC Press LLC, Boca Raton (2004)
6. Gupta, V., Chung, T.H., Hassibi, B., Murray, R.M.: On a stochastic sensor selection algorithm with applications in sensor scheduling and sensor coverage. Automatica 42(2), 251–260 (2006)
7. Joshi, S., Boyd, S.: Sensor Selection via Convex Optimization. Submitted to IEEE Transactions on Signal Processing (2007)
8. Liu, X., Goldsmith, A.J.: Kalman Filtering with Partial Observation Losses. In: IEEE Conference on Decision and Control (2004)
9. Matveev, A., Savkin, A.: The problem of state estimation via asynchronous communication channels with irregular transmission times. IEEE Trans. on Automat. Contr. 48(4), 670–676 (2003)
10. Misic, J., Shafi, S., Misic, V.B.: Performance of a Beacon Enabled IEEE 802.15.4 Cluster with Downlink and Uplink Traffic. IEEE Transactions on Parallel and Distributed Systems 17(4), 361–376 (2006)
11. Moura, J.: Sensor Andrew: Ubiquitous wide campus sensing exploitation. In: International Conference on Networked Sensing Systems (2008)
12. Mutambara, A.G.O.: Decentralized Estimation and Control for multisensor systems. CRC Press, Boca Raton (1998)
13. Polastre, J., Hill, J., Culler, D.: Versatile Low Power Media Access for Wireless Sensor Networks. In: Proceeding of SenSys 2004, Baltimore, Maryland, November 3-5 (2004)
14. Pottie, G., Kaiser, W.: Wireless integrated network sensors. Comm. of the ACM 43(5) (2000)
15. Sinopoli, B., Schenato, L., Franceschetti, M., Poolla, K., Jordan, M.I., Sastry, S.S.: Kalman filtering with intermittent observations. IEEE Trans. on Automat. Contr. 49(9), 1453–1464 (2004)
16. Standard for Part 15.4: Wireless Medium Access Control (MAC) and Physical Layer (PHY) Specifications for Low Rate Wireless Personal Area Networks (WPAN). In: IEEE Std 802.15.4. IEEE, New York
17. Weimer, J.E., Sinopoli, B., Krogh, B.H.: A Relaxation Approach to Dynamic Sensor Selection in Large-ScaleWireless Networks. In: The 28th International Conference on Distributed Computing Systems Workshops (2008)
18. Wells, A., Hammack, R., Veloski, G., Diehl, J., Strazisar, B.: Monitoring, mitigation, and verification at sequestration sides: Sequre technologies and the challenge for geophysical detection. Technical report, NETL, Pittsburgh, PA (2006)
19. Xu, Y., Hespanha, J.P.: Communication logics for networked control systems. In: Proc. of the American Control Conference, vol. 1, pp. 572–577 (2004)
20. Xu, Y., Hespanha, J.P.: Estimation under uncontrolled and controlled communications in networked control systems. In: IEEE Conference on Decision and Control (2005)
21. Yook, J.K., Tilbury, D.M., Soparkar, N.R.: Trading computation for bandwidth: Reducing communication in distributed control systems using state estimators. EEE Trans. Contr. Syst. Technol. 10(4), 503–518 (2002)
22. Zhu, B., Sinopoli, B., Poolla, K., Sastry, S.S.: Estimation over Wireless Sensor Networks. In: Proceedings of American Control Conference, New York, NY (2007)

# Chapter 9
# Growing Fully Distributed Robust Topologies in a Sensor Network

Andrea Gasparri, Sandro Meloni, and Stefano Panzieri

**Abstract.** Wireless Sensor Networks (WSN) are at the forefront of emerging technologies due to the recent advances in Micro-Electro-Mechanical Systems (MEMS) technology. WSN are considered to be unattended systems with applications ranging from environmental sensing, structural monitoring, and industrial process control to emergency response and mobile target tracking.The distributed nature and the limited hardware capabilities of WSN challenge the development of effective applications. The strength of a sensor network, which turns out to be also its weakness, is the capability to perform inter-node processing while sharing data across the network. However, the limited reliability of a node, due to the low-cost nature of the hardware components, drastically constrains this aspect. For this reason, the availability of a mechanism to build distributed robust connectivity topologies, where robustness is meant against random failures of nodes and intentional attacks of nodes, is crucial. The complex network theory along with the percolation theory provides a suitable framework to achieve that. Indeed, topologies such as multi-modal and scale free ones, show interesting properties which might be embedded into a sensor network to significantly increase its robustness. In this work, a mechanisms to build robust topologies in a distributed fashion is proposed, its effectiveness is analytically investigated and results are validated through simulations.

---

Andrea Gasparri
University of "Roma Tre", Via della Vasca Navale 79
e-mail: `gasparri@dia.uniroma3.it`

Sandro Meloni
University of "Roma Tre", Via della Vasca Navale 79
e-mail: `sandro@dia.uniroma3.it`

Stefano Panzieri
University of "Roma Tre", Via della Vasca Navale 79
e-mail: `panzieri@uniroma3.it`

## 9.1 Introduction

A sensor network consists of a collection of nodes deployed in an environment that cooperate to perform a task. Each node, which is equipped with a radio transceiver, a micro-controller and a set of sensors, shares data to reach the common objective. Sensor networks provide a framework in which, exploiting the collaborative processing capabilities, several problems can be faced and solved in a new way. However, it comes along with several challenges such as limited processing, storage and communication capabilities as well as limited energy supply and bandwidth. Performing a partial computation locally on each node, and exploiting inter-node cooperation, is the ideal way to use sensor networks. Unfortunately, this *modus-operandi* is highly constrained by the reduced hardware capabilities as well as by the limited energy resources that makes communication extremely unreliable as well as expensive in terms of node life-time. As a consequence, the availability of a mechanism to build distributed robust connectivity topologies, where robustness is meant against random node failures and intentional node attacks, is crucial.

Sensor networks can be of interest to different areas of application, ranging from environmental monitoring [5, 38], civil infrastructures [13, 25], medical care [31, 28] to home and office applications [32, 15]. In each field, the deployment of a sensor network has provided interesting advantages. For instance, in the context of environmental monitor the introduction of a sensor network made it possible to keep environments intrinsically threatening for human beings [38] under surveillance, or in the context of medical care it made it possible to remotely monitor the health condition of patients by continuously extracting clinical relevant information [28].

Regardless to the specific application, for a sensor network in oder to properly operate, information must be shared across the network allowing for data dissemination and data aggregation. Indeed, a big effort has been done by the research community to develop efficient topology discovery and control algorithms able to achieve that. Strictly speaking, the topology discovery aims to infer the topological structure of the network for management purpose, while the topology control aims to maintain some desired network properties in order to improve the performance of networking services such as routing. In particular, regarding the topology control problem, the majority of works available in literature address this problem in terms of per-node transmission power in oder to increase the life-time of the sensor network [4, 24, 2]. Some contributions focus their attention on the fault-tollerance aspects in terms of network deployment or power assignment [21, 12, 35].

In this work, a novel topology control algorithm is proposed. The main idea is to design a robust connectivity topology by exploiting the complex network along with the percolation theory. Indeed, complex networks such as the scale-free networks show interesting properties which might be embedded into a sensor network to significantly increase its robustness. These properties can be further refined by exploiting the percolation theory which turns out to be a very suitable framework for analysis purposes. In detail, a mechanisms to build an arbitrary topology over a geographical environment is proposed. In addition, a robust distribution against random failures and intentional attacks has been exploited, its effectiveness is analytically investigated according to [34] and results are validated through simulations.

It must be mentioned that the use of the percolation theory is not new to this field. Indeed, it was already applied to analyze the connectivity of ad-hoc networks in [11, 10] and, to develop topology control algorithms for wireless sensor network that are large-scale and lack a centralised authority in [29]. The main novelty introduced by this work is that the main results of percolation theory applied to the complex networks have been extended to provide a better tolerance to random failures of sensor network nodes. This idea to use the percolation theory not as an evaluation tool but instead to build a topology with specific properties is new, up to the knowledge of the authors, to the field of sensor networks.

The rest of the chapter is organized as follows. In Section 9.2 a summary of the state of the art is presented. In Section 9.3 an overview of the theoretical background is provided. In Section 9.4 the proposed algorithm is described. In Section 9.5 a numerical analysis to corroborate the analytical results is given. Finally, in Section 9.6 conclusions are drawn and future work is discussed.

## 9.2 Related Work

In the last years, a great effort has been devoted by the research community to the design of energy-efficient and fault-tolerant algorithms for sensor networks, e.g., routing algorithms [8, 23, 22]. This objective is very challenging in particular due to the dynamic nature of the sensor networks. For this reason, the topology control has become a very attractive field of research. An interesting survey of this family of algorithms is given in [30]. According to it, a possible classification of topology control techniques can be drawn either in regard of constrains on the power-range assignment or with respect to topological properties of the connectivity graph. As far as the power-assignment is concerned, a distinction can be made between between homogeneous techniques [17, 27] and non-homogenous techniques [14, 20]. In the first case nodes are assumed to have the same transmitting range while in the second case nodes are allowed to have a different transmitting range. Moreover, non-homogeneous approaches can be classified into location-based [19, 20], direction based [4, 16] and neighbor based [39, 14]. For instance, location based techniques use information regarding the location of a node, while direction based techniques exploit the relative distance between nodes and finally neighbor based techniques use only information regarding the ID of a node, neither location nor distance information is available. Alternatively, topology control techniques can be classified with respect to the topological properties of the connectivity graph resulting from their applications. In particular, a significant amount of works presented in literature are concerned with building and maintaining a connected network topology which allows data to be shared across the network. In particular, some authors have considered the problem of building k-connected network topologies with the aim of improving the fault-tolerance [1, 18, 12, 35]. Some other authors instead have focused on topology control schemas where nodes alternate between active and sleeping time while maintaining connectivity of the whole network [9, 40].

## 9.3 Theoretical Background

### *9.3.1 Complex Network*

The discovery of small-world and scale-free properties of many natural and artificial complex networks has originated a big interest in investigating the fundamental organizing principles of various complex networks. In the context of network theory, a complex network is a network (graph) with non-trivial topological features, i.e., features that do not occur in simple networks such as lattices or random graphs. Indeed, a remarkable number of systems in nature present non-trivial topological features which can be properly modeled by exploiting the complex network theory. Airlines routing maps for instance are neither random graphs or regular lattices. Nodes of this network are airports connected by direct flights among them. In particular, there are a few hubs on the airline routing map representing the major cities from which flights depart to almost all other airports. Instead, the majority of airports are tiny, appearing as nodes with one or a few links connecting them to one or several hubs. The same argument can be adopted to describe the Internet as well. In fact, also in this case, there are a few hubs constituting the backbone of the network, while the majority of nodes are connected to it by a limited number of links. Another interesting example is given by how diseases are transmitted through social networks. In fact, the way in which a disease rapidly spreads can be hardly motivated by describing the social network as a random graph or a lattice. In the same way, the protein-to-protein interaction networks (PINs) at the base of any biological process show non-trivial topological features which cannot be embedded into random graphs or lattices.

In the last decades several quantities have been investigated to characterize the properties of a complex network. Thus far, three concepts, i.e., the characteristic path length, the clustering coefficient, and the degree distribution, turned out to play a crucial role in the recent study and development of complex networks theory. The characteristic path length $\langle d \rangle$ of the network is the mean distance between two nodes, averaged over all pairs of nodes, where the distance between two nodes is defined as the number of the edge along the shortest path connecting them. The cluster coefficient $\langle c \rangle$ of the network is the average of $\langle c_i \rangle$ over all nodes $i$, where the coefficient $\langle c_i \rangle$ of node $i$ is the average fraction of pairs of neighbors of the node $i$ that are also neighbors of each other. The degree distribution of the network is the distribution function $P(k)$ describing the probability that a randomly selected node has exactly degree $k$, that is the number of links a node owns.

Indeed, the original attempt of Watts and Strogatz in their work on small-world networks [37] was to construct a network model with small average path length as a random graph and relatively large clustering coefficient as a regular lattice, which evolved to become a new network model as it stands today. On the other hand, the discovery of scale-free networks was based on the observation that the degree distributions of many real networks have a power-law form, albeit power- law distributions have been investigated for a long time in physics for many other systems and processes. For a comprehensive overview of the basic concepts and significant results concerning the complex network theory the reader is referred to [36].

## 9.3.2 Percolation Theory

The percolation theory provides a suitable framework to analytically investigate the robustness of a network, i.e., the ability of a network to properly operate even when a fraction of its components is damaged [3].

Strictly speaking, the percolation theory is a general mathematical theory of connectivity and transport in geometrical complex system. Percolation is of particular interest to physicists as it can be considered the simplest model of a disordered system capable of experiencing a phase transition. A remarkable aspect of percolation is that many results can be often encapsulated in a small number of simple algebraic relationships. For a comprehensive introduction to the percolation theory the reader is referred to [33].

A standard percolation process can be, in general, of two types: site or bond. Site percolation on a given graph means that the vertices are empty with a given probability $f$ (or occupied with a probability $p = 1 - f$ ), while bond percolation refers to the existence or not of an edge between two arbitrarily chosen nodes. Once the random deletion (or placement) of nodes or edges is done, several quantities allow the characterization of the network properties. In particular, it is possible to look at the existence and size of the giant component as a function of $f$, and at the average size and fluctuation in the size of finite components. In this way, it can be defined a critical probability $f_c$ below which the network percolates, i.e., it has a giant component, and a set of critical exponents characterizing the phase transition. The exact value of such a threshold $f_c$ depends on which kind of grid (graph) is considered and its dimension. Percolation theory gives an analytical framework for the study of failures or attacks on a network in general. During the last decades some exacts results have been proposed for special types of graphs such as one and two-dimensional lattices, Cayley trees and a general criterion for study networks robustness. In 1998 Molloy and Reed [26] defined a criterion for the appearance of the giant component in a graph with generic degree distribution $P(k)$ only analyzing its first $\langle k \rangle$ and second moment $\langle k^2 \rangle$. The Molloy and Reed criterion has been used by Cohen et al.[7][6] to give a general form for the percolation threshold $f_c$ both for random failures and intentional attacks

The study of random failures for a sensor network can be exactly mapped into an inverse percolation problem. More precisely, given an adjacency graph $A_o$ describing the connectivity topology of a sensor network, the inverse percolation problem consists of finding the critical fraction $f_c$ of the links of $A_o$ for which the giant component disappears. Obliviously, such a breakdown of the connectivity topology drastically influences the capability of a sensor network to share data across and properly operate.

## 9.4 The Proposed Algorithm

In this work, a way of reproducing an arbitrary degree distribution $P(k)$ on a geographical space where nodes are characterized by limited visibility is proposed. The underlying idea is that well-known techniques in the field of complex networks robustness can be suitably applied to a geographical environment. In particular, a degree distribution with properties of robustness against both faults and attacks is sought . Indeed, the multi-modal distribution proposed by [34] provides these properties. As a result, a robust topology for a sensor network can be designed.

The following scenario is considered:

- Nodes are uniformly distributed in a closed 2-Dimensional plane of side $L$ and area $A = L^2$.
- Nodes have a limited radius of interaction $r$ defined as a fraction of $L$,

Now, given a sensor network consisting of $N$ nodes, the number of neighbors (degree) of a generic node $i$ is $\langle k_i \rangle = \rho \pi r^2$, where $\rho = N/L^2$ is the density of nodes deployment.

According to the given scenario the proposed algorithm works as follow: i) $N$ nodes are distributed uniformly on a square of side $L$ ii) as each node $i$ starts operating, it extracts a integer $k$ from a selected distribution iii) then $i$ tries to make $k$ connections with the nodes in its visibility radius $r$, iv) to assure the full connectivity of all the nodes an additional connectivity maintenance step is introduced. Note that, a node might not be able to establish the desired number of connections, due to the limited radius of visibility $r$ with respect to the density of deployment $\rho$. Nonetheless, a good approximation of the distribution can always be reached for reasonable values of $\rho$ and $r$. Indeed, this is the case for a realistic sensor network scenario.

Regarding the connectivity maintenance step, the idea is to exploit a consensus algorithm by which nodes share their ID within their visibility neighborhood, i.e., nodes within its range of visibility. From an algorithmic perspective, each node broadcasts its ID to its neighbors, if a node receives a lower ID it starts sharing the received lower number. Periodically, each node check IDs within visibility neighborhood. If one of these nodes $k$ holds a lower ID, then node $i$ creates a new connection to $k$ and starts sharing $k$'s ID. This step permits to obtain a connected network only adding few links to the original distribution, and if executed periodically to readapt network topology to failures and damages. Note that, even though the connectivity maintenance step is required, from a practical standpoint this can be avoided by performing a proper choice of $\rho$ and $r$.

Regarding the distributed nature of the algorithm, it is worthwhile to mention that a few parameters concerning the degree distribution $P(k)$ are required for the algorithm in order to properly operate. Moreover, as these parameters are fixed, they can be easily hardcoded into each node.

At this point, being a technique for constructing an arbitrary distribution over a geographical space available, the analytical evaluation of the best form for the $P(k)$ is faced.

### 9.4.1 *Optimal Degree Distribution for Random Failures and Attacks*

As previously mentioned, a degree distribution which provides interesting properties of robustness has been proposed in [34]. In this work, the authors show that a network which maximizes the value of the threshold $f_T$, defined as $f_T = f_r + f_a$ with $f_r$ the percolation threshold for random node removal and $f_a$ the threshold for targeting node removal, can be obtained by exploiting the following functional form:

$$P(k) = \sum_{i=1}^{m} r_i \delta(k - k_i) = \sum_{i=1}^{m} r_1 a^{-(i-1)} \delta(k - k_i) \tag{9.1}$$

where $k_i = k_1 b^{-(i-1)}$ with $k_1$ min degree of the network and $\delta(x)$ Dirac delta function. In detail, the model is characterized by three different quantities: $a$ that represents the fraction of nodes having different degrees (with $a > 1$), $b$ that controls the values of the degrees, and $k_1$ that is the smallest degree in the network. The two remaining parameters $r_1$ and $r_m$ can be obtained from the following normalization condition:

$$\sum_{i=1}^{m} r_i = r_1 \sum_{i=1}^{m} a^{-(i-1)} = 1 \tag{9.2}$$

as follows:

$$r_1 = \frac{1 - a^{-1}}{1 - a^{-m}} \quad \text{or} \quad r_m = \frac{a - 1}{a^m - 1} \tag{9.3}$$

and

$$\frac{a - 1}{a^m - 1} = \frac{q}{N} \tag{9.4}$$

$$r_m = \frac{q}{N} = N^{\alpha - 1} \text{ with } 0 < \alpha < 0.25 \tag{9.5}$$

with $q$ the number of nodes with the highest degree $k_m$. In addition, the authors evidence an inter-dependency among all the parameters of Eq. 9.1 which leads to a model that is only a function of $N$ and $m$, i.e, the number of nodes and the number of distinct modes in the distribution respectively. They also demonstrate that the mean degree $\langle k \rangle$ is:

$$\langle k \rangle = \sum_{i=1}^{m} k_i r_i = k_1 r_1 \sum_{i=1}^{m} (ab)^{-(i-1)} \tag{9.6}$$

leading to the following general form of the parameters $a$ and $b$:

$$ab \sim N^{(1/2 - \alpha)/(m-1)} \tag{9.7}$$

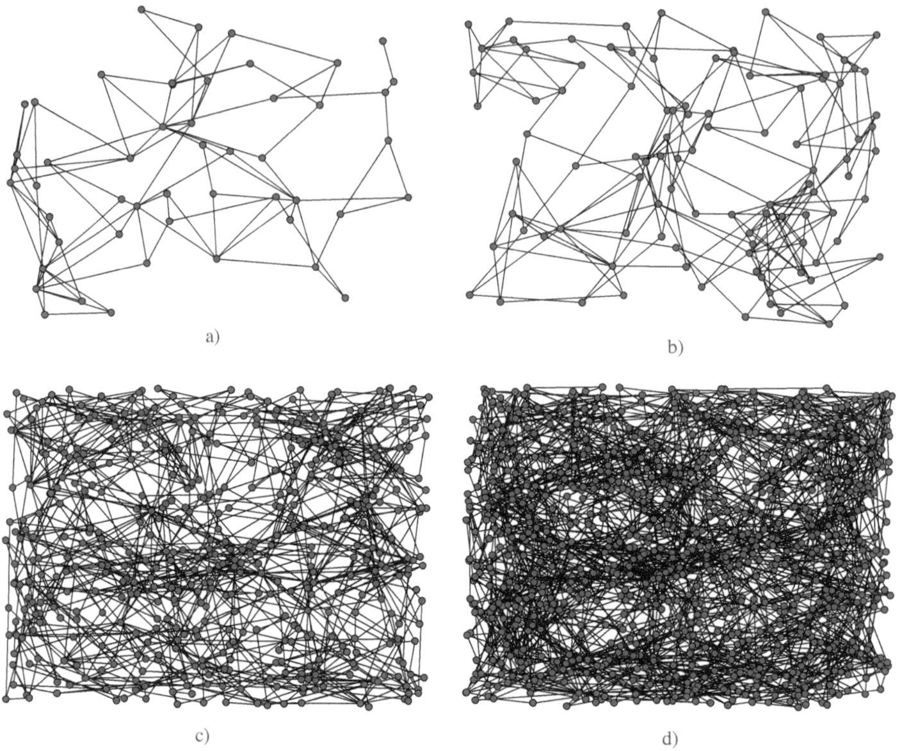

**Fig. 9.1** Four examples of network produced by the proposed algorithm with different number of nodes $N$, a) $N = 50$, b) $N = 100$, c) $N = 500$ and d) $N = 1000$, with $m = 3$, $k_1 = 3$ and $\alpha = 0$.

from which:

$$a \sim N^{(1-\alpha)/(m-1)} \tag{9.8}$$

and

$$b \sim N^{1/2(m-1)} \tag{9.9}$$

At this point, for a given number of nodes $N$, the two parameters $a$ and $b$ depend only on $m$ and $\alpha$. This leads to a two parameters model, for which it is possible to compute the optimal value of the percolation threshold $f_T^{opt}$ by assuming $\alpha = 0$ and $m = 2$, where $f_T^{opt} = f_a^{opt} + f_r^{opt}$ is defined as the sum of the two percolation thresholds for nodes attacks and random failures respectively. Note that, as the $f_T^{opt}$ is a linear combination of two factors, a slightly different behavior, i.e., higher robustness to random node failures or higher robustness to intentional node attacks, can be obtained with a proper choice of the two parameters $\alpha$ and $m$.

An example of network topology created with the proposed algorithm when exploiting the multi-modal distribution described so far is given in Fig. 9.1. It can be noticed that the obtained topologies are characterized by a high number of triangles which guarantee robustness. At the same time, the degree of the most connected nodes is kept sufficiently low which allows to both mitigate the impact of intentional attacks and limit the effect of random failures.

## 9.5 Numerical Analysis

The proposed algorithm have been thoroughly investigated through simulations. Two aspects of interest have been investigated: the robustness to random node failures and the robustness to node attacks. The first aims to evaluate the capability of the sensor network to properly operate even when suddenly some nodes stop working, while the second investigates the resistance of the network when in presence of organized attack aiming to destabilize the normal operating conditions.

The following indexes of quality have been considered: i) the number of components ii) the size of the giant component iii) the percentage of network disconnected. The first index gives an information about the overall connectivity of the network, the second one gives an idea about the remaining operability, while the last one gives an information about the number of nodes still functioning.

Moreover, a comparison against a null-model has been performed. Such a null-model is built starting from the network produced by the proposed algorithm by keeping the same constraints on the number of nodes, the visibility radius $r$ but introducing a randomized version of the link connections leading to a Poisson degree distribution. As a result, a random network topology is achieved.

Several network configurations have been analyzed. In the following only results regarding a network composed by 2500 nodes deployed in an geographical space with density $\rho = 0.5$ are shown.

Fig. 9.2-a) shows the degree distribution $P(k)$ obtained for the proposed model with the parameter $m = 3$. It can be noticed the presence of three peaks, respectively for $k = 3, 22, 33$, representing the three modes of the distribution. The two remaining spare peaks can be explained by the limited visibility $r$ of nodes. Indeed, these two peaks would tend to the closest ones on the line if the radius $r$ were sufficiently large. Note that for $m \rightarrow \infty$ the distribution $P(k)$ tends to a scale-free distribution [34]. On the other hand, Fig. 9.2-b) describes the degree distribution $P(k)$ obtained for the null-model, which is, as expected, a Poisson distribution.

Table 14.1 gives a synoptical overview of the conducted analysis. In particular, it can be noticed that when considering two networks with a comparable number of nodes and links the proposed model turns out to be more robust. This can be explained by the higher value of the clustering coefficient $\langle c \rangle$ leading to an higher number of triangles in the network that are known to be the most robust structure against random failures. Moreover, another interesting aspect can be pointed out: both the characteristic path length $\langle d \rangle$ and diameter $d_{max}$ are lower for the proposed model. Indeed, this is a good property for a sensor network as it implies a lower consumption to spread data over the network.

Fig. 9.3 shows the number of connected components (CC) for both the proposed model (circles) and the null-model (squares) when varying the fraction of removed links. In detail, Fig. 9.3-a) represents the behavior of the models against random

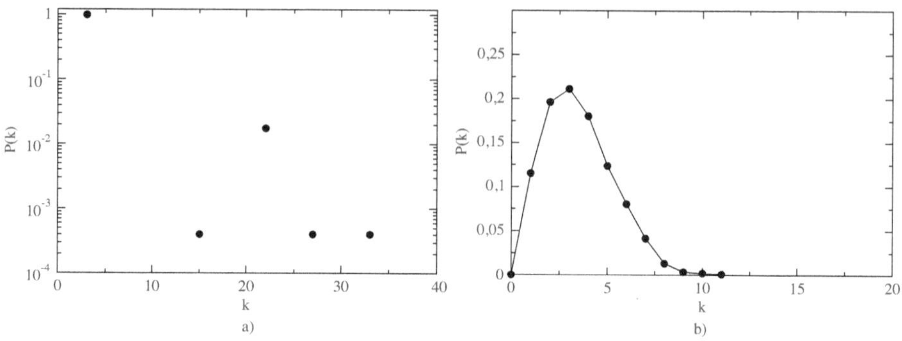


**Fig. 9.2** Degree distribution of the proposed model a), and of the null-model b). Parameters setting: $N = 2500, m = 3, k_1 = 3, \alpha = 0$.

**Table 9.1** Principal topological features of the proposed model and the null-model. In detail, $N$ is the number of nodes in the network, $E$ the numerb of links, $\langle k \rangle$ the mean degree, $k_{max}$ the highest degree, $\langle d \rangle$ the characteristic path length, $d_{max}$ network diameter and $\langle c \rangle$ the mean clustering coefficient.

| Model | N | E | $\langle k \rangle$ | $k_{max}$ | $\langle d \rangle$ | $d_{max}$ | $\langle c \rangle$ |
|---|---|---|---|---|---|---|---|
| Proposed Model | 2500 | 4201 | 3.36 | 33 | 5.99 | 13 | 0.004766 |
| Null Model | 2500 | 4277 | 3.42 | 11 | 7.24 | 15 | 0.001279 |

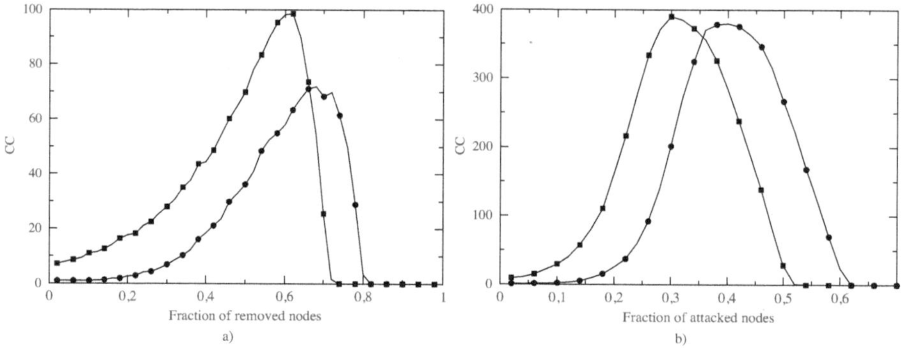


**Fig. 9.3** Number of connected components (CC) vs. fraction of removed nodes for the proposed model (circles) and the null-model (squares) in case of random node failures a) and intentional node attacks b). Parameters setting: $N = 2500, m = 3, k_1 = 3, \alpha = 0$.

node failures, while Fig. 9.3-b) depicts the same behavior against intentional attacks. In both cases, the proposed model outperforms the null-model, i.e., the network starts to break down after a higher fraction of node (approx. 20%). Note that, isolated nodes are not counted as components.

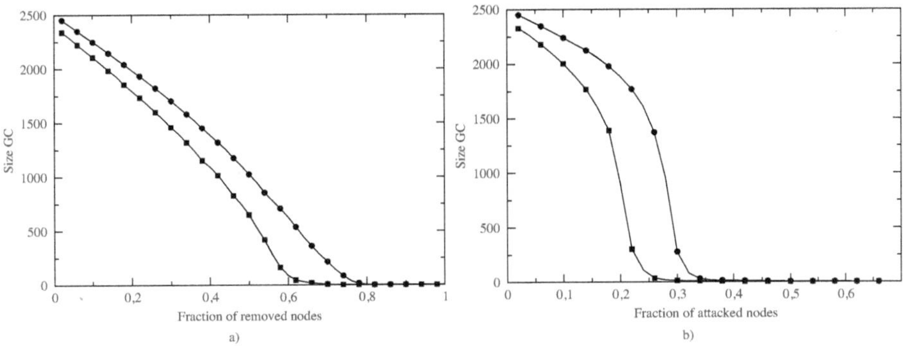


**Fig. 9.4** Size of the giant component (Size GC) vs. fraction of removed nodes for the proposed model (circles) and the null-model (squares) in case of random node failures a) and intentional node attacks b). Parameters setting: $N = 2500, m = 3, k_1 = 3, \alpha = 0$.

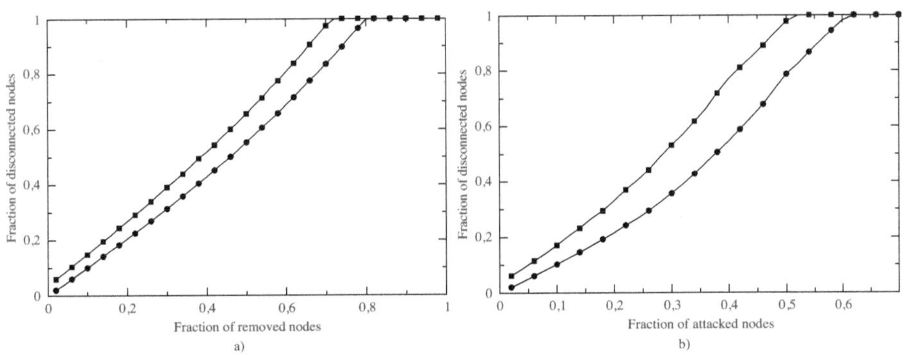


**Fig. 9.5** Fraction of disconnected nodes vs. fraction of removed nodes for the proposed model (circles) and the null-model (squares) in case of random node failures a) and intentional node attacks b). Parameters setting: $N = 2500, m = 3, k_1 = 3, \alpha = 0$.

Fig. 9.4 shows the size of the giant component for both the proposed model (circles) and the null-model (squares) when varying the fraction of removed links. Also in this case, the proposed model outperforms the null-model. In particular, the size of the biggest component decreases almost linearly with the fraction of removed nodes in the case of random nodes removal.

Finally, Fig. 9.5 shows the fraction of disconnected nodes for both the model (circles) and the null-model (squares) when varying the fraction of removed links. As before, the performance of the proposed model is significantly better than the null-model.

An additional analysis of the behavior of proposed technique has been successively carried out. In particular the following aspects have been investigated: the rate of growth of the number of links with the respect to the number of nodes, the fraction of isolated nodes resulting from the removal of a fraction of nodes and the

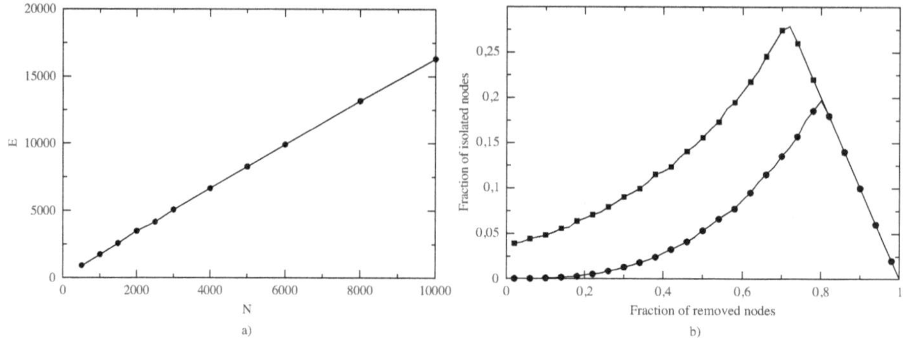


**Fig. 9.6** a) Growth rate of the number of links vs. the number of nodes. b) Fraction of isolated nodes resulting from the removal of a fraction of nodes in case of random node failures.

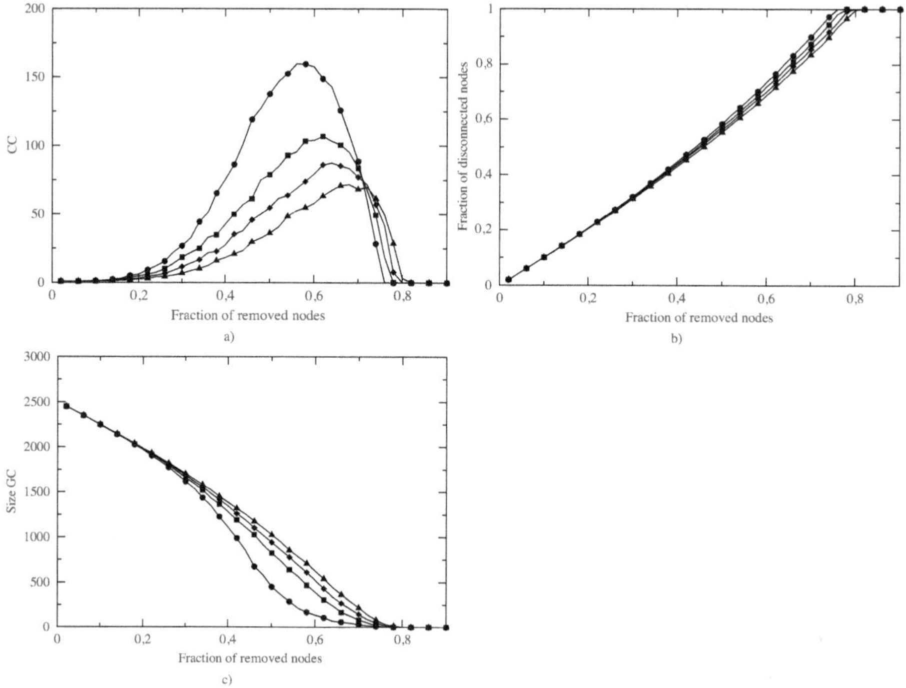


**Fig. 9.7** a) Number of connected components, b) Size of the giant component and c) Fraction of isolated nodes vs fraction of removed nodes in the proposed model with different values of $m$. $m = 3$ (circles), $m = 5$ (squares), $m = 7$ (diamonds), $m = 10$ (triangles). Parameters setting: $N = 2500$, $k_1 = 3$, $\alpha = 0$.

variation of the tree indexes previously introduced, i.e., the number of components, the size of the giant component and the percentage of network disconnected, when varying the value of the parameter $m$.

Fig. 9.6-a) shows the rate of growth of the number of links with respect to the number of nodes. it can be easily noticed that the number of links increases linearly with the number of nodes. This is indeed a good property of the algorithm as the higher is the number of links the higher is the power consumption of the network leading to a good scalability. Fig. 9.6-b) illustrates the fraction of isolated nodes resulting as a consequence of the removal of a fraction of nodes. This is another interesting property of the algorithm. In fact, it points out that only a negligible percentage of nodes are affected by the removal of other nodes. In other words, by removing a node the connectivity of its neighbors is not significantly influenced.

Figs. 9.7 shows how the tree indexes change when varying the value of the parameter $m$. Note that, this result is referred to the proposed model against random node failures. According to the theoretical results, the higher is the value of the parameter $m$ the better is the performance against random failures as the scale-free characterization of the degree distribution becomes more and more notable. Indeed, this is in agreement with the results obtained in [34] as the percolation threshold is not influenced by the variation of the parameter $m$, but at the same time other characteristics, such as the number of connected components, are positively influenced in the case of random node failures.

## 9.6 Conclusions

In this work, a novel topology control algorithm has been proposed. The idea is to take advantage of the complex networks along with the percolation theory to design robust topologies. Indeed, the availability of a connectivity topology algorithm able to properly operate even when in presence of random failures of nodes drastically increases the robustness as well as the operability of a sensor network.

In detail, an algorithm to build an arbitrary topology over a geographical environment is proposed. In addition, a robust degree distribution against random failures and intentional attacks has been exploited [34]. The properties of the resulting model have been analytically characterized by exploiting the percolation theory and the related results have been corroborated by numerical simulations. In particular, three different indexes of quality have been investigated, namely the number of connected components, the size of the giant component and the fraction of disconnected nodes in the network. Moreover, a comparison against a randomized version of the network (null-model) has been performed. According to these results, the proposed topology control technique has turned out to be very effective as it always outperforms the null-model in terms of connectivity maintenance against both random node failures and intentional node attacks.

To conclude, the proposed algorithm is very simple, distributed and easy to implement on-board each node. It requires a limited number of messages in order to build the topology and the number of links scales linearly with the size of the network. Moreover, even though the algorithm has been implemented only in a 2-dimensional plane, there is no additional cost to extend it to a n-dimensional space, as the topology construction relies only on the Euclidian distance.

Several challenges still remain for future work. An extension where a node independently sets its radius of visibility $r$ might be investigated in order to reduce the energy consumption. In addition, a dynamical network rewiring process able to reconnect the network anytime two or more components arise might be considered. Finally, an enhanced scenario where mobility is taken into account for some nodes might be of interest.

## References


1. Bettstetter, C.: On the minimum node degree and connectivity of a wireless multihop network. In: MobiHoc 2002: Proceedings of the 3rd ACM international symposium on Mobile ad hoc networking & computing, pp. 80–91. ACM Press, New York (2002)
2. Blough, D.M., Leoncini, M., Resta, G., Santi, P.: The k-neighbors approach to interference bounded and symmetric topology control in ad hoc networks. IEEE Trans. Mob. Comput. 5(9), 1267–1282 (2006)
3. Boccaletti, S., Latora, V., Moreno, Y., Chavez, M., Hwang, D.-U.: Complex networks: Structure and dynamics. Physics Reports 424(4-5), 175–308 (2006)
4. Borbash, S.A., Jennings, E.H.: Distributed topology control algorithm for multihop wireless networks. In: Proc. 2002 World Congress on Computational Intelligence (WCCI 2002), pp. 355–360 (2002)
5. Cerpa, A., Elson, J., Hamilton, M., Zhao, J., Estrin, D., Girod, L.: Habitat monitoring: application driver for wireless communications technology. In: SIGCOMM LA 2001: Workshop on Data communication in Latin America and the Caribbean, pp. 20–41. ACM Press, New York (2001)
6. Cohen, R., ben Avraham, D., Havlin, S.: Percolation critical exponents in scale-free networks. Phys. Rev. E 66(3), 036113 (2002)
7. Cohen, R., Erez, K., ben Avraham, D., Havlin, S.: Resilience of the internet to random breakdowns. Phys. Rev. Lett. 85(21), 4626–4628 (2000)
8. Datta, M.-A.: A fault-tolerant protocol for energy-efficient permutation routing in wireless networks. IEEE Trans. Comput. 54(11), 1409–1421 (2005)
9. Deb, B., Nath, B.: On the node-scheduling approach to topology control in ad hoc networks. In: MobiHoc 2005: Proceedings of the 6th ACM international symposium on Mobile ad hoc networking and computing, pp. 14–26. ACM Press, New York (2005)
10. Dousse, O., Thiran, P., Hasler, M.: Connectivity in ad-hoc and hybrid networks. In: Twenty-First Annual Joint Conference of the IEEE Computer and Communications Societies. IEEE INFOCOM 2002, pp. 1079–1088 (2002)
11. Gupta, P., Kumar, P.R.: Critical power for asymptotic connectivity. In: Proceedings of the 37th IEEE Conference on Decision & Control, pp. 1106–1110 (1998)
12. Hajiaghayi, M.T., Immorlica, N., Mirrokni, V.S.: Power optimization in fault-tolerant topology control algorithms for wireless multi-hop networks. IEEE/ACM Trans. Netw. 15(6), 1345–1358 (2007)
13. Kim, S., Pakzad, S., Culler, D., Demmel, J., Fenves, G., Glaser, S., Turon, M.: Health monitoring of civil infrastructures using wireless sensor networks. In: IPSN 2007: Proceedings of the 6th international conference on Information processing in sensor networks, pp. 254–263. ACM Press, New York (2007)
14. Leoncini, M., Resta, G., Santi, P.: The k-neighbors approach to interference bounded and symmetric topology control in ad hoc networks. IEEE Transactions on Mobile Computing 5(9), 1267–1282 (2006); Senior Member-Douglas M. Blough

15. Lesser, V., Atighetchi, M., Benyo, B., Horling, B., Raja, A., Vincent, R., Wagner, T., Ping, X., Zhang, S.X.Q.: The intelligent home testbed. In: Proceedings of the Autonomy Control Software Workshop (Autonomous Agent Workshop) (January 1999)
16. Li, L., Halpern, J.Y., Bahl, P., Wang, Y.-M., Wattenhofer, R.: A cone-based distributed topology-control algorithm for wireless multi-hop networks. IEEE/ACM Trans. Netw. 13(1), 147–159 (2005)
17. Li, N., Hou, J.C., Sha, L.: Design and analysis of an mst-based topology control algorithm. In: Twenty-Second Annual Joint Conference of the IEEE Computer and Communications Societies. IEEE INFOCOM 2003, vol. 3, pp. 1702–1712 (2003)
18. Li, N., Hou, J.C.: Flss: a fault-tolerant topology control algorithm for wireless networks. In: MobiCom 2004: Proceedings of the 10th annual international conference on Mobile computing and networking, pp. 275–286. ACM Press, New York (2004)
19. Li, N., Hou, J.C.: Localized topology control algorithms for heterogeneous wireless networks. IEEE/ACM Trans. Netw. 13(6), 1313–1324 (2005)
20. Li, X.-Y., Song, W.-Z., Wang, Y.: Localized topology control for heterogeneous wireless sensor networks. ACM Trans. Sen. Netw. 2(1), 129–153 (2006)
21. Li, X.-Y., Wan, P.-J., Wang, Y., Yi, C.-W.: Fault tolerant deployment and topology control in wireless networks. In: MobiHoc 2003: Proceedings of the 4th ACM international symposium on Mobile ad hoc networking & computing, pp. 117–128. ACM Press, New York (2003)
22. Liang, C., Huang, X., Deng, J.: A fault tolerant and energy efficient routing protocol for urban sensor networks. In: InfoScale 2007: Proceedings of the 2nd international conference on Scalable information systems, ICST, Brussels, Belgium, Belgium, pp. 1–8. ICST (Institute for Computer Sciences, Social-Informatics and Telecommunications Engineering) (2007)
23. Liang, Q., Wang, L., Ren, Q.: Fault and tolerant and energy efficient cross-layer design for wireless sensor networks. Int. J. Sen. Netw. 2(3/4), 248–257 (2007)
24. Liu, J., Li, B.: Distributed topology control in wireless sensor networks with asymmetric links. In: Global Telecommunications Conference, 2003. GLOBECOM 2003, vol. 3, pp. 1257–1262 (2003)
25. Mehta, V., El Zarki, M.: A bluetooth based sensor network for civil infrastructure health monitoring. Wirel. Netw. 10(4), 401–412 (2004)
26. Molloy, M., Reed, B.: The size of the giant component of a random graph with a given degree sequence. Combin. Probab. Comput. 7, 295–305 (1998)
27. Nayebi, A., Sarbazi-Azad, H.: Lifetime analysis of the logical topology constructed by homogeneous topology control in wireless mobile networks. In: International Conference on Parallel and Distributed Systems, vol. 2(5), pp. 1–8 (2007)
28. Patel, S., Lorincz, K., Hughes, R., Huggins, N., Growdon, J.H., Welsh, M., Bonato, P.: Analysis of feature space for monitoring persons with parkinson's disease with application to a wireless wearable sensor system. In: Proceedings of the 29th IEEE EMBS Annual International Conference, Lyon, France (August 2007)
29. Raghavan, U.N., Kumara, S.R.T.: Decentralised topology control algorithms for connectivity of distributed wireless sensor networks. Int. J. Sen. Netw. 2(3/4), 201–210 (2007)
30. Santi, P.: Topology control in wireless ad hoc and sensor networks. ACM Comput. Surv. 37(2), 164–194 (2005)
31. Shnayder, V., Chen, B.r., Lorincz, K., Thaddeus, R.F., Jones, F., Welsh, M.: Sensor networks for medical care. In: SenSys 2005: Proceedings of the 3rd international conference on Embedded networked sensor systems, p. 314. ACM Press, New York (2005)
32. Srivastava, M.B., Muntz, R.R., Potkonjak, M.: Smart kindergarten: sensor-based wireless networks for smart developmental problem-solving enviroments. In: Mobile Computing and Networking, pp. 132–138 (2001)

33. Stauffer, D., Aharony, A.: Introduction to percolation theory. CRC Press, Boca Raton (1998)
34. Tanizawa, T., Paul, G., Havlin, S., Stanley, H.E.: Optimization of the robustness of multimodal networks. Phys. Rev. E 74(1), 016125 (2006)
35. Thallner, B., Moser, H., Schmid, U.: Topology control for fault-tolerant communication in wireless ad hoc networks. In: Wireless Networks (2008)
36. Wang, X.F., Chen, G.: Complex networks: small-world, scale-free and beyond. Circuits and Systems Magazine, IEEE 3(1), 6–20 (2003)
37. Watts, D.J., Strogatz, S.H.: Collective dynamics of 'small-world' networks. Nature 393(6684), 440–442 (1998)
38. Werner-Allen, G., Lorincz, K., Welsh, M., Marcillo, O., Johnson, J., Ruiz, M., Lees, J.: Deploying a wireless sensor network on an active volcano. IEEE Internet Computing 10(2), 18–25 (2006)
39. Xue, F., Kumar, P.R.: The number of neighbors needed for connectivity of wireless networks. Wirel. Netw. 10(2), 169–181 (2004)
40. Yuanyuan, Z., Medidi, M.: Sleep-based topology control for wakeup scheduling in wireless sensor networks. In: 4th Annual IEEE Communications Society Conference on Sensor, Mesh and Ad Hoc Communications and Networks, 2007. SECON 2007, June 2007, pp. 304–313 (2007)

# Part IV
# System Science

# Chapter 10
# Topological Properties in Identification and Modeling Techniques

Giacomo Innocenti and Donatello Materassi

**Abstract.** This contribution deals with the problem of finding models and dependencies within a large set of time series or processes. Nothing is assumed about their mutual influences and connections. The problem can not be tackled efficiently, starting from a classical system identification approach. Indeed, the general optimal solution would provide a large number of models, since it would consider every possible interdependence. Then a suboptimal approach will be developed. The proposed technique will present interesting modeling properties which can be interpreted in terms of graph theory. The application of this procedure will also be exploited as a tool to provide a clusterization of the time series. Finally, we will show that it turns out to be a dynamical generalization of other techniques described in literature.

## 10.1 Introduction

In a variety of scientific fields the problem of deriving information from data has widely been studied and a large amount of results have been produced, especially in the engineering, physics, biology and economy areas [10, 5, 8, 16]. In this respect, one of the most relevant contribution is represented by the Identification theory. Such an approach provides a theoretical framework to develop mathematical models attempting to describe the dynamics of the observed processes. In

---

Giacomo Innocenti
Università di Firenze, Dipartimento di Sistemi e Informatica,
via S. Marta 3, 50139 Firenze, Italy
e-mail: `giacomo.innocenti@gmail.com`

Donatello Materassi
University of Minnesota, Department of Electrical and Computer Engineering,
200 Union St SE, 55455, Minneapolis (MN)
e-mail: `mater013@umn.edu`

particular the Auto-Regressive Moving-Average (ARMAX) and Box-Jenkins techniques are widely employed tools in the linear framework [7, 1]. Identification theory is a typical approach when understanding the evolution of the system is crucial to the formulation of reliable forecasts. The clusterization problem, instead, has a quite different aim. It deals with the analysis of a multivariate data set and its objective concerns the similarities and relations hidden inside the original values, attempting to divide them into homogeneous subsets [2, 6]. A common tool in this framework is represented by the artificial neural network approach. Such an approach is based on the employment of a suitable learning algorithm, that is responsible for the detection of analogies and differences in the observed data set [13]. Data mining techniques, instead, provides a variety of methodologies to order and sort into connected layers the original data assuming that no a priori information is available. Though Identification theory and clusterization tools were developed in different fields and with different objectives, it is remarkable how "close" to each other they often result.

In this contribution we exploit the modeling approach to develop a clusterization technique based on frequency analysis. Considering just the linear component of the dynamical connections among the original values, a simplified model explaining their mutual relations is first derived. Then, such a result is interpreted in terms of graph theory and some clusterization properties are derived as well. Moreover, the proposed approach is compared to some results in literature. It is remarkable that it turns out to be the dynamical extension of a well-known technique successfully employed in the economic field.

## 10.2 Problem Formulation

Let us assume that the observations of a system are represented by a set of $N$ scalar time series, namely $\{S_j\}_{j=1,\dots,N}$. Our aim is to derive for each of them a dynamical model describing the possible connections and the mutual influences with the others. Besides, exploiting the related modeling errors to quantify the "distances" among the time series, we also provide a technique to group them into connected networks representing the clusters of the original data set.

Suppose that any deterministic component can be removed from the observed time series in order to obtain $N$ stochastic processes $\{X_j\}_{j=1,\dots,N}$ which are wide sense stationary and with zero mean [15]. We are interested in describing each stochastic process $X_j$ as the superposition of linear dynamical transformations of the other processes' outputs, i.e. as

$$X_j(t) = e_j(t) + \sum_{j=1, j\neq i}^{N} W_{ji}(z) X_i(t), \tag{10.1}$$

where $W_{ji}(z)$ are suitable transfer functions and $e_j$ is the modeling error. In a general framework, the objective boils down to find the set $\{W_{ji}(z)\}_{i,j=1\dots N, i\neq j}$ of the transfer functions best describing the time series according to the least squares criterion

$$\min \sum_j E\left[(Q_j(z)e_j)^2\right], \qquad (10.2)$$

where $Q_j(z)$ are dynamical weight functions. The set of models (10.1) obtained by the minimization of the cost functions (10.2) can be exploited to provide a dynamical model of the original system. Moreover, the analysis of such a representation is useful to detect (linear) connections among the observed processes and to divide them into homogeneous clusters on this basis.

Unfortunately, the complexity of the resulting model may turn out to be very high. Indeed, the amount of transfer functions to be estimated is $N(N-1)$ and in the worst case all of them could be not null, even if their influence could be almost negligible. Therefore, the development of a suboptimal strategy turns out to be a more suitable approach in order to reduce the complexity of the final model. To this aim, the most natural idea is to consider only those transfer functions which provide the "most significant" reduction of the cost. Hence, hereafter we will consider the case of just a unique transfer function per process, i.e., for each time series we will detect the process that provides its best (linear) description.

**Problem 10.1.** *For each observed process $X_j$ find the process $X_i$ and the transfer function $W_{ji}(z)$ providing the model*

$$X_j = W_{ji}(z)X_i + e_j \qquad (10.3)$$

*that minimize the quadratic cost*

$$E\left[e_j^2\right]. \qquad (10.4)$$

Even though this is the simplest case, it still provides useful insights about the connection topology and interesting theoretical properties.

## 10.3 The Frequency Approach

In order to develop our approach, let us first recall some notions of System Identification theory.

Given two stochastic processes $X_i$ and $X_j$, let us consider the problem of evaluating the transfer function $\hat{W}(z)$ that minimizes the quadratic cost

$$E\left[(\varepsilon_Q)^2\right], \qquad (10.5)$$

where

$$\varepsilon_Q \doteq Q(z)\left(X_j - W(z)X_i\right)$$

and $Q(z)$ is an arbitrary stable and causally invertible function. The minimization of the cost (10.5) is a well-known problem in scientific literature and its solution is referred to as the Wiener filter [7]. The following statement summarizes this result.

**Proposition 10.2 (Wiener filter).** *The Wiener filter modeling $X_j$ by $X_i$ is the linear stable filter $\hat{W}_{ji}(z)$ minimizing the filtered quantity* (10.5). *Its expression is given by*

$$\hat{W}_{ji}(z) = \frac{\Phi_{X_i X_j}(z)}{\Phi_{X_i}(z)} \tag{10.6}$$

*and it does not depend upon $Q(z)$. Moreover, the minimized cost is equal to*

$$\min E\left[(Q(z)\varepsilon)^2\right] = \frac{1}{2\pi}\int_{-\pi}^{\pi} |Q(\omega)|^2 \left(\Phi_{X_j}(\omega) - |\Phi_{X_j X_i}(\omega)|^2 \Phi_{X_i}^{-1}(\omega)\right) d\omega .$$

*Proof.* See for example [7].

**Remark 10.3.** *The stable implementation of the Wiener filter $\hat{W}_{ji}(z)$ is in general non-causal [7]. That is, the output of the filtered system $\hat{W}_{ji}(z)X_i$ depends on both past and future values of the input process $X_i$.*

We want to stress that concerning this work the Wiener filter turns out to be interesting just from an information and modeling point of view, but in order to make predictions a causal filter is needed.

Since the weighting function $Q(z)$ does not affect the Wiener filter but only the energy of the filtered error, we choose $Q(z)$ equal to $F_j(z)$, the inverse of the spectral factor of $\Phi_{X_j}(z)$, that is

$$\Phi_{X_j}(z) = F_j^{-1}(z)(F_j^{-1}(z))^* \tag{10.7}$$

with $F_j(z)$ stable and causally invertible [14]. In such a case the minimum cost assumes the value

$$\min E[\varepsilon_{F_j}^2] = \int_{-\pi}^{\pi} \left(1 - \frac{|\Phi_{X_j X_i}(\omega)|^2}{\Phi_{X_i}(\omega)\Phi_{X_j}(\omega)}\right) d\omega. \tag{10.8}$$

This peculiar choice of $Q(z)$ makes the cost depend explicitly on the coherence function of the two processes

$$C_{X_i X_j}(\omega) \doteq \frac{|\Phi_{X_j X_i}(\omega)|^2}{\Phi_{X_i}(\omega)\Phi_{X_j}(\omega)}, \tag{10.9}$$

which turns out to be non negative and symmetric with respect to $\omega$. It is also well-known that the Cross-Spectral Density satisfies the Schwartz Inequality. Hence, the coherence function is limited between 0 and 1. The choice $Q(z) = F_j(z)$ can be now understood as motivated by the necessity to achieve a dimensionless cost function not depending on the power of the signals as in (10.8).

Let us consider, now, a set $\Theta$ of time-discrete zero mean and wide sense stationary stochastic processes. The cost obtained by the minimization of the error $\varepsilon_{F_j}$ using the Wiener filter as before allows us to define on $\Theta$ the binary function

$$d(X_i,X_j) \doteq \left[\frac{1}{2\pi}\int_{-\pi}^{\pi}\left(1-C_{X_iX_j}(\omega)\right)d\omega\right]^{1/2} \qquad \forall X_i,X_j \in \Theta\,. \tag{10.10}$$

**Proposition 10.4.** *The function $d(\cdot,\cdot)$ defined as in* (10.10) *satisfies the following properties:*

1. $d(X_1,X_2) \geq 0$
2. $d(X_1,X_2) = 0 \Leftrightarrow X_1 = X_2$
3. $d(X_1,X_2) = d(X_2,X_1)$
4. $d(X_1,X_3) \leq d(X_1,X_2) + d(X_2,X_3)$

*for all $X_1,X_2,X_3 \in \Theta$. Hence, $d(\cdot,\cdot)$ is a metric on $\Theta$.*

*Proof.* The only non trivial property to prove is the forth, i.e. the triangle inequality (or sublinear additivity). Let us consider the processes $X_i,X_j \in \Theta$ and let $\hat{W}_{ji}(z)$ and $e_{ji}$ be respectively the Wiener filter and the modeling error computed according to (10.6). Then, the following relations hold:

$$X_3 = \hat{W}_{31}(z)X_1 + e_{31}\,, \quad X_3 = \hat{W}_{32}(z)X_2 + e_{32}\,, \quad X_2 = \hat{W}_{21}(z)X_1 + e_{21}\,.$$

Observe that at any frequency the Wiener $\hat{W}_{31}(z)$ filter between the two processes $X_1$ and $X_3$ performs better than any other linear filter, such as $\hat{W}_{32}(z)\hat{W}_{21}(z)$, for instance. Therefore, we have

$$\begin{aligned}\Phi_{e_{31}}(\omega) &\leq \Phi_{e_{32}}(\omega) + |\hat{W}_{32}(\omega)|^2\Phi_{e_{21}}(\omega) + \Phi_{e_{32}e_{21}}(\omega)\hat{W}_{32}^*(\omega) + \hat{W}_{32}(\omega)\Phi_{e_{21}e_{32}}(\omega) = \\ &\leq \Phi_{e_{32}}(\omega) + |\hat{W}_{32}(\omega)|^2\Phi_{e_{21}}(\omega) + 2|\Phi_{e_{32}e_{21}}(\omega)||\hat{W}_{32}(\omega)| \leq \\ &\leq \left(\sqrt{\Phi_{e_{32}}(\omega)} + |\hat{W}_{32}(\omega)|\sqrt{\Phi_{e_{21}}(\omega)}\right)^2 \quad \forall\, \omega \in \mathbb{R}\,.\end{aligned}$$

For the sake of simplicity we neglect to explicitly write the argument $\omega$ in the following passages. Normalizing with respect to $\Phi_{X_3}$, we find

$$\frac{\Phi_{e_{31}}}{\Phi_{X_3}} \leq \frac{1}{\Phi_{X_3}}\left(\sqrt{\Phi_{e_{32}}} + |\hat{W}_{32}|\sqrt{\Phi_{e_{21}}}\right)^2$$

and exploiting the 2-norm properties we obtain

$$\begin{aligned}\left(\int_{-\pi}^{\pi}\frac{\Phi_{e_{31}}}{\Phi_{X_3}}d\omega\right)^{\frac{1}{2}} &\leq \left(\int_{-\pi}^{\pi}\frac{\Phi_{e_{32}}}{\Phi_{X_3}}d\omega\right)^{\frac{1}{2}} + \left(\int_{-\pi}^{\pi}|\hat{W}_{32}|^2\frac{\Phi_{e_{21}}}{\Phi_{X_3}}d\omega\right)^{\frac{1}{2}} = \\ &= \left(\int_{-\pi}^{\pi}\frac{\Phi_{e_{32}}}{\Phi_{X_3}}d\omega\right)^{\frac{1}{2}} + \left(\int_{-\pi}^{\pi}\frac{|\Phi_{X_3X_2}|^2}{\Phi_{X_3}\Phi_{X_2}}\frac{\Phi_{e_{21}}}{\Phi_{X_2}}d\omega\right)^{\frac{1}{2}}\end{aligned}$$

where we have substituted the expression of $\hat{W}_{32}$. Therefore, considering that

$$0 \leq \frac{|\Phi_{X_3X_2}|^2}{\Phi_{X_3}\Phi_{X_2}} \leq 1,$$

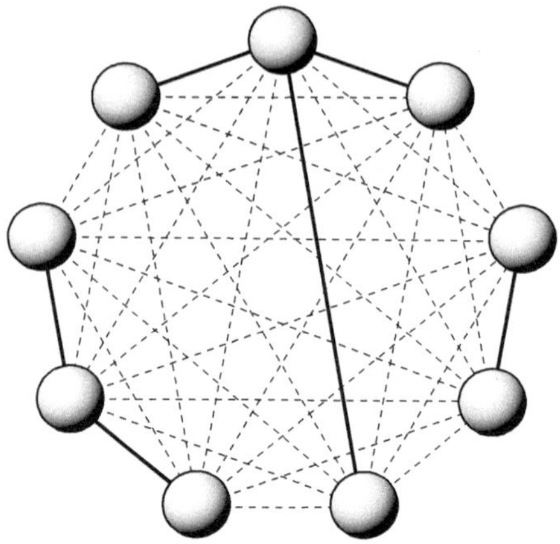

**Fig. 10.1** The figures illustrates in a nine-node network all the possible connections between two nodes (dashed lines). The solid lines depict a forest as if it were the result after the application of the `Algorithm A`.

we find

$$d(X_1,X_3) \leq d(X_1,X_2) + d(X_2,X_3).$$

According to the problem formulation of Section 10.2, we are interested in modeling each process $X_j$ as the output of a SISO linear system, whose input is the process $X_i$ that provides the largest reduction of the cost function (10.2). Hence, exploiting the previous results, we can solve such a problem just employing the following strategy:

`Algorithm A:`

1. initialize the set $A = \emptyset$
2. for every process $X_j$ ($j = 1, ..., N$)
   - 2a. for every $i = 1, ..., N$, $i \neq j$
     compute the distance $d_{ij} \doteq d(X_i, X_j)$;
   - 2b. define the set $M(j) \doteq \{k | d_{kj} = \min_i d_{ij}\}$
   - 2c. choose, if possible, $m(j) \in M(j)$ such that $(m(j), j) \notin A$
   - 2d. choose the model
     $X_j = \hat{W}_{jm(j)}(z) X_{m(j)} + e_{jm(j)}$
   - 2e. add the unordered couple $(j, m(j))$ to $A$.

It is worth observing that the above procedure admits a graph theory interpretation [4]. This is obtained by representing each process $X_j$ as the node of a graph and, then, for each couple $(j, m(j)) \in A$ connecting the $j$-th and $m(j)$-th nodes by an arc whose weight is $d_{m(j)j}$. A scheme to exemplify this representation is illustrated in Figure 10.1.

**Proposition 10.5.** *The graph resulting from the* `Algorithm A` *satisfies the following properties:*

- *on every node there is at least an incident arc*
- *if there is a cycle, then all the arcs of the cycle have the same weight*
- *there are at least $\lceil N/2 \rceil$ and at most $N$ arcs.*

*Proof.* The proof of the first property is straightforward because for every node the algorithm considers an incident arc. Let us suppose there is a cycle and be $k$ the number of nodes $n_1,...,n_k$ and arcs $a_1,...,a_k$ of such a cycle. Every arc $a_1,...,a_k$ has been chosen at the step `2e` when the algorithm was taking into account one of the nodes $n_1,...,n_k$. Conversely, every node $n_1,...,n_k$ is also responsible for one of the arcs $a_1,...,a_k$. Indeed, if a node $n_i$ causes the selection of an arc $\hat{a} \notin \{a_1,...,a_k\}$, then we are left with the $k$ arcs which cannot all be chosen by $k-1$ nodes.

Let us consider the node $n_1$. Without loss of generality assume that it is responsible for the selection of the arc $a_1$ with weight $d_1$ and linking it to the node $n_2$. According to the previous results, $n_2$ can not be responsible for the choice of $a_1$. Let $a_2$ be the arc selected because of $n_2$ with weight $d_2$ and connecting it to $n_3$. Observe that necessarily $d_2 \leq d_1$. We may repeat this process till the node $n_{k-1}$. Hence, we obtain that every node $n_i$ is connected to $n_{i+1}$ by the arc $a_i$ whose cost is $d_i \leq d_{i-1}$, for $i = 2,...,k-1$. Finally consider $n_k$. It must be responsible for $a_k$ which has to connect it to $n_1$ with cost $d_k \leq d_{k-1}$. Since $d_k$ is incident to $n_1$ it holds that $d_1 \leq d_k$ Therefore $d_1 \leq d_k \leq d_{k-1} ... \leq d_2 \leq d_1$ and we have the assertion of the second property.

About the third property, the upper bound $N$ follows from the consideration that every node causes the choice of at most a new arc. In step `2c` of the algorithm, it may happen at most $\lfloor N/2 \rfloor$ times that we are forces to pick up an arc which is already in $A$. So we have at least $N - \lceil N/2 \rceil = \lfloor N/2 \rfloor$ arcs.

We want to underline that the presence of cycles in the resulting graph is a pathological situation. Indeed, a necessary condition of existence for a cycle is the presence of more than two nodes with common multiple minimum cost arcs. Then, a mild sufficient condition in order to avoid cycles in the graph is to assume that every node has a unique minimum cost arc. Therefore, if the costs of the arcs are obtained by estimation from real data the probability to obtain a cycle is zero almost everywhere [15]. In such a case the expected link structure of the graph is a forest, i.e. a collection of one or more disjoined subgraphs with tree topology.

It is worth observing that the connectivity property may be lost only during the step `2c`, when it is not possible to choose a new arc that does not belong to $A$, i.e. when the (undirected) arc $(m(j), j)$ is chosen during the estimation of both $X_j$ and $X_{m(j)}$. This observation highlights a situation that may result in a over-clusterization of the original set of processes. Assume that the processes $X_i$ and $X_j$ have a strong (linear) connection, that is one of them casts a relevant (linear) influence on the other. Moreover, suppose that both of them have several very weak (linear) influences with other nodes. In such a case it is very likely that processing the `Algorithm A` the node $X_i$ would provide the choice of the arc to $X_j$ and vice versa. Therefore, if all the processes linked to them have just a slightly stronger connection to a different process, then the couple $(X_i, X_j)$ will be set aside, forming a single cluster of only two elements. To deal with the problem of the over-clusterization of the original data set the most natural strategy is the application of a grouping procedure to the clusters resulting from `Algorithm A`:

`Algorithm B:`

1. let $\{C_j\}_{j=1,\dots,M}$ be $M$ clusters resulting from the application of `Algorithm A` to the original $N$ processes $\{X_i\}_{i=1,\dots,N}$
2. initialize the set $A^* = \emptyset$
3. for every cluster $C_j$ $(j = 1,\dots,M)$
   3a. for every $i = 1,\dots,M$, $i \neq j$ compute
   $$d^*_{ij} \doteq \min_{\substack{X_r \in C_i \\ X_s \in C_j}} d(X_r, X_s)$$
   3b. define the set $M^*(j) \doteq \{k | d^*_{kj} = \min_i d^*_{ij}\}$
   3c. choose, if possible, $m(j) \in M^*(j)$ such that $(m(j), j) \notin A^*$
   3d. add the unordered couple $(j, m(j))$ to $A^*$.

The above procedure provides a set of arcs linking the clusters originally detected by `Algorithm A` by means of the minimum weighted links among all the connections between two clusters. However, those connections are not related to any linear dynamical model. Therefore, `Algorithm B` contribution is useful only to deal with the over-clusterization issue, but not to improve the performance of the (linear) dynamical model describing the original set of processes. Observe that when the data set is known to be naturally divided into certain groups, the refining procedure depicted in `Algorithm B` can be replaced by techniques designed just to ensure that number of clusters. Nonetheless, the dynamical model is still provided by the only `Algorithm A`. The previous observations are summarized in the following remark.

**Remark 10.6.** *The procedure described in* `Algorithm A` *provides for each process a (linear) dynamical model of the form* (10.3)*, such that the overall cost* (10.4) *is minimized. Moreover, the resulting connections among the processes boils down in a clusterization of the data set. However, since* `Algorithm A` *has no connectivity constraints, i.e. it does not consider a maximum number of clusters, it may result in a over-clusterization of the data. To this aim several strategies may be considered, such as the one proposed in* `Algorithm B`*, that groups the clusters without affecting the nature of the final model.*

It is worth observing that, unfortunately, the adoption of non causal filters prevents us from employing the resulting model to make predictions. To this aim, however, a causal approach could be developed.

## 10.4 Static Models Case

In the previous section we have considered the modeling problem for a set of random processes $\Theta$. The best models (in the mean square sense) have been found in the space of the linear transfer functions. In particular, we have chosen to limit this work to the model (10.3), i.e. we have developed a procedure to describe each process detecting an other process and a linear transfer function which minimize the cost function (10.4). We have observed that in such scenario an apt choice for the error function allows one to define a metric on $\Theta$. Moreover, some clusterization properties of the final model have been highlighted as well.

In [10] and [11] a technique to derive topological information from a set of random processes is described. There, $N$ realizations of $N$ random processes $X_i$ are considered. First, an estimation of the correlation index $\rho_{ij}$ related to every couple $X_i$, $X_j$ is computed, along with the associated distances

$$d_{ij} \doteq \sqrt{2(1-\rho_{ij})} \,. \tag{10.11}$$

Then, a graph is defined where every node represents a random process and the arcs are weighted according to (10.11). Eventually, the minimum spanning tree is extracted by the graph in order to provide a quantitative and topological analysis of the time series. This procedure has been successfully employed especially in the economic field (see e.g. [9], [16] and [12]). It is worth observing that such a technique can be derived from the procedure described in the previous sections. In particular, it turns out to be analogous to the case where only static gains are employed as models and the connectivity constraint is forced, i.e. the data set is allowed to have just one cluster.

Consider the problem of modeling a process $X_j$ by scaling another process $X_i$ by a suitable constant $\alpha_{ji}$. If we choose

$$\alpha_{ji} = \sqrt{\frac{E[X_j^2]}{E[X_i^2]}} \,, \tag{10.12}$$

the following result holds:

$$\begin{aligned} E\left[(X_j - \alpha_{ji}X_i)^2\right] &= E[X_j^2] \cdot E\left[\left(\frac{X_j}{\sqrt{E[X_j^2]}} - \frac{X_i}{\sqrt{E[X_i^2]}}\right)^2\right] = \\ &= 2\,E[X_j^2](1-\rho_{ij}) \,. \end{aligned}$$

Hence, the distance (10.11) can be interpreted as a properly normalized modeling error in the case that the simple gain (10.12) is employed to describe $X_j$ by $X_i$. It is worth observing that the choice of (10.12) is not optimal even among the constant gains. Indeed, it is immediate to check (see e.g. [7]) that the best choice of $\alpha_{ji}$ in the sense of the least square error is given by

$$\hat{\alpha}_{ji} = \frac{R_{X_j X_i}}{R_{X_i}} \tag{10.13}$$

and that the relative quadratic error amounts to

$$E[e_{ji}^2] = R_{X_j} - \frac{R_{X_jX_i}^2}{R_{X_i}} . \tag{10.14}$$

Note that the gain model (10.13) turns out to be the result of the approach depicted in Section 10.3 once that the linear model space has been narrowed to the constant gains only. Indeed, let us define a dimensionless cost function not depending on the signal powers as illustrated in Section 10.3. To this aim, considering the square root of the normalized mean square error, we can define

$$d(X_i, X_j) \doteq \sqrt{\frac{E[e_{ji}^2]}{R_{X_j}}} = \sqrt{1 - \frac{R_{X_iX_j}^2}{R_{X_i}R_{X_j}}} = \sqrt{1 - \rho_{X_iX_j}^2} . \tag{10.15}$$

Since (10.15) satisfies the same properties of (10.10), it is a distance exactly as (10.11).

**Proposition 10.7.** *The function* (10.15) *represents a distance on* $\Theta$.

*Proof.* The only non trivial property to show is the triangle inequality. Consider the following relations involving the optimal gains $\hat{\alpha}_{31}$, $\hat{\alpha}_{32}$, $\hat{\alpha}_{21}$

$$X_3 = \hat{\alpha}_{31}X_1 + e_{31} \quad X_3 = \hat{\alpha}_{32}X_2 + e_{32} \quad X_2 = \hat{\alpha}_{21}X_1 + e_{21} .$$

Since $\hat{\alpha}_{31}$ is the best constant model, we have that it must perform better than any other constant model (in particular $\hat{\alpha}_{32}\hat{\alpha}_{21}$)

$$\begin{aligned} R_{X_3} - \frac{R_{X_3X_1}^2}{R_{X_1}} &\leq E[(e_{32} + \hat{\alpha}_{32}e_{21})^2] = \\ &= E[e_{32}^2] + \hat{\alpha}_{32}^2 E[e_{21})^2] + 2\hat{\alpha}_{32}E[e_{32}e_{21}] \leq \\ &\leq \left(\sqrt{E[e_{32}^2]} + |\hat{\alpha}_{32}|\sqrt{E[e_{21})^2]}\right)^2 . \end{aligned}$$

Normalize with respect to $R_{X_3}$ and consider the square root

$$\begin{aligned} \sqrt{1 - \rho_{X_1X_3}^2} &\leq \sqrt{\frac{1}{R_{X_3}}\left(\sqrt{E[e_{32}^2]} + |\hat{\alpha}_{32}|\sqrt{E[e_{21}^2]}\right)^2} \leq \\ &\leq \sqrt{\frac{E[e_{32}^2]}{R_{X_3}}} + |\hat{\alpha}_{32}|\sqrt{\frac{E[e_{21}^2]}{R_{X_3}}} = \\ &= \sqrt{\frac{E[e_{32}^2]}{R_{X_3}}} + \rho_{X_2X_3}\sqrt{\frac{E[e_{21}^2]}{R_{X_2}}} \end{aligned}$$

Since $|\rho_{X_2X_3}| \leq 1$, we have the assertion.

In [10], each process is represented as a node of a graph and all the possible links are weighted according to (10.11). Then, the minimum spanning tree is derived from the resulting structure. This turns out to be equivalent to the definition of a hierarchical structure for the time series relying on the linear gain (10.12) as model for the processes. We stress that (10.12) represents a normalization of the random process energies, before applying the clusterization algorithm. On the other hand the choice of (10.13) allows one to set up the best gain model in the sense of least squares. From a System theory point of view, it can be said that both the approaches are static. Indeed, the models do not have a state, thus they do not have any dynamics. They simply capture a direct relation between two process samples at the same time instant. Conversely, the use of Wiener filters we propose allows one to take into account more complex behaviors such as the presence of delays or even auto-regressive moving average dynamics. Hence, the more general procedure developed in this contribution is expected to capture a larger amount of information. In particular, due to the filtering action, it is likely that the high frequency behaviour could be more accurately described (for an alternative approach see [3]).

## 10.5 Numerical Example

In this section we illustrate the results obtained by the application of the modeling approach described in Section 10.3 to several data sets numerically generated according to the following procedure.

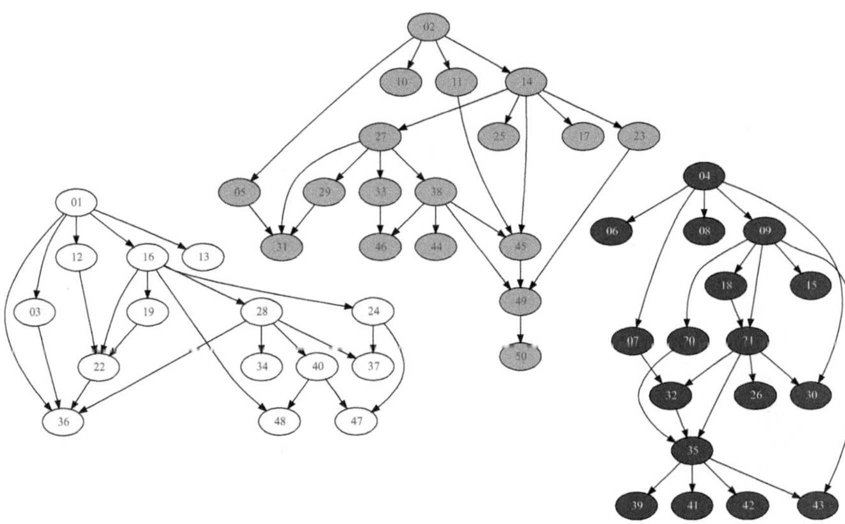


**Fig. 10.2** The 50 nodes network employed to generate the data set. The three root nodes represent three processes obtained by a pseudo-random generator. The other time series are computed by adding a pseudo white noise and the filtered signals coming from the other processes, according to the link orientations. The observation horizon is set to 1500 time steps.

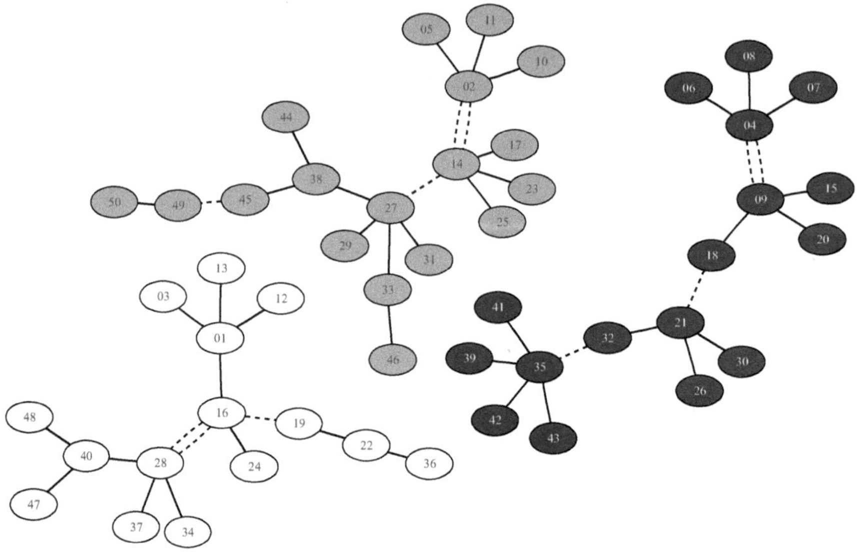


**Fig. 10.3** The dynamical connections detected by `Algorithm A` (solid lines) and `Algorithm B` (dotted lines). The first procedure ensures the linear models describing the generation of the time series. However, it boils down to a over-clusterization of the data set. This situation is recovered by the refining procedure of `Algorithm B`, that, nonetheless, does not provide any further contribution to the model. It is worth underlining that no a priori knowledge about the number of cluster is provided to the algorithm. The groups of the original processes are exactly reconstructed.

The data set counts 50 processes randomly grouped in 3 clusters. The link structure inside each group has been initially modeled as a (random) tree. The choice of such a topology is functional to the introduction of a partial ordering as described hereafter. For each cluster a process has been randomly chosen to act as the "root" of the tree and its time evolution has been supposed not affected by any other process. In particular, every root time series has been generated by a pseudo-random generator so to behave as zero-mean white processes. The rooted tree topology provides the "parent-child" and "ancestor-descendant" relationships among the processes of the same group [4]. To compute the samples of the other processes the linear model (10.3) has been initially considered, assuming that the $i$-th time series is the father of the $j$-th one, according to the rooted tree topology of the cluster. Again, the noise $e_j$ has been obtained by a pseudo-random generator. Then, the connections inside each cluster have been made more complex by randomly adding new arcs to the starting tree topology. Such links have been modeled as additive contributions and their flow has been chosen to prevent the formation of algebraic loops, i.e. we have avoided that a process could cast its influence on his own ancestors. Therefore, for each time series but the roots the final generating model has the form:

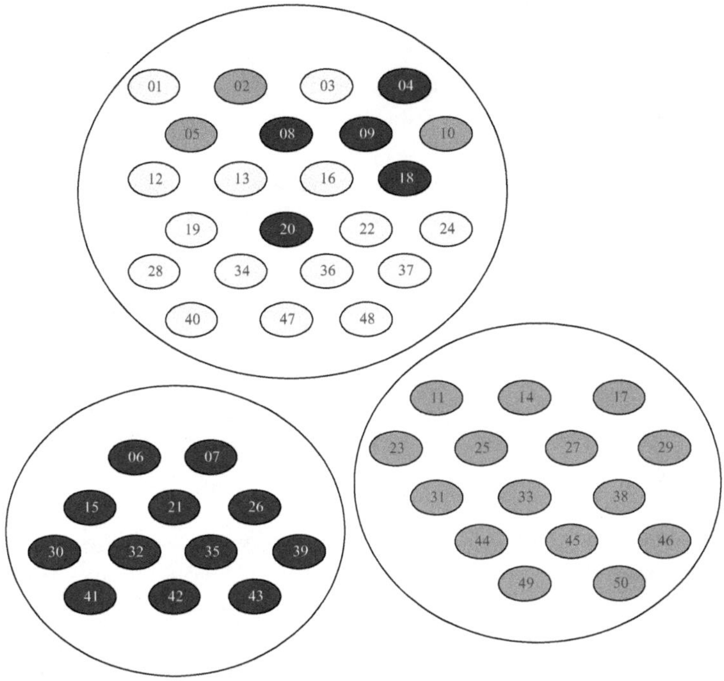


**Fig. 10.4** The clusters obtained by the best application of the standard tools provided by the Matlab environment. The `pdist` function has been first employed to compute the correlation distances among every couple of processes. Then, the `linkage` function with the "complete" method (i.e. the one based on the furthest distance between the nodes of two chosen clusters) has been applied. The result has finally been passed to the `cluster` function constrained to detect a maximum of 3 groups. Nonetheless, it is important to underline that in many real world situations the number of clusters is not a priori known. Observe that eight of the fifty original processes are wrongly placed.

$$X_j = e_j + \sum_{i \in H_j} W_{ji}(z) X_i \,,$$

where $H_j$ is a set of indexes such that $X_i$ is not a descendant of $X_j$ for each $i \in H_j$. For the sake of simplicity the transfer functions $W_{ji}(z)$ have been randomly generated to be at most of the second order. The number of samples considered for each process is 1500.

Several data sets computed according to the above procedure have been analyzed through the approach of Section 10.3. Since the qualitative results turns out to be very similar for every data set, hereafter we are reporting just one of the considered cases. In Figure 10.2 a generating network is depicted, while Figure 10.3 illustrates the link structure obtained by the application of `Algorithm A` (the solid lines) and then refined by `Algorithm B` (the dotted arcs). Even if the number of the clusters initially appointed by `Algorithm A` is eleven, the application of `Algorithm B` correctly recovers such situation of over-clusterization. We underline, however, that only the connections provided by `Algorithm A` corresponds to dynamical models that can be employed to describe the evolution of the time series. The reliability of the complete procedure is remarkable. In particular it is worth observing that no wrong connections are detected.

In Figure 10.4 the best result obtained by using the standard clusterization tools provided by the Matlab environment (i.e. the functions `pdist`, `linkage` and `cluster`) is presented. The group division can be considered fairly satisfactory, but several wrong matches are present in the realization of one of the clusters. Moreover, we stress that this latter procedure does not provide any model to justify the connection among the processes and it is expected to perform worse as the order of the original transfer functions $W_{ji}(z)$ is increased.

## 10.6 Conclusions

In this contribution, we have approached the problem of describing dynamical relations within a large set of time series or processes. No assumptions has been made about the relative influences and dependencies. In such a scenario the general optimal solution in the sense of the Identification theory would provide a too large number of models since it would consider every possible connection. Hence, a suboptimal approach has been taken into account: every process has been described as the output of a SISO linear system, suitably driven by an other process so to provide the "best" model in the sense of the least squares. The procedure has been developed from the classic modeling/identification point of view via the frequency approach, but an interpretation in terms of graph theory has been provided to derive remarkable clusterization properties as well. Such an interpretation has allowed us to compare the technique with other ones in literature. It particular it has been shown that it generalizes certain procedures, which have already been successfully employed especially in the economic and financial fields. It is worth underlining that the main and novel contribution of our approach is in the fact that such a technique attempts to capture a topological structure describing not only static, but also dynamic relations among the observed processes.

## References

1. Alexander, C.: Market Models: A Guide to Financial Data Analysis. Wiley, Chichester (2001)
2. Anderberg, M.: Cluster Analysis for Applications. Academic Press, New York (1973)
3. Bonanno, G., Lillo, F., Mantegna, R.N.: High frequency cross-correlation in a set of stocks. Quant Fin 1, 96–104 (2001)
4. Diestel, R.: Graph Theory. Springer, Berlin (2006)
5. Eisen, M.B., Spellman, P.T., Brown, P.O., Botstein, D.: Cluster analysis and display of genome-wide expression patterns. Proc. Natl. Acad. Sci. USA 95(25), 14, 863–868 (1998)

6. Gan, G., Ma, C., Wu, J.: Data clustering: theory, algorithms and applications. SIAM, Philadelphia (2007)
7. Kailath, T., Sayed, A.H., Hassibi, B.: Linear Estimation. Prentice Hall, Upper Saddle River (2000)
8. Ljung, L.: System identification: theory for the user, 2nd edn. Prentice-Hall, Inc., Upper Saddle River (1999)
9. Mantegna, R.N.: Hierarchical structure in financial markets. Eur. Phys. J. B 11, 193–197 (1999)
10. Mantegna, R.N., Stanley, H.E.: Scaling behaviour in the dynamics of an economic index. Nature 376, 46–49 (1995)
11. Mantegna, R.N., Stanley, H.E.: Turbulence and financial markets. Nature (Scientific Correspondence) 383, 587–588 (1996)
12. Naylora, M.J., Roseb, L.C., Moyle, B.J.: Topology of foreign exchange markets using hierarchical structure methods. Physica A 382, 199–208 (2007)
13. Rojas, R.: Neural networks: a systematic introduction. Springer, New York (1996)
14. Sayed, A.H., Kailath, T.: A survey of spectral factorization methods. Numerical Linear Algebra with Applications 8, 467–469 (2001)
15. Shiryaev, A.N.: Probability. Springer, New York (1995)
16. Tumminello, M., Coronnello, C., Lillo, F., Micciche, S., Mantegna, R.N.: Spanning trees and bootstrap reliability estimation in correlation-based networks. Int. J. of Bifurcations & Chaos 17, 2319–2329 (2007)

# Chapter 11
# Network Abstract Linear Programming with Application to Cooperative Target Localization*

Giuseppe Notarstefano and Francesco Bullo

**Abstract.** We identify a novel class of distributed optimization problems, namely a networked version of abstract linear programming. For such problems we propose distributed algorithms for networks with various connectivity and/or memory constraints. Finally, we show how a suitable target localization problem can be tackled through appropriate linear programs.

## 11.1 Introduction

This paper focuses on a class of distributed computing problems and on its applications to cooperative target localization in sensor networks. To do so, we study abstract linear programming, that is, a generalized version of linear programming that was introduced by Matousěk, Sharir and Welzl in [1] and extended by Gärtner in [2]. Abstract linear programming is applicable also to some geometric optimization problems, such as the minimum enclosing ball, the minimum enclosing stripe and the minimum enclosing annulus. These geometric optimization problems are relevant in the design of efficient robotic algorithms for minimum-time formation control problems as shown in [3].

Giuseppe Notarstefano
Department of Engineering, University of Lecce, Via per Monteroni,
73100 Lecce, Italy
e-mail: giuseppe.notarstefano@unile.it

Francesco Bullo
Center for Control, Dynamical Systems and Computation,
University of California at Santa Barbara, Santa Barbara, CA 93106, USA
e-mail: bullo@engineering.ucsb.edu

* This material is based upon work supported in part by ARO MURI Award W911NF-05-1-0219 and ONR Award N00014-07-1-0721. The research leading to these results has received funding from the European Community's Seventh Framework Programme (FP7/2007-2013) under grant agreement no. 224428 (CHAT Project).

Linear programming and its generalizations have received widespread attention in the literature. The following references are most relevant in our treatment. The earliest (deterministic) algorithm that solves a linear program in a fixed number of variables subject to $n$ linear inequalities in time $O(n)$ is given in [4]. An efficient randomized incremental algorithm for linear programming is proposed in [1], where a linear program in $d$ variables subject to $n$ linear inequalities is solved in expected time $O(d^2n + e^{O(\sqrt{d \log d})})$; the expectation is taken over the internal randomizations executed by the algorithm. An elegant survey on randomized methods in linear programming is [5]; see also [6]. The survey [7], see also [8], discusses the application of abstract linear programming to a number of geometric optimization problems. Regarding parallel computation approaches to linear programming, we only note that linear programs with $n$ linear inequalities can be solved [9] by $n$ parallel processors in time $O((\log\log(n))^d)$. The approach in [9] and the ones in the references therein are, however, limited to parallel random-access machines (usually denoted PRAM), where a shared memory is readable and writable to all processors. In this paper, we focus on networks described by arbitrary graphs.

The problem of target localization has been widely investigated and the related literature is therefore quite rich. Recently, the interest in sensor networks and distributed computation has lightened the attention on this problem from this new perspective. A good reference for localization and tracking in sensor networks is [10]. Two approaches may be used to tackle target localization, a stochastic and a deterministic one. As regards the deterministic approach (the one we use in this paper), a set membership estimation technique was proposed in [11]. Recently, a sensor selection problem for target tracking was studied in [12]. References for target localization and tracking in sensor networks, even by use of a stochastic approach, may be found therein.

The contributions of this paper are three-fold. First, we identify a class of distributed optimization problems that appears to be novel and of intrinsic interest. Second, we propose a novel simple algorithmic methodology to solve these problems in networks with various connectivity and/or memory constraints. Specifically, we propose three algorithms, prove their correctness and establish halting conditions. Finally, we illustrate how these distributed computation problems are relevant for distributed target localization in sensor networks. Specifically, we cast the target localization problem in the problem of approximating the intersection of convex polytopes by using a small number of halfplanes (among all those defining the polytopes). We show that suitable linear programs running in parallel, in fact, solve the problem, so that the proposed distributed algorithms may be used.

The paper is organized as follows. Section 11.2 introduces abstract linear programs. Section 11.3 introduces network models. Section 11.4 contains the definition of network abstract linear programs and the proposed distributed algorithms. Section 11.5 shows the relevance of the proposed distributed computing algorithms in the context of cooperative target localization.

Notation

We let $\mathbb{N}$, $\mathbb{N}_0$, and $\mathbb{R}_+$ denote the natural numbers, the non-negative integer numbers, and the positive real numbers, respectively. For $f, g : \mathbb{N} \to \mathbb{R}$, we say that $f \in O(g)$ if there exist $n_0 \in \mathbb{N}$ and $k \in \mathbb{R}_+$ such that $|f(n)| \le k|g(n)|$ for all $n \ge n_0$.

## 11.2 Abstract Linear Programming

In this section we present an abstract framework that captures a wide class of optimization problems including linear programming and various geometric optimization problems. These problems are known as *abstract linear programs* (or *LP-type problems*). They can be considered a generalization of linear programming in the sense that they share some important properties. A comprehensive analysis of these problems may be found for example in [7].

### *11.2.1 Abstract Framework*

We consider optimization problems specified by a pair $(H, \omega)$, where $H$ is a finite set, and $\omega : 2^H \to \Omega$ is a function with values in a linearly ordered set $(\Omega, \le)$; we assume that $\Omega$ has a minimum value $-\infty$. The elements of $H$ are called *constraints*, and for $G \subset H$, $\omega(G)$ is called the *value* of $G$. Intuitively, $\omega(G)$ is the smallest value attainable by a certain objective function while satisfying the constraints of $G$. An optimization problem of this sort is called *abstract linear program* if the following two axioms are satisfied:

(i) *Monotonicity*: if $F \subset G \subset H$, then $\omega(F) \le \omega(G)$;
(ii) *Locality*: if $F \subset G \subset H$ with $-\infty < \omega(F) = \omega(G)$, then, for all $h \in H$,

$$\omega(G) < \omega(G \cup \{h\}) \implies w(F) < w(F \cup \{h\}).$$

A set $B \subset H$ is *minimal* if $\omega(B) > \omega(B')$ for all proper subsets $B'$ of $B$. A minimal set $B$ with $-\infty < \omega(B)$ is a *basis*. Given $G \subset H$, a *basis of $G$* is a minimal subset $B \subset G$, such that $-\infty < \omega(B) = \omega(G)$. A constraint $h$ is said to be *violated* by $G$, if $\omega(G) < \omega(G \cup \{h\})$.

The *solution* of an abstract linear program $(H, \omega)$ is a minimal set $B_H \subset H$ with the property that $\omega(B_H) = \omega(H)$. The *combinatorial dimension* $\delta$ of $(H, \omega)$ is the maximum cardinality of any basis. Finally, an abstract linear program is called *basis regular* if, for any basis with $\text{card}(B) = \delta$ and any constraint $h \in H$, every basis of $B \cup \{h\}$ has the same cardinality of $B$. We now define two important primitive operations that are useful to solve abstract linear programs.

(i) *Violation test*: given a constraint $h$ and a basis $B$, it tests whether $h$ is violated by $B$; we denote this primitive by $\texttt{Viol}(B, h)$;
(ii) *Basis computation*: given a constraint $h$ and a basis $B$, it computes a basis of $B \cup \{h\}$; we denote this primitive by $\texttt{Basis}(B, h)$.

*Remark 11.1 (Examples of abstract linear programs).* We present three useful geometric examples; see Figure 11.1.

(i) *Smallest enclosing ball:* Given $n$ points in $\mathbb{R}^d$, compute the center and radius of the ball of smallest volume containing all the points. This problem has combinatorial dimension $d+1$.
(ii) *Smallest enclosing stripe:* Given $n$ points in $\mathbb{R}^2$ in generic positions, compute the center and the width of the stripe of smallest width containing all the points. This problem has combinatorial dimension 5.
(iii) *Smallest enclosing annulus:* Given $n$ points in $\mathbb{R}^2$, compute the center and the two radiuses of the annulus of smallest area containing all the points. This problem has combinatorial dimension 4.

More examples are discussed in [1, 2, 5, 7]. □

### 11.2.2 Randomized Sub-exponential Algorithm

A randomized algorithm for solving abstract linear programs has been proposed in [1]. Such algorithm has linear expected running time in terms of the number of constraints, whenever the combinatorial dimension $\delta$ is fixed, and subexponential in $\delta$. The algorithm, called `SUBEX_lp`, has a recursive structure and is based on the two primitives introduced above, i.e., the violation test and the basis computation primitives. For simplicity, we assume here that such primitives may be implemented in constant time, independent of the number of constraints. Given a set of constraints $G$ and a candidate basis $C \subset G$, the algorithm is as follows.

**function** `SUBEX_lp`$(G,C)$
**if** $G = C$, **then** return $C$
**else**
choose a random $h \in G \setminus C$
$B :=$ `SUBEX_lp`$(G \setminus \{h\}, C)$
**if** `Viol`$(B,h)$, i.e., $h$ is violated by $B$,
return `SUBEX_lp`$(G,$ `Basis`$(B,h))$
**else** return $B$
**end if**
**end if**

For the abstract linear program $(H,\omega)$, the routine is invoked with `SUBEX_lp`$(H,B)$, given any initial candidate basis $B$.

In [1] the expected completion time for the `SUBEX_lp` algorithm in conjunction with Clarkson's algorithms was shown to be in $O(d^2 n + e^{O(\sqrt{d \log d})})$ for basis regular

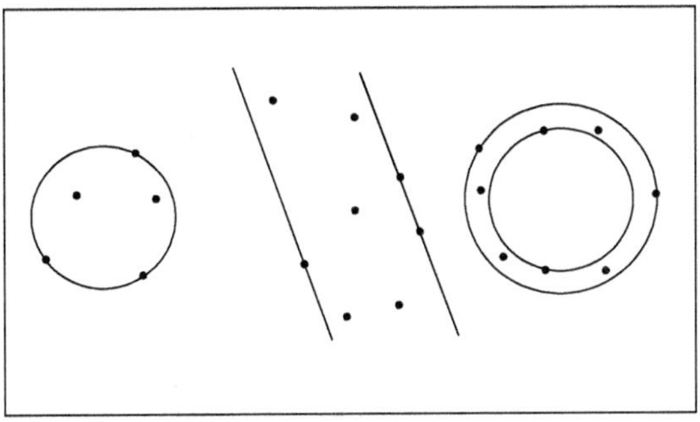

**Fig. 11.1** Smallest enclosing ball, stripe and annulus

abstract linear programs. In [5] the result was extended to problems that are not basis regular.

## 11.3 Network Models

Following [13], we define a synchronous network system as a "collection of *computing elements* located at nodes of a directed network graph." These computing elements are sometimes called *processors*.

### *11.3.1 Digraphs and Connectivity*

We let $I = \{1,\dots,n\}$ and let $\mathscr{G} = (I,E)$ denote a directed graph, where $I$ is the set of nodes and $E \subset I \times I$ is the set of edges. For each node $i$ of $\mathscr{G}$, the number of edges going out from (coming into) node $i$ is called *out-degree* (*in-degree*) and is denoted $\mathtt{outdeg}^{[i]}$ ($\mathtt{indeg}^{[i]}$). The set of outgoing (incoming) neighbors of node $i$ are the set of nodes to (from) which there are edges from (to) $i$. They are denoted $\mathscr{N}_O(i)$ and $\mathscr{N}_I(i)$, respectively. A directed graph is called *strongly connected* if, for every pair of nodes $(i,j) \in I \times I$, there exists a path of directed edges that goes from $i$ to $j$. In a strongly connected digraph, the minimum number of edges between node $i$ and $j$ is called the *distance from i to j* and is denoted $\text{dist}(i,j)$. The maximum $\text{dist}(i,j)$ taken over all pairs $(i,j)$ is the *diameter* and is denoted $\text{diam}(\mathscr{G})$. Finally, we consider time-dependent directed graphs of the form $t \mapsto \mathscr{G}(t) = (I,E(t))$. The time-dependent directed graph $\mathscr{G}$ is *jointly strongly connected* if, for every $t \in \mathbb{N}_0$,

$$\cup_{\tau=t}^{+\infty} \mathscr{G}(\tau) \text{ is strongly connected.}$$

Moreover, the time-dependent directed graph $\mathscr{G}$ is *uniformly strongly connected* if, there exists $S > 0$ s.t. for every $t \in \mathbb{N}_0$

$$\cup_{\tau=t}^{t+S} \mathscr{G}(\tau) \text{ is strongly connected.}$$

### 11.3.2 Synchronous Networks and Distributed Algorithms

Strictly speaking, a *synchronous network* is a directed graph $\mathscr{G} = (I, E_{\text{cmm}})$ where the set $I = \{1,\ldots,n\}$ is the set of *identifiers* of the computing elements, and the time-dependent map $E_{\text{cmm}} : \mathbb{N}_0 \to 2^{I\times I}$ is the *communication edge map* with the following property: an edge $(i,j)$ belongs to $E_{\text{cmm}}(t)$ if and only if processor $i$ can communicate to processor $j$ at time $t$.

**Definition 11.1 (Distributed algorithm).** Let $\mathscr{G} = (I, E_{\text{cmm}})$ be a synchronous network. A distributed algorithm consists of the sets

- $W$, set of "logical" states $w^{[i]}$, for all $i \in I$;
- $W_0 \subset W$, subset of allowable initial values;
- $M$, message alphabet, including the `null` symbol;

and the maps

- $\text{msg} : W \times I \to M$, message-generation function;
- $\text{stf} : W \times M^n \to W$, state-transition function. □

Execution of the network begins with all processors in their start states and all channels empty. Then the processors repeatedly perform the following two actions. First, the $i$th processor sends to each of its outgoing neighbors in the communication graph a message (possibly the `null` message) computed by applying the message-generation function to the current value of $w^{[i]}$. After a negligible period of time, the $i$th processor computes the new value of its logical variables $w^{[i]}$ by applying the state-transition function to the current value of $w^{[i]}$, and to the incoming messages (present in each communication edge). The combination of the two actions is called a *communication round* or simply a round.

In this execution scheme we have assumed that each processor executes all the calculations in one round. If it is not possible to upper bound the execution-time of the algorithm, we may consider a slightly different network model that allows the state-transition function to be executed in multiple rounds. When this happens, the message is generated by using the logical state at the previous round.

The last aspect to consider is the *algorithm halting*, that is a situation such that the network (and therefore each processor) is in a idle mode. Such status can be used to indicate the achievement of a prescribed task. Formally we say that a distributed algorithm is in *halting status* if the logical state is a fixed point for the state-transition function (that becomes a self-loop) and no message (or equivalently the `null` message) is generated at each node.

## 11.4 Network Abstract Linear Programming

In this section we define a *network abstract linear program* and propose novel distributed algorithms to solve it.

### 11.4.1 Problem Statement

Informally we can say that a *network abstract linear program* consists of three main elements: a network, an abstract linear program and a mapping that associates to each constraint of the abstract linear program a node of the network. A more formal definition is the following.

**Definition 11.2.** A network abstract linear program (NALP) is a tuple $(\mathcal{G}, (H, \omega), \mathcal{B})$ consisting of

(i) $\mathcal{G} = (I, E_{\text{cmm}})$, a communication digraph;
(ii) $(H, \omega)$, an abstract linear program;
(iii) $\mathcal{B} : H \to I$, a surjective map called *constraint distribution map*. □

The *solution* of the network abstract linear program is attained when all processors in the network have computed a solution to the abstract linear program.

*Remark 11.2.* Our definition allows for various versions of network abstract linear programs. Regarding the constraint distribution map, the most natural case to consider is when the constraint distribution map is bijective. In this case one constraint is assigned to each node. More complex distribution laws are also interesting depending on the computation power and memory of the processors in the network. In what follows, we assume $\mathcal{B}$ to be bijective. □

### 11.4.2 Distributed Algorithms

Next we define three distributed algorithms that solve network abstract linear programs. First, we describe a synchronous version that is well suited for time-dependent networks whose nodes have bounded computation time and memory, but also bounded in-degree or equivalently arbitrary in-degree, but also arbitrary computation time and memory. Then we describe two variations that take into account the problem of dealing with arbitrary in-degree versus short computation time and small memory. The second version of the algorithm is suited for time-dependent networks that have arbitrary in-degree and bounded computation time, but are allowed to store arbitrarily large amount of information, in the sense that the number of stored messages may depend on the number of nodes of the network. The third algorithm considers the case of time-independent networks with arbitrary in-degree and bounded computation time and memory.

In the algorithms we consider a uniform network $\mathcal{S}$ with communication digraph $\mathcal{G} = (I, E_{\text{cmm}})$ and a network abstract linear program $(\mathcal{G}, (H, \omega), \mathcal{B})$. We assume $\mathcal{B}$ to be bijective, that is, the set of constraints $H$ has dimension $n$, $H = \{h_1, \cdots, h_n\}$. The combinatorial dimension is $\delta$.

Here is an informal description of what we shall refer to as the *FloodBasis* algorithm:

*[Informal description]* Each processor has a logical state of $\delta+1$ variables taking values in $H$. The first $\delta$ components represent the current value of the basis to compute, while the last element is the constraint assigned to that node. At the start round the processor initializes every component of the basis to its constraint, then, at each communication round, performs the following tasks: (i) it acquires from its neighbors (a message consisting of) their current basis; (ii) it executes the `SUBEX_lp` algorithm over the constraint set given by the collection of its and its neighbors' basis and its constraint (that it maintains in memory), thus computing a new basis; (iii) it updates its logical state and message using the new basis obtained in (ii).

In the second scenario we work with a time-dependent network with no bounds on the in-degree of the nodes and on the memory size. In this setting the execution of the `SUBEX_lp` may exceed the communication round length. In order to deal with this problem, we slightly change the network model as described in Section 11.3, so that each processor may execute the state transition function "asynchronously", in the sense that the time-length of the execution may take multiple rounds. If that happens, the message generation function in each intermediate round is called using the logical state of the previous round. Here is an informal description of what we shall refer to as the *FloodBasisMultiRound* algorithm:

*[Informal description]* Each processor has the same message alphabet and logical state as in *FloodBasis* and also the same state initialization. At each communication round it performs the following tasks: i) it acquires the messages from its in-neighbors; ii) if the execution of the `SUBEX_lp` at the previous round was over it starts a new instance, otherwise it keeps executing the one in progress; iii) if the execution of the `SUBEX_lp` ends it updates the logical state and runs the message-generation function with the new state, otherwise it generates the same message as in the previous round.

In the third scenario we work with a time-independent network with no bounds on the in-degree of the nodes. We suppose that each processor has limited memory capacity, so that it can store at most $D$ messages. The memory is dimensioned so to guarantee that the `SUBEX_lp` is always solvable during two communication rounds. The memory constraint is solved by processing only part of the incoming messages at each round and cycling in a suitable way in order to process all the messages in multiple rounds.

Here is an informal description of what we shall refer to as the *FloodBasisCycling* algorithm:

*[Informal description]* The first $\delta+1$ components of the logical state are the same as in *FloodBasis* and are initialized in the same way. A further component is added. It is simply a counter variable that keeps trace of the current round. At each communication round each processor performs the following tasks: (i) it acquires from its neighbors (a message consisting of) their current basis; (ii) it chooses $D$ messages according to a scheduled protocol, e.g., it labels its in-neighboring edges with natural numbers from 1 up to $\texttt{indeg}^{[i]}$ and cycles over them in increasing order; (iii) it executes the `SUBEX_lp` algorithm over the constraint set given by the collection of the $D$ messages plus its basis and its constraint (that it maintains in memory), thus computing a new basis; (iv) it updates its logical state and message using the new basis obtained in (iii).

*Remark 11.3.* For the algorithm to converge it is important that each agent keeps in memory its constraint and thus implements the SUBEX_lp on the bases received from its neighbors together with its constraint. This requirement is important because of the following reason: no element of a basis $B$ for a set $G \subset H$ needs to be an element in the basis of $G \cup \{h\}$ for any $h \in H \setminus G$. □

We are now ready to prove the algorithms' correctness.

**Proposition 11.1 (Correctness of *FloodBasis*).** *Let $\mathcal{S}$ be a synchronous time-dependent network with communication digraph $\mathcal{G} = (I, E_{\mathrm{cmm}})$ and let $(\mathcal{G}, (H, \omega), \mathcal{B})$ be a network abstract linear program. If $\mathcal{G}$ is jointly strongly connected, then the* FloodBasis *algorithm solves $(\mathcal{G}, (H, \omega), \mathcal{B})$, that is, in a finite number of rounds each node acquires a copy of the solution of $(H, \omega)$, i.e., the basis $B$ of $H$.*

*Proof.* In order to prove correctness of the algorithm, observe, first of all, that each law at every node converges in a finite number of steps. In fact, using axioms from abstract linear programming and finiteness of $H$, each sequence $\omega(B^{[i]}(t))$, $t \in \mathbb{N}_0$, is monotone nondecreasing, upper bounded and can assume a finite number of values. Then we proceed by contradiction to prove that all the laws converge to the same $\omega(B)$ and that it is exactly $\omega(B) = \omega(H)$. Suppose that for $t > t_0 > 0$ all the nodes have converged to their limit basis and that there exist at least two nodes, call them $i$ and $j$, such that $\omega(B^{[i]}(t)) = \omega(B^{[i]}) \neq \omega(B^{[j]}) = \omega(B^{[j]}(t))$, for all $t \geq t_0$. For $t = t_0 + 1$, for every $k_1 \in \mathcal{N}_O(i)$, $B^{[i]}$ does not violate $B^{[k_1]}$, otherwise they would compute a new basis thus violating the assumption that they have converged. Using the same argument at $t = t_0 + 2$, for every $k_2 \in \mathcal{N}_O(k_1)$, $B^{[k_1]}$ does not violate $B^{[k_2]}$. Notice that this does not imply that $B^{[i]}$ does not violate $B^{[k_2]}$, but it implies that $\omega(B^{[i]}) \leq \omega(B^{[k_2]})$. Iterating this argument we can show that for every $S > 0$, every $k$ connected to $i$ in the graph $\cup_{t=t_0}^{t_0+S} \mathcal{G}(t)$ must have a basis $B^{[k]}$ such that $\omega(B^{[i]}) \leq \omega(B^{[k]})$. However, using the joint connectivity assumption, there exists $S_0 > 0$ such that $\cup_{t=t_0}^{t_0+S_0} \mathcal{G}(t)$ is strongly connected and therefore $i$ is connected to $j$, thus showing that $\omega(B^{[i]}) \leq \omega(B^{[j]})$. Repeating the same argument by starting from node $j$ we obtain that $\omega(B^{[j]}) \leq \omega(B^{[i]})$, that implies $\omega(B^{[i]}) = \omega(B^{[j]})$, thus giving the contradiction. Now, the basis at each node satisfies, by construction, the constraints of that node. Since the basis is the same for each node, it satisfies all the constraints, then $\omega(B) = \omega(H)$. ■

*Remark 11.4.* Correctness of the other two versions of the *FloodBasis* algorithm may be established along the same lines. For example, it is immediate to establish that the basis at each node reaches a constant value in finite time. It is easy to show that this constant value is the solution of the abstract linear program for the *FloodBasisMultiRound* algorithm. For the *FloodBasisCycling* algorithm we note that the procedure used to process the incoming data is equivalent to considering a time-dependent graph whose edges change with that law. □

**Proposition 11.2 (Halting condition).** *Consider a network $\mathcal{S}$ with time-independent, strongly connected digraph $\mathcal{G}$ where the* FloodBasis *algorithm is* *running. Each processor can halt the algorithm execution if the value of its basis has not changed after* $2\,\mathrm{diam}(\mathscr{G})+1$ *communication rounds.*

*Proof.* First, notice that, for all $t \in \mathbb{N}_0$ and for every $(i,j) \in E_{\mathrm{cmm}}$,

$$\omega(B^{[i]}(t)) \leq \omega(B^{[j]}(t+1)). \tag{11.1}$$

This holds by simply noting that $B^{[j]}(t+1)$ is not violated by $B^{[i]}(t)$ by construction of the *FloodBasis* algorithm. Assume that node $i$ satisfies $B^{[i]}(t) = B$ for all $t \in \{t_0, \dots, t_0 + 2\,\mathrm{diam}(\mathscr{G})\}$, and pick any other node $j$. Without loss of generality assume that $t_0 = 0$. Because of equation (11.1), if $k_1 \in \mathscr{N}_O(i)$, then $\omega(B^{[k_1]}(1)) \geq \omega(B)$ and, recursively, if $k_2 \in \mathscr{N}_O(k_1)$, then $\omega(B^{[k_2]}(2)) \geq \omega(B^{[k_1]}(1)) \geq \omega(B)$. Iterating this argument $\mathrm{dist}(i,j)$ times, the node $j$ satisfies $\omega(B^{[j]}(\mathrm{dist}(i,j))) \geq \omega(B)$. Now, consider the out-neighbors of node $j$. For every $k_3 \in \mathscr{N}_O(j)$, it must hold that $\omega(B^{[k_3]}(\mathrm{dist}(i,j)+1)) \geq \omega(B^{[j]}(t))$. Iterating this argument $\mathrm{dist}(j,i)$ times, the node $i$ satisfies $\omega(B^{[i]}(\mathrm{dist}(i,j)+\mathrm{dist}(j,i))) \geq \omega(B^{[j]}(\mathrm{dist}(i,j)))$. In summary, because $\mathrm{dist}(i,j)+\mathrm{dist}(j,i) \leq 2\,\mathrm{diam}(\mathscr{G})$, we know that $B^{[i]}(\mathrm{dist}(i,j)+\mathrm{dist}(j,i)) = B$ and, in turn, that

$$\omega(B) \geq \omega(B^{[j]}(\mathrm{dist}(i,j))) \geq \omega(B).$$

This shows that, if basis $i$ does not change for a duration $2\,\mathrm{diam}(\mathscr{G})+1$, then it will never change afterwards because all bases $B^{[j]}$, for $j \in \{1,\dots,n\}$, have cost equal to $\omega(B)$ at least as early as time equal to $\mathrm{diam}(\mathscr{G})+1$. Therefore, node $i$ can safely stop after a $2\,\mathrm{diam}(\mathscr{G})+1$ duration. ∎

## 11.5 Distributed Computation of the Intersection of Convex Polytopes for Target Localization

In this section we discuss an application of network abstract linear programming to sensor networks, namely a distributed solution for target localization. We consider a set of (fixed) sensors $\{1,\dots,n\}$ deployed on a plane. These sensors have to detect a target located at position $x \in \mathbb{R}^2$. Each sensor $i$ detects a region of the plane, $m^{[i]}(x) \subset \mathbb{R}^2$, containing the target; we assume that this region, possibly unbounded, can be written as the intersection of a finite number of half-planes. For $i \in \{1,\dots,n\}$, let $c_i$ denote the number of half-planes defining the sensing region of sensor $i$. An example scenario with $c_i = 2$ for $i \in \{1,\dots,n\}$ is illustrated in Figure 11.2. From now on, in order to simplify the notation, we assume that $c_i = c$ for all $i \in \{1,\dots,n\}$, so that the number of half-planes (and thus the number of constraints) is $nc$.

The intersection of the regions detected by each sensor provides the best estimate of the target location, $M(x) = \cap_{i\in\{1,\dots,n\}} m^{[i]}(x)$. It easy to see that $M(x)$ is a non empty convex set, since it is the finite intersection of convex sets all containing the position $x$ of the target.

Here, we are interested in approximating the intersection of convex polytopes by means of a "small" number of halfplanes. We consider the following approximation problem. Given a finite collection of convex polytopes with nonempty intersection,

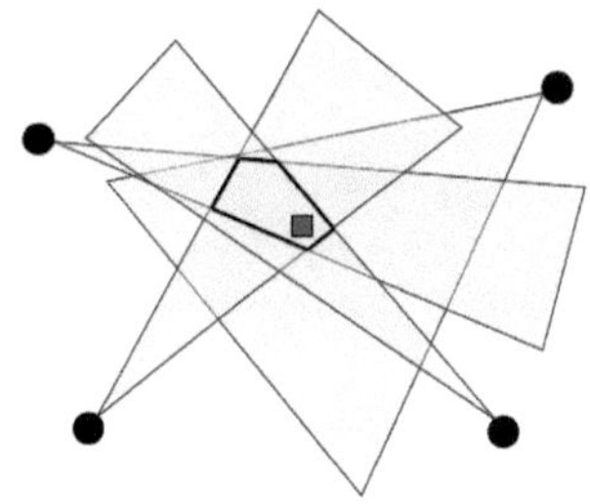

**Fig. 11.2** Target localization: set measurements

find the smallest axis-aligned rectangle that contains the intersection. We refer this rectangle as the "bounding rectangle."

The bounding rectangle has two important features. First, the rectangle provides bounds for the coordinates of the target. That is, let $p_o = (p_o^1, p_o^2)$ be the center of the rectangle with sides of length $a$ and $b$ respectively. For any $p = (p^1, p^2) \in M(x)$, then $|p^1 - p_o^1| \leq a/2$ and $|p^2 - p_o^2| \leq b/2$. Second, the bounding rectangle is characterized by at most four points of the polytope or, equivalently, by at most eight halfplanes.

It can be easily shown that computing the bounding rectangle is equivalent to solving four linear programs respectively in the positive and negative directions of each reference axis. More formally, let $v_\theta \in S^1$ be a vector forming an angle $\theta$ with the first reference axis. Given a set of half-planes $H = \{h_1, \ldots, h_n\}$, $h_i \subset \mathbb{R}^2$ for $i \in \{1, \ldots, n\}$, we denote $(H, \omega_\theta)$ the linear program

$$\begin{aligned} &\min v_\theta^T x \\ &\text{subj. to } a_i^T x \leq b_i, \quad i \in \{1, \ldots, n\} \end{aligned}$$

where $h_i = \{x \in \mathbb{R}^2 \mid a_i^T x \leq b_i, a_i \in \mathbb{R}^2 \text{ and } b_i \in \mathbb{R}\}$. The bounding rectangle may be computed by solving the linear programs $(H, \omega_\theta)$, $\theta \in \{0, \pi/2, \pi, 3\pi/2\}$. An example is depicted in Figure 11.3

*Remark 11.5.* (i) A tighter approximation of the intersection may be obtained by choosing a finer grid for the angle $\theta$. Choosing angles at distance $2\pi/k$, $k \geq 4$, the intersection is approximated by a $k$-polytope.

(ii) An inner approximation to a polytope in dimension $d$ with $n_p$ facets can also be computed via the largest ball contained in the polytope. This center and radius of this ball are referred to as the incenter and inradius of the polytope. It is known [14] that the incenter and the inradius may be computed by solving a linear program of dimension $d + 1$ with $n_p$ constraints.

(iii) The computation of a bounding $k$-polytope is a sensor selection problem in which one wants to select a few representative among a large set of sensors by optimizing some appropriate criterion. Indeed, the $2k$ sensors solving the problem are the only ones needed to localize the target. □

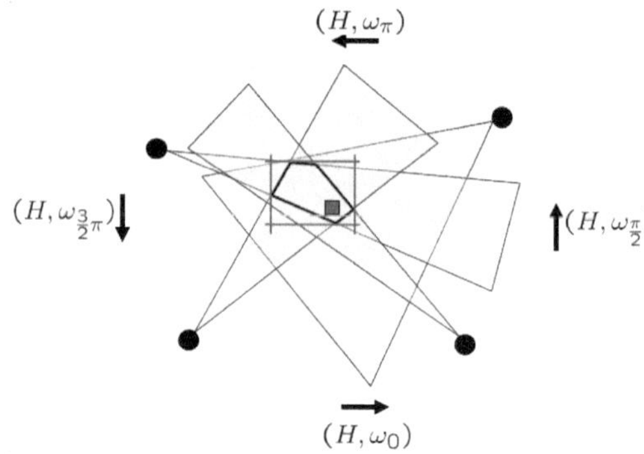


**Fig. 11.3** Target localization: bounding rectangle

In the following we want to design a distributed algorithm running on a network to approximate the intersection of planar convex polytopes. We assume the sensor network may be described by the mathematical model introduced in Section 11.3. Let $\mathcal{G} = (I, E_{\mathrm{wsn}})$ be the associated communication graph, where the set $I = \{1,\ldots,n\}$ is the set of identifiers of the sensors and $E_{\mathrm{wsn}}$ is the communication edge map. We assume that $\mathcal{G}$ is a fixed undirected connected graph. We consider the network (abstract) linear programs defined by $(\mathcal{G},(H,\omega_\theta),\mathcal{B})$ where $(H,\omega_\theta)$ are the linear programs defined above and $\mathcal{B}$ is the mapping associating to each node the $c$ constraints describing its sensing region. Therefore, the distributed algorithm to compute the bounding rectangle ($k$-polytope) consists of 4 ($k$) instances of *FloodBasis* running in parallel (one for each linear program). We denote *FloodRect* (*FloodPoly*) the algorithm consisting of the 4 ($k$) instances of *FloodBasis* running in parallel and the routine to compute the bounding rectangle (polytope).

Next, we are interested in estimating the position of a moving target. The proposed algorithm can be generalized according to the set membership approach, described for example in [11]. The idea is to track the target position by means of a prediction and "measurement update" iteration. The idea may be summarized as follows. We consider the sensor network described above, but with the objective of tracking a moving target. The sensors can measure the position of the target every $T \in \mathbb{N}$ communication rounds, so that during the $T$ rounds they can perform a distributed computation in order to improve the estimate of the target. The target moves in the sensing area with bounded velocity $|v| \leq v_{\max}$. That is, given its position $x(t)$, the position after $\tau \in \mathbb{N}$ communication rounds may be bounded by $x(t+\tau) \in B(x(t), v_{\max}\tau)$, where we have assumed the inter-communication interval to be of unit duration.

In order to simplify notation we assume that each sensor can measure only one halfplane containing the target, i.e., we set $c = 1$. Also, we suppose that the sensors may keep in memory $k$ past measures and use them to improve the estimate.

We begin the algorithm description by discussing the *prediction step*. Let $h^{[i]}(t) = \{x \in \mathbb{R}^2 \mid a^{[i]}(t)^T x \leq b^{[i]}(t)\}$ be the halfplane (containing the target) measured by

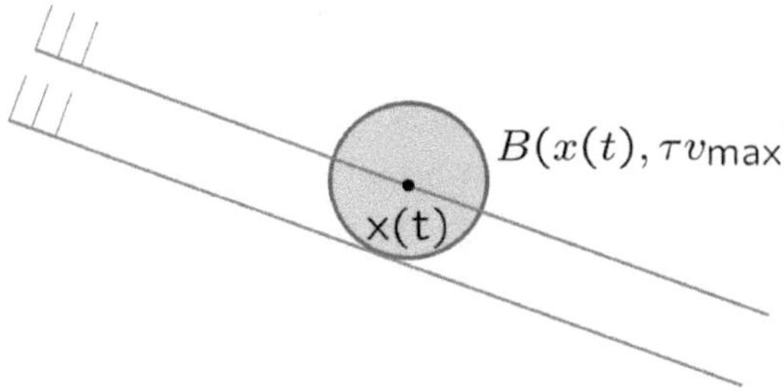


**Fig. 11.4** Constraint after the prediction step

sensor $s^{[i]}$ at time $t$. Since the target moves with bounded velocity, it follows easily that at instant $t+\tau$ the target will be contained in the halfplane $h^{[i]}(t+\tau|t) = \{x \in \mathbb{R}^2 \mid a^{[i]}(t)^T(x - v_{\max}\tau a^{[i]}(t)/\|a^{[i]}(t)\|) \leq b^{[i]}(t)\}$. The idea is illustrated in Figure 11.4.

We are now ready to describe the estimation procedure. Informally, between two measurement instants each sensor runs a *FloodRect* algorithm such that each *FloodBasis* instance has constraints given by the current measured halfplane and the prediction at time $t$ of the latest $k$ measures. More formally, each node solves the network linear programs $(\mathcal{G}, (H(t), \omega_\theta), \mathcal{B})$, $\theta \in \{0, \pi/2, \pi, 3\pi/2\}$, where $H(t) = \{H^{[1]}(t), \ldots, H^{[n]}(t)\}$, with

$$H^{[i]}(t) = \{h^{[i]}(t|t-kT), h^{[i]}(t|t-(k-1)T), \ldots, h^{[i]}(t|t-T), h^{[i]}(t)\}.$$

Then each node computes the corresponding bounding rectangle. The following result follows directly.

**Proposition 11.3.** *Let $t \in \mathbb{N}$ be a time instant when a new measure arrives. We denote Rect$^{[i]}(t+\tau)$ the estimate of the bounding rectangle at instant $t+\tau$ obtained by running the* FloodRect *algorithm for the network linear programs $(\mathcal{G}, (H(t), \omega_\theta), \mathcal{B})$, $\theta \in \{0, \pi/2, \pi, 3\pi/2\}$. Then*

*(i) For any $\tau \in [0, T]$ and any $i \in \{1, \ldots, n\}$, the rectangle Rect$^{[i]}(t+\tau)$ is a subset of the rectangle Rect$^{[i]}(t)$ and it contains the target.*
*(ii) For sufficiently large $T$, there exists $\tau_0 \in [0, T]$ such that Rect$^{[i]}(t+\tau) =$ Rect$^{[i]}(t+\tau_0) =$ Rect$_0$ for all $\tau_0 \leq \tau \leq T$, where Rect$_0$ solves the network linear programs.*

*Proof.* To prove statement i) first note that for any $i \in \{1, \ldots, n\}$ each constraints in $H^{[i]}(t)$ contains the target. Therefore each estimate of the bounding rectangle will contain the target. The monotonicity property of Rect$^{[i]}(t+\tau)$ follows by the monotonicity property of each *FloodBasis* algorithm solving the corresponding network linear program.

Statement ii) follows easily by the fact that each *FloodBasis* algorithm running in parallel solves the respective network linear program. ∎

## 11.6 Conclusions

In this paper we have shown how to solve a class of optimization problems, namely abstract linear programs, over a network in a distributed way. We have proposed distributed algorithms to solve such problems in network with various connectivity and/or memory constraints. The proposed methodology has been used to compute an outer approximation of the intersection of convex polytopes in a distributed way. In particular, we have shown that a set approximation problem of this sort may be posed to perform cooperative target localization in sensor networks.

## References

1. Matousek, J., Sharir, M., Welzl, E.: A subexponential bound for linear programming. Algorithmica 16(4/5), 498–516 (1996)
2. Gärtner, B.: A subexponential algorithm for abstract optimization problems. SIAM Journal on Computing 24(5), 1018–1035 (1995)
3. Notarstefano, G., Bullo, F.: Network abstract linear programming with application to minimum-time formation control. In: IEEE Conf. on Decision and Control, New Orleans, December 2007, pp. 927–932 (2007)
4. Megiddo, N.: Linear programming in linear time when the dimension is fixed. Journal of the Association for Computing Machinery 31(1), 114–127 (1984)
5. Gärtner, B., Welzl, E.: Linear programming - randomization and abstract frameworks. In: Puech, C., Reischuk, R. (eds.) STACS 1996. LNCS, vol. 1046, pp. 669–687. Springer, Heidelberg (1996)
6. Goldwasser, M.: A survey of linear programming in randomized subexponential time. SIGACT News 26(2), 96–104 (1995)
7. Agarwal, P.K., Sen, S.: Randomized algorithms for geometric optimization problems. In: Pardalos, P., Rajasekaran, S., Reif, J., Rolim, J. (eds.) Handbook of Randomization. Kluwer Academic Publishers, Dordrecht (2001)
8. Agarwal, P.K., Sharir, M.: Efficient algorithms for geometric optimization. ACM Computing Surveys 30(4), 412–458 (1998)
9. Ajtai, M., Megiddo, N.: A deterministic $\text{poly}(\log\log n)$-time $n$-processor algorithm for linear programming in fixed dimension. SIAM Journal on Computing 25(6), 1171–1195 (1996)
10. Zhao, F., Guibas, L.: Wireless Sensor Networks: An Information Processing Approach. Morgan Kaufmann, San Francisco (2004)
11. Garruli, A., Vicino, A.: Set membership localization of mobile robots via angle measurements. IEEE Transactions on Robotics and Automation 17(4), 450–463 (2001)
12. Isler, V., Bajcsy, R.: The sensor selection problem for bounded uncertainty sensing models. IEEE Transactions on Automation Sciences and Engineering 3(4), 372–381 (2006)
13. Lynch, N.A.: Distributed Algorithms. Morgan Kaufmann, San Francisco (1997)
14. Bullo, F., Cortés, J., Martínez, S.: Distributed Control of Robotic Networks. Applied Mathematics Series. Princeton University Press, Princeton (2008) (manuscript under contract), http://www.coordinationbook.info

# Chapter 12
# On the Effect of Packet Acknowledgment on the Stability and Performance of Networked Control Systems*

Emanuele Garone, Bruno Sinopoli, and Alessandro Casavola

**Abstract.** This work concerns discrete-time Linear Quadratic Gaussian (LQG) optimal control of a remote plant that communicates with the control unit by means of a packets dropping channel. Namely, the output measurements are sent to the control unit through an unreliable network and the actions decided by the control unit are sent to the plant actuator via the same network. Sensor and control packets may be randomly lost according to a Bernoulli process. In this work we focus on the importance of acknowledgments in the communication between the control unit and the actuators. In the literature two extreme cases have been considered: either guaranteed acknowledgment or complete lack of it. Although very common in practice, the case where the acknowledgment packets can be lost has not been dealt with a sufficient level of detail. In this work we focus on such a case by assuming that also the acknowledgment packets can be lost according to a Bernoulli process. We can show how the partial loss of acknowledgements yields a non classical information

Emanuele Garone
Dipartimento di Elettronica, Informatica e Sistemistica,
Università degli Studi della Calabria, Via Pietro Bucci, Cubo 42-c, Rende (CS), 87036, Italy
e-mail: egarone@deis.unical.it

Bruno Sinopoli
Department of Electrical and Computer Engineering, Carnegie Mellon University,
Pittsburgh, PA 15213, USA
e-mail: brunos@ece.cmu.edu

Alessandro Casavola
Dipartimento di Elettronica, Informatica e Sistemistica,
Università degli Studi della Calabria, Via Pietro Bucci, Cubo 42-c, Rende (CS), 87036, Italy
e-mail: casavola@deis.unical.it

* This research was supported in part by CyLab at Carnegie Mellon under grant DAAD19-02-1-0389 from the Army Research Office, and grant CNS-0509004 from the National Science Foundation. The views and conclusions contained here are those of the authors and should not be interpreted as necessarily representing the official policies or endorsements, either express or implied, of ARO, CMU, NSF, or the U.S. Government or any of its agencies.

pattern [1], making the optimal control law a nonlinear function of the information set. For the special case each observation packet contains the complete state information, we can prove linearity of the optimal controller. Furthermore, we can compute the control law in closed form and show that the stability range increases monotonically with the arrival rate of the acknowledgement packets.

## 12.1 Introduction

The ubiquity of networking together with technological advances in embedded sensing, computing and wired/wireless communication is today enabling the use of shared general purpose networks for control system applications, in stark contrast with the traditional use of dedicated connections [2]. Due to inherent unreliability that such a paradigm introduces, the control system community has focused his attention to the implications of using imperfect communication channels to "close the loop". Several aspects (see [3] and [4] for a survey) concerning Networked Control Systems have been analyzed with special attention to limited bandwidth ([5]- [7]), transmission delays ([8]-[12]) and packet dropouts. In this work we will focus on the effect of packet loss on system stability and performance.

Here we consider a generalized formulation of the Linear Quadratic Gaussian (LQG) optimal control problem where sensors and actuators communicate with a remote control unit via a lossy network. Observation and control packet drops can be modeled with independent Bernoulli's processes of parameters $\bar{\gamma}$ and, respectively, $\bar{\nu}$. Please note that the latter description presents two ambiguities which must be solved in order to correctly define the control problem.

First, we need to specify the control signal applied by the actuators in the case that some control packets are lost. The most common approaches presented in literature consist of either providing a zero input [13] when the communication fails or using a zero-holder [14] to maintain the previous applied input. Recently, in [15] it has been shown that, in general, none of the two approaches can be claimed superior to the other. In the rest of the chapter the zero input approach will be used.

The second ambiguity to be addressed is the specification of the information available at each time instant to the control unit. This ambiguity resides in the actuator's communication channel. In fact, because the control unit is the transmitter of the packet, it could either have or not knowledge of the successful command transmission, the latter depending on the presence of an acknowledgment protocol.

Usually in literature the two extreme cases are considered: situations where a perfect acknowledgment is guaranteed at each time instant or where no acknowledgment is provided. Following the framework proposed by Imer [16], approaches related to the first case are usually referred to as TCP-like and UDP-like for the second. Previous work [17, 18, 19] has shown the existence of a critical domain of values for the parameters of the Bernoulli arrival processes, $\overline{\nu}$ and $\overline{\gamma}$, outside of which a transition to instability occurs and the optimal controller fails to stabilize the system. In particular, it has been proven that under TCP-like protocols the

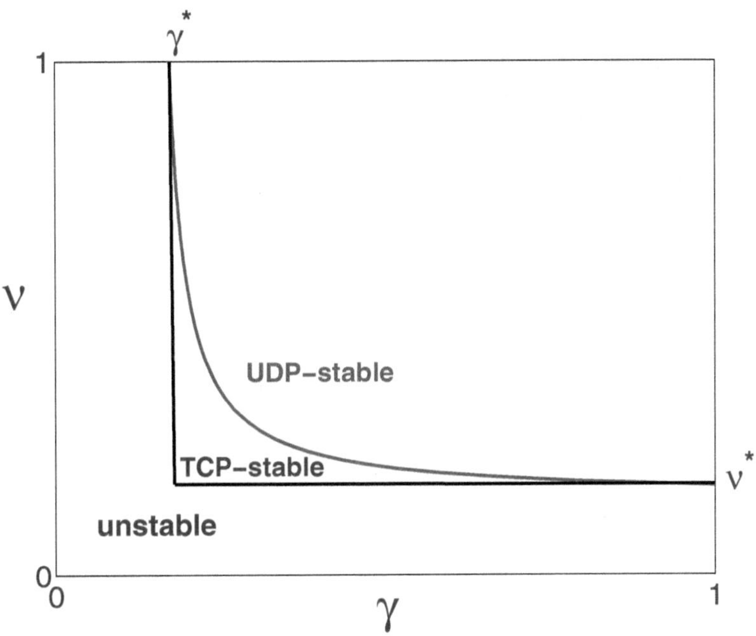


**Fig. 12.1 Stability regions for TCP-like and UDP-like systems** (special case with perfect observations).

critical arrival probabilities for the control and observation channels are independent each other. A more involved situation regards UDP-like protocols.

We have also shown that in the TCP-like case the classic separation principle holds, and consequently the controller and estimator can be designed independently. Moreover, the optimal controller is a linear function of the state. In sharp contrast, in the UDP-like case, the optimal controller is in general non-linear. In this case the absence of an acknowledgement structure generates a nonclassical information pattern [1]. Because of the importance of UDP protocols for wireless sensor networks, we have analyzed a special case when the arrival of a sensor packet provides complete knowledge of the state and, despite the lack of acknowledgements, the optimal control design problem yields a linear controller [19]. In this case, the critical arrival probabilities for the control and observation channels are coupled. The stability domain and the performance of the optimal controller degrade considerably as compared with TCP-like protocols as shown in Figure 12.1.

Also, for the general case, a sub-optimal solution was provided in [20], by designing the optimal linear static regulator, composed by constant gains for both the observer and the controller. This is particularly attractive for sensor networks, where the simplicity of implementation is highly desirable and the complexity issues are a primary concern. Recently Epstein et al. [21] proposed, in the context of UDP-like control, to estimate not only the state of the system, but also a binary variable which indicates wether the previous control packet has been received or not. Such strategy, improves closed-loop performance at the price of a somewhat larger computational complexity.

In this work, similarly to the analysis carried out in [22], we drop the assumption of guaranteed deterministic acknowledgement and assume only that the acknowledgment packets can be lost accordingly to a Bernoulli's process. It is shown that loss of acknowledgement leads once again to a nonclassical information pattern, and we are able to prove that in general the optimal control law is a nonlinear function of the information set. By restricting ourselves to the complete observability case, we are able to solve the LQG problem. We show that probabilistic acknowledgements increase the stability range of the system with respect to UPD-like controls. Furthermore, we can also show that such domains converge to the TCP-like one as the erasure probability for the acknowledgement channel tends to zero.

The remainder of the paper is organized as follows. In Section 2 we provide the problem formulation; in Section 3, we derive the estimator equations; in Section 4, we consider the control problem in the general case; in Section 5 we consider the special case of complete observability; in Section 6 we present an illustrative example and in Section 7 we provide conclusions and directions for future work.

## 12.2 Problem and Formulation

Consider the following linear stochastic system with intermittent observation and control packets:

$$\begin{aligned} x_{k+1} &= Ax_k + Bu_k^a + \omega_k \\ u_k^a &= \nu_k u_k^c + [1-\nu_k]\, u_k^l \\ y(k) &= \gamma_k C x_k + v_k \end{aligned} \tag{12.1}$$

where $u_k^a$ is the control input to the actuator, $u_k^c$ is the desired control input computed by the controller, $(x_0, \omega_k, v_k)$ are Gaussian, uncorrelated, white, with mean $(\overline{x_0}, 0, 0)$ and covariance $(P_0, Q, R)$ respectively, and $\gamma_k$ and $\nu_k$ are i.i.d. Bernoulli random variable with $P(\gamma_k = 1) = \overline{\gamma_k}$ and $P(\nu_k = 1) = \overline{\nu_k}$. $u_k^l$ is the signal it is locally provided to the actuators in the case $\nu_k = 0$ (the control packet packet to the actuators is lost). In this work we will consider $u_k^l = 0$.

A key point towards the design of any control strategy is the definition of the *Information Set* available to the controller at each time instant. It is usual in literature (see [16]) to consider the following two Information Sets

$$I_k = \begin{cases} F_k = \{\gamma_k y_k, \gamma_k, \nu_{k-1} | k = 0, \ldots, t\} & TCP-like \\ G_k = \{\gamma_k y_k, \gamma_k | k = 0, \ldots, t\} & UDP-like \end{cases}$$

As depicted in Figure 12.2, the difference between the two Information Sets is the availability of the control packet acknowledgement i.e. $\nu_{k-1}$. While for the "TCP-like" case, several useful and important results (separation principles, linear quadratic gaussian optimal control, etc...) are known, it is well known from networks and computer science literature that guaranteeing a deterministic "perfect" acknowledgement is in general a very difficult task and, in the case the

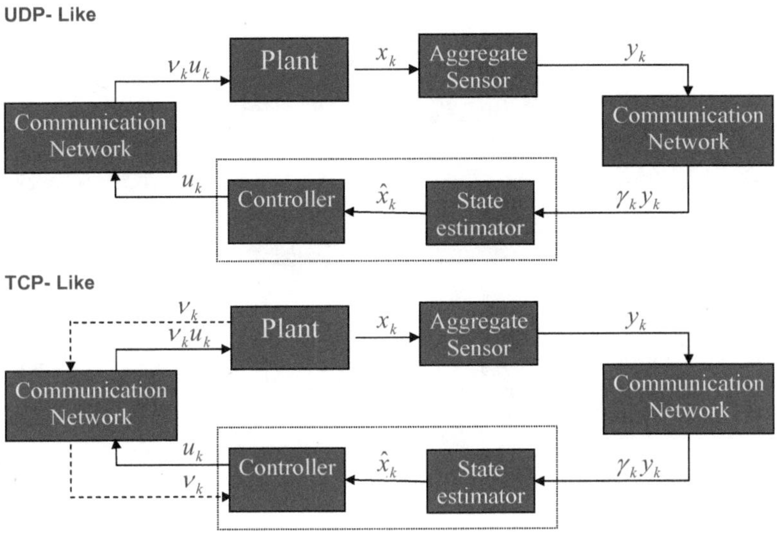


**Fig. 12.2 TCP-like and UDP-like Framework:** Bernoulli variables $\gamma_k, \nu_k$ model the packets arrival through the network. Dashed lines represent the communication from the actuators to the control unit. Note that in the UDP case no acknowledgment traffic exists. On the contrary, in the TCP case, all the acknowledgment packets sent from the actuators are supposed to arrive at the control unit.

acknowledgement packets use unreliable channels, theoretically impossible as it constitutes a particular case of the *two-armies problem* (see [23]).

On the other hand it is extremely difficult [19] to design optimal estimators and controllers under the information set $G_k$, since the separation principle does not hold and it can be shown that the optimal control is not linear. Moreover, performance and stability regions can be highly affected, due to the fact that no "real" information on the actual input is exploited.

In many practical cases, it is reasonable to use communication channels where acknowledgements are provided although they can be dropped with a probability which depends on both the channel reliability and the protocol employed. This means that, during each process, we have a non-zero probability to lose the acknowledgment packet. In order to formalize this assumption, let us denote by $\theta_k$ the Bernoulli variable which models the acknowledgment arrivals and by $\overline{\theta}$ its arrival probability. The new Information Set can then be defined as follows

$$E_k = \{\gamma_k y_k, \gamma_k, \theta_{k-1}, \theta_{k-1}\nu_{k-1} | k = 0, \dots, t\} \tag{12.2}$$

and the overall closed loop system is depicted in Figure 12.3.

Let us now define $u^{N-1} = \{u_0, u_1, \dots, u_{N-1}\}$ as the set of all the input values between time instants 0 and $N-1$. In this work we will consider the LQG control problem, i.e. we will look for a control input sequence $u^{N-1*}$, function of the information set $E_k$, that solves the following optimization problem

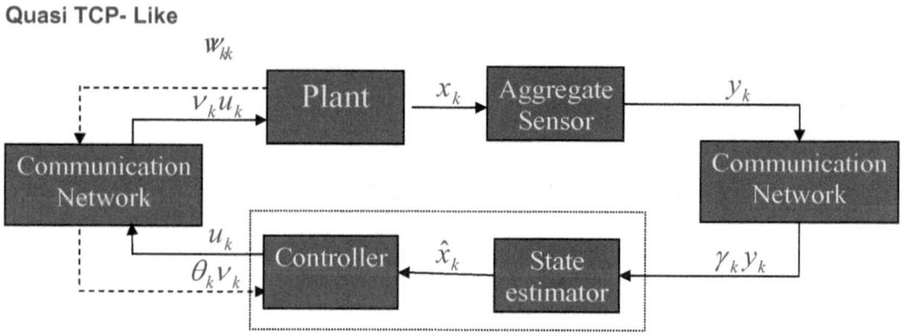


**Fig. 12.3 Quasi TCP-like Framework:** Dashed lines represent the communication from the actuators to the control unit. Note that in this case the acknowledgment packets sent from the actuators can be dropped according to the random Bernoulli variable $\theta_k$.

$$J_N^*(\bar{x}_0, P_0) = min_{u_k = g_k(E_k)} J_N(u^{N-1}, \bar{x}_0, P_0), \tag{12.3}$$

where the cost function $J_N\left(u^{N-1}, \bar{x}_0, P_0\right)$ is defined as follows:

$$J_N\left(u^{N-1}, \bar{x}_0, P_0\right) = E\left[x_N^T W_N x_N + \sum_{k=0}^{N-1} x_k^T W_k x_k + u_k^{aT} U_k u_k^a \middle| u^{N-1}, \bar{x}_0, P_0\right]. \tag{12.4}$$

## 12.3 Estimator Design

By the knowledge of the information set (12.2), the one-step prediction can be easily written as follows:

$$\hat{x}_{k+1|k} = A\hat{x}_{k|k} + \theta_k \nu_k B u_k + (1-\theta_k)\bar{\nu} B u_k. \tag{12.5}$$

Using (12.5), it is possible to rewrite the prediction error as follows:

$$\begin{aligned} e_{k+1|k} &= x_{k+1} - \hat{x}_{k+1|k} = \\ &= Ax_k + \nu_k B u_k + \omega_k - A\hat{x}_{k|k} + \theta_k \nu_k B u_k - (1-\theta_k)\bar{\nu} B u_k = \\ &= Ae_{k|k} + (\nu_k - \theta_k \nu_k - (1-\theta_k)\bar{\nu}) B u_k + \omega_k. \end{aligned} \tag{12.6}$$

We can then compute the associated covariance of the one-step prediction error:

$$\begin{aligned} P_{k+1|k} &= E\left[e_{k+1|k} e_{k+1|k}^T | E_k, \theta_k, \theta_k \nu_k\right] = \\ &= E\left[Ae_{k|k} e_{k|k}^T A | E_k\right] + E\left[\omega_k \omega_k^T | E_k\right] + \\ &\quad + E\left[(\nu_k - \theta_k \nu_k - (1-\theta_k)\bar{\nu})^2 | E_k, \theta_k, \theta_k \nu_k\right] B u_k u_k^T B^T, \end{aligned}$$

obtaining

$$P_{k+1|k} = AP_{k|k}A^T + Q + (1-\theta_k)(1-\bar{\nu})\bar{\nu}\left[Bu_k u_k^T B^T\right]. \tag{12.7}$$

Equations (12.5), (12.6) and (12.7) represent the predictions of the Kalman Filter for the system (12.1). The correction steps, instead, are the classical ones reported in ([32]):

$$\hat{x}_{k+1|k+1} = \hat{x}_{k+1|k} + \gamma_{k+1} K_{k+1} \left( y_{k+1} - C x_{k+1|k} \right) \quad (12.8)$$

$$P_{k+1|k+1} = P_{k+1|k} - \gamma_{k+1} K_{k+1} C P_{k+1|k} \quad (12.9)$$

$$K_{k+1} = P_{k+1|k} C^T \left( C P_{k+1|k} C^T + R \right)^{-1} \quad (12.10)$$

*Remark 12.1.* Note that:

$$\theta_k = 1 \Rightarrow P_{k+1|k} = A P_{k|k} A + Q$$
$$\theta_k = 0 \Rightarrow P_{k+1|k} = A P_{k|k} A + Q + \overline{\nu}(1-\overline{\nu}) \left[ B u_k u_k^T B^T \right].$$

This implies that, at each time $k$, the prediction switches between the "TCP-like" predictions and the "UDP-like" ones, depending on the instantaneous value of $\theta_k$.

## 12.4 Optimal Control - General Case

In this section we will show that, in the presence of stochastic acknowledgements, the optimal control law is not a linear function of the state and that the estimation and control designs cannot be treated separately. In order to prove such a claim, it is sufficient to consider the following simple counterexample.

**Example 1.** Consider a simple scalar discrete-time Linear Time-Invariant (LTI) system with a single sensor and a single actuator, i.e. A=B=C=$W_N$=$W_k$=R=1,$U_k$=Q=0. We can define

$$V(N) = E\left[ x_N^T W_N x_N | E_N \right] = E\left[ x_N^2 | E_N \right].$$

For $k = N-1$ we have:

$$\begin{aligned} V_{N-1}(x_{N-1}) &= \min_{u_N} E\left[ x_{N-1}^2 + V_N(x_N) | E_{N-1} \right] = \\ &= \min_{u_N} E\left[ x_{N-1}^2 + x_N^2 | E_{N-1} \right] = \\ &= \min_{u_N} E\left[ x_{N-1}^2 + (x_{N-1} + \nu_{N-1} u_{N-1})^2 | E_{N-1} \right], \end{aligned} \quad (12.11)$$

and then finally

$$V_{N-1}(x_{N-1}) = E\left[ 2 x_{N-1|N-1}^2 | E_{N-1} \right] + \min_{u_N} \overline{\nu} \left( u_{N-1}^2 + 2 \hat{x}_{N-1|N-1} u_{N-1} \right). \quad (12.12)$$

If we differentiate the latter, we obtain the following optimal input :

$$u^*_{N-1} = -\hat{x}_{N-1|N-\hat{1}} \tag{12.13}$$

If we substitute (12.13) in (12.11) the cost becomes:

$$\begin{aligned} V_{N-1}(x) &= E\left[2x^2_{N-1}|E_{N-1}\right] - \overline{\nu}\,\hat{x}^2_{N-1|N-1} = \\ &= (2-\overline{\nu})E\left[x^2_{N-1}|E_{N-1}\right] - \overline{\nu}P_{N-1|N-1}. \end{aligned} \tag{12.14}$$

Let us focus now on the covariance matrix:

$$\begin{aligned} P_{N-1|N-1} &= P_{N-1|N-2} - \gamma_{N-1}\frac{P^2_{N-1|N-2}}{(P_{N-1|N-2}+1)} = \\ &= P_{N-1|N-2} - \gamma_{N-1}\left(P_{N-1|N-2} - 1 + \frac{1}{(P_{N-1|N-2}+1)}\right) \end{aligned} \tag{12.15}$$

Because

$$P_{N-1|N-2} = P_{N-2|N-2} + (1-\theta_{N-2})(1-\overline{\nu})\overline{\nu}u^2_{N-2} \tag{12.16}$$

one finds that

$$\begin{aligned} &E\left[P_{N-1|N-1}|E_{N-2}\right] = P_{N-2|N-2} + (1-\overline{\theta})(1-\overline{\nu})\overline{\nu}u^2_{N-2} + \\ &-\overline{\gamma}\Big(P_{N-2|N-2} + (1-\overline{\theta})(1-\overline{\nu})\overline{\nu}\,u^2_{N-2} - 1 + \overline{\theta}\tfrac{1}{P_{N-2|N-2}} + \\ &(1-\overline{\theta})\tfrac{1}{P_{N-2|N-2}+(1-\overline{\nu})\overline{\nu}u^2_{N-2}}\Big). \end{aligned} \tag{12.17}$$

Finally, we get

$$\begin{aligned} V_{N-2}(x) &= \min_{u_{N-2}} E\left[x^2_{N-2} + V_{N-1}(x_{N-1})|E_{N-2}\right] = \\ &= (3-\overline{\nu})E\left[x^2_{N-1}|E_{N-2}\right] + \min_{u_{N-2}} P_{N-2|N-2} + \\ &\quad + (1-\overline{\theta})(1-\overline{\nu})\overline{\nu}u^2_{N-2} - \overline{\gamma}\Big(P_{N-2|N-2} + \\ &\quad + (1-\overline{\theta})(1-\overline{\nu})\overline{\nu}u^2_{N-2} - 1 + \overline{\theta}\tfrac{1}{P_{N-2|N-2}} + \\ &\quad + (1-\overline{\theta})\tfrac{1}{P_{N-2|N-2}+(1-\overline{\nu})\overline{\nu}u^2_{N-2}}\Big) \end{aligned} \tag{12.18}$$

The first terms within the last parenthesis in (12.18) are convex quadratic functions of the control input $u_{N-2}$, however the last term is not. Then, the optimal control law is, in general, a nonlinear function of the information set $E_k$. □

There are only two cases where the optimal control is linear. The first case is when $\bar{\theta} = 1$ (TCP-Case). This corresponds to the TCP-like case studied in [32].

The second case is when the measurement noise covariance is zero ($R = 0$) and any delivered packet contains full state information, i.e. $Rank(C) = n$. In fact, this would mean that, at each time instant $k$, $\gamma_k C$ is either zero or full column-rank. If it is zero, the dependence is linear. If $\gamma_k C$ is full column-rank, it is equivalent to having an exact measurement of the actual state (see [32]). However, it is important to remark that in such a second case the separation principle still does not hold, since the control input affects the estimator error covariance. These results can be summarized in the following Theorem.

**Theorem 12.1.** *Let us consider the stochastic system defined in (12.1) with horizon $N \geq 2$. Then:*

- *if $\overline{\theta} < 1$, the separation principle does not hold*
- *The optimal control feedback $u_k = f_k^*(E_k)$ that minimizes the cost functional defined in Equation (12.4) is, in general, a nonlinear function of information set $E_k$*
- *The optimal control feedback $u_k = f_k^*(E_k)$ is a linear function of the estimated state if and only if one of the following conditions holds true:*
  - $\overline{\theta} = 1$
  - *$Rank(C) = n$ and $R = 0$* □

In the next Section we will focus on the case $Rank(C) = n$, and $R = 0$. In particular, we will compute the optimal control and we will show that, in the infinite horizon scenario, the optimal state-feedback gain is constant, i.e. $L_k^* = L^*$ and can be computed as the solution of a convex optimization problem.

## 12.5 Optimal Control – Rank(C)=n, R=0 Case

Without loss of generality we can assume $C = I$. Because of the hypothesis of no measurement noise, i.e. $R = 0$, it is possible to simply measure the state $x_k$ when a packet is delivered. The estimator equations then simplify in the following way:

$$K_{k+1} = I \tag{12.19}$$

$$P_{k+1|k} = AP_{k|k}A + Q + (1-\theta_k)(1-\overline{\nu})\overline{\nu}\left[Bu_k u_k^T B^T\right] \tag{12.20}$$

$$\begin{aligned} P_{k+1|k+1} &= (1-\gamma_{k+1})P_{k+1|k} \\ &= (1-\gamma_{k+1})\left(AP_{k|k}A + Q + \right. \\ &\quad \left. + (1-\theta_k)(1-\overline{\nu})\overline{\nu}\left[Bu_k u_k^T B^T\right]\right) \end{aligned} \tag{12.21}$$

$$\begin{aligned} E\left[P_{k+1|k+1}|E_k\right] &= (1-\overline{\gamma})\left(AP_{k|k}A + Q \right. \\ &\quad \left. + (1-\overline{\theta})(1-\overline{\nu})\overline{\nu}\left[Bu_k u_k^T B^T\right]\right). \end{aligned} \tag{12.22}$$

In the last equation the independence of $E_k, \gamma_{k+1}, \theta_k$ is exploited. By following the classical dynamic programming approach to solve optimal control problems, it can be seen that the value function $V_k^*(x_k)$ can be written as follows:

$$V_k(x_k) = \hat{x}_{k|k}^T S_k \hat{x}_{k|k} + trace\left(T_k P_{k|k}\right) + trace\left(D_k Q\right) = \qquad (12.23)$$
$$= E\left[x_{k|k}^T S_k x_{k|k}\right] + trace\left(H_k P_{k|k}\right) + trace\left(D_k Q\right)$$

for each $k = N,\ldots,0$ where $H_k = T_k - S_k$. This is clearly true for $k = N$. In fact, we have

$$V_N(x_N) = E\left[x_N^T W_N x_N | E_N\right]$$
$$= \hat{x}_{N|N}^T W_N \hat{x}_{N|N} + trace\left(W_N P_{N|N}\right),$$

and the statement is satisfied by $S_N = T_N = W_N, D_N = 0$. Let us suppose that (12.23) holds true for $k+1$ and let us show by induction that it holds true for $k$ as well

$$V_k(x_k) = \min_{u_k} E\left[x_k^T W_k x_k + \nu_k u_k^T U_k u_k + V_{k+1}(x_{k+1}) | E_k\right] =$$
$$\min_{u_k} E\left[x_k^T W_k x_k | E_k\right] + \overline{\nu} u_k^T U_k u_k + E\left[x_{k+1}^T S_{k+1} x_{k+1} | E_k\right] +$$
$$trace\left(H_{k+1} P_{k+1|k+1}\right) + trace\left(D_{k+1} Q\right) =$$
$$= \min_{u_k} E\left[x_k^T W_k x_k | E_k\right] + \overline{\nu}_k u_k^T U_k u_k + trace\left(D_{k+1} Q\right) +$$
$$+ trace\left(H_{k+1}\left((1-\overline{\gamma})\left(A P_{k|k} A + Q + (1-\theta_k)\overline{\nu}(1-\overline{\nu})\left[B u_k u_k^T B^T\right]\right)\right)\right)$$
$$+ E\left[\left(A x_{k|k} + \theta_k \nu_k B u_k + (1-\theta_k)\overline{\nu} B u_k\right)^T S_{k+1}\right.$$
$$\left.\left(A x_{k|k} + \theta_k \nu_k B u_k + (1-\theta_k)\overline{\nu} B u_k\right) \middle| E_k\right].$$

Further manipulation yields:

$$V_k(x_k) = \min_{u_k} E\left[x_k^T W_k x_k + \overline{\nu}_k u_k^T U_k u_k + \left(x_{k|k}^T A^T S_{k+1} A x_{k|k}\right) + \right.$$
$$+ \left(\theta_k \nu_k u_k^T B^T B u_k\right) + \left((1-\theta_k)\overline{\nu} u_k^T B^T B u_k^T\right) + 2\theta_k \nu_k x_{k|k}^T A^T S_{k+1} B u_k +$$
$$\left. + 2(1-\theta_k)\overline{\nu} x_{k|k}^T A^T S_{k+1} B u_k | E_k\right] + trace\left(D_{k+1} Q\right) +$$
$$+ trace\left(H_{k+1}\left((1-\overline{\gamma})\left(A P_{k|k} A + Q + (1-\overline{\theta})\overline{\nu}(1-\overline{\nu})\left[B u_k u_k^T B^T\right]\right)\right)\right) =$$
$$= E\left[x_{k|k}^T \left(W_k + A^T S_{k+1} A\right) x_{k|k}\right] + (1-\overline{\gamma}) trace\left(H_{k+1}\left(\left(A P_{k|k} A + Q\right)\right)\right) +$$
$$+ \min_{u_k} \overline{\nu}\left(u_k^T \left(U_k + B^T \left(S_{k+1} + (1-\overline{\theta})(1-\overline{\nu})\overline{\nu} H_{k+1}\right) B\right) u_k^T\right) +$$
$$+ 2\overline{\nu}\left(x_{k|k}^T A^T S_{k+1} B u_k\right) + trace\left(D_{k+1} Q\right).$$

Since $V_k(x_k)$ is a convex quadratic function w.r.t. $u_k$, the minimizer is the solution of $\partial V_k(x_k)/\partial u_k = 0$, given by:

$$u_k^* = -\left(U_k + B^T\left(S_{k+1} + \overline{\alpha} H_{k+1}\right) B\right)^{-1}\left(B^T S_{k+1} A x_{k|k}\right) = L_k x_{k|k}, \qquad (12.24)$$

where $\overline{\alpha} = (1-\overline{\gamma})\left(1-\overline{\theta}\right)(1-\overline{\nu})\overline{\nu}$. The optimal control is a linear function of the estimated state $x_{k|k}$. Substituting back (12.24) into the value function we get:

$$V_k(x_k) = trace\left((1-\overline{\gamma})H_{k+1}\left((AP_{k|k}A)\right)\right) + trace\left(((1-\overline{\gamma})T_{k+1} + D_{k+1})Q\right) + \\ +E\left[x_{k|k}^T(W_k + A^T S_{k+1}A)x_{k|k}\right] - \overline{\nu}x_{k|k}^T(A^T S_{k+1}BL_k)x_{k|k},$$

which becomes

$$\begin{aligned} V_k(x_k) = {} & trace\left((1-\overline{\gamma})H_{k+1}\left((AP_{k|k}A)\right)\right) + \\ & +E\left[x_{k|k}^T(W_k + A^T S_{k+1}A)x_{k|k} + \left(\overline{\nu}\left(x_{k|k}^T A^T S_{k+1}B\right)L_k x_{k|k}\right)\right] + \\ & +trace((D_{k+1} + (1-\overline{\gamma})T_{k+1})Q) - trace\left((\overline{\nu}A^T S_{k+1}BL_k P_{k|k})\right). \end{aligned}$$

Finally, we obtain

$$\begin{aligned} V_k(x_k) = {} & trace\left((D_{k+1} + (1-\overline{\gamma})H_{k+1})Q\right) + \\ & +E\left[x_{k|k}^T(W_k + A^T S_{k+1}A)x_{k|k} + \left(\overline{\nu}\left(x_{k|k}^T A^T S_{k+1}B\right)L_k x_{k|k}\right)\right] + \\ & +trace\left((((1-\overline{\gamma})A^T H_{k+1}A - \overline{\nu}A^T S_{k+1}BL_k)P_{k|k})\right). \end{aligned}$$

From the last equation we see that the value function can be written as in (12.23) if and only if the following equations are satisfied:

$$S_k = W_k + A^T S_{k+1}A + \overline{\nu}\left(A^T S_{k+1}B\right)L_k \tag{12.25}$$
$$T_k = (1-\overline{\gamma})A^T T_{k+1}A + W_k + \overline{\gamma}A^T S_{k+1}A \tag{12.26}$$
$$D_k = D_{k+1} + (1-\overline{\gamma})T_{k+1} + \overline{\gamma}S_{k+1}. \tag{12.27}$$

*Remark 12.2.* Notice that, if $\overline{\theta} \to 0$, the control design system soon regresses to the UDP-like case.

The optimal minimal cost for the finite horizon, $J_N^* = V_0(x_0)$ is then given by:

$$J_N^* = \bar{x_0}^T S_0 x_0 + trace(S_0 P_0) + trace(D_0 Q).$$

For the infinite horizon optimal controller, necessary and sufficient conditions for the average minimal cost $J_\infty^* = \lim_{N\to\infty} \frac{1}{N} J_N^*$ to be finite, are that the coupled recurrent equations (12.26) and (12.25) converge to a finite value $S_\infty$ and $T_\infty$ as $N \to \infty$.

**Theorem 12.2.** *Consider system (12.1) and the problem of minimizing the cost function (12.4) within the class of admissible policies $u_k = f(E_k)$. Assume also that $R = 0$ and $rank C = n$. Then:*

*(i) The optimal estimator gain is constant and in particular* $K_k = I$ *if* $C = I$.
*(ii) The infinite horizon optimal control exists if and only if there exist positive definite matrices* $S_\infty, T_\infty > 0$ *such that* $S_\infty = \Phi_S(S_\infty, T_\infty)$ *and* $T_\infty = \Phi_T(S_\infty, T_\infty)$, *where* $\Phi_S$ *and* $\Phi_T$ *are:*

$$\Phi_S(S_k, W_k) = W_k + A^T S_k A - \overline{\nu}(A^T S_k B) \left(U_k + B^T((1-\overline{\alpha})S_{k+1} + \overline{\alpha}T_{k+1})B\right)^{-1}(B^T S_{k+1} A) \tag{12.28}$$

$$\Phi_T(S_k, T_k) = (1-\overline{\gamma})A^T T_{k+1} A + W_k + \overline{\gamma} A^T S_{k+1} A. \tag{12.29}$$

*(iii) The infinite horizon optimal controller gain is constant:* $\lim_{k\to\infty} L_k = L_\infty$

$$L_\infty = -\left(U + B^T((1-\overline{\alpha})S_\infty + \overline{\alpha}T_\infty)B\right)^{-1}(B^T S_\infty A). \tag{12.30}$$

*(iv) A necessary condition for existence of* $S_\infty, T_\infty > 0$ *is*

$$\begin{aligned} & 1 - |A|^2 \left(1 - \frac{\overline{\nu}}{(1-\overline{\alpha}) + \overline{\alpha}\frac{\overline{\gamma}|A|^2}{1-(1-\overline{\gamma})|A|^2}}\right) \geq 0 \\ & \overline{\gamma} > 1 - \frac{1}{|A|^2}, \end{aligned} \tag{12.31}$$

*where* $|A| = \max_i |\lambda_i(A)|$ *is the largest eigenvalue of the matrix A. This condition is also sufficient if B is square and invertible.*
*(v) The expected minimum cost for the infinite horizon scenario converges to:*

$$J_\infty^* = \lim_{k\to\infty} \frac{1}{N} J_N^* = trace\left(((1-\overline{\gamma})T_k + \overline{\gamma}S_k)Q\right). \tag{12.32}$$

*Proof.* Proof is omitted for space limitations. Please refer to [33] for a complete proof.

## 12.6 Example 2 - The Batch Reactor

The goal of this example is to show how stability and control performance are affected by the acknowledgment packet arrival probability. Let us consider an unstable batch reactor (see [34], pp. 62) with full sensing. Due to physical distance between sensors and actuators it may be convenient to connect sensors and actuators through an (eventually wireless) channel. The linearized process model is:

$$\dot{x}(t) = \begin{pmatrix} 1.3800 & -0.2077 & 6.715 & -5.676 \\ -0.5814 & -4.29 & 0 & 0.675 \\ 1.067 & 4.273 & -6.654 & 5.893 \\ 0.048 & 4.273 & 1.343 & -2.104 \end{pmatrix} x(t) + \begin{pmatrix} 0 & 0 \\ 5.679 & 0 \\ 1.136 & -3.146 \\ 1.136 & 0 \end{pmatrix} u(t)$$

$$y(t) = x(t).$$

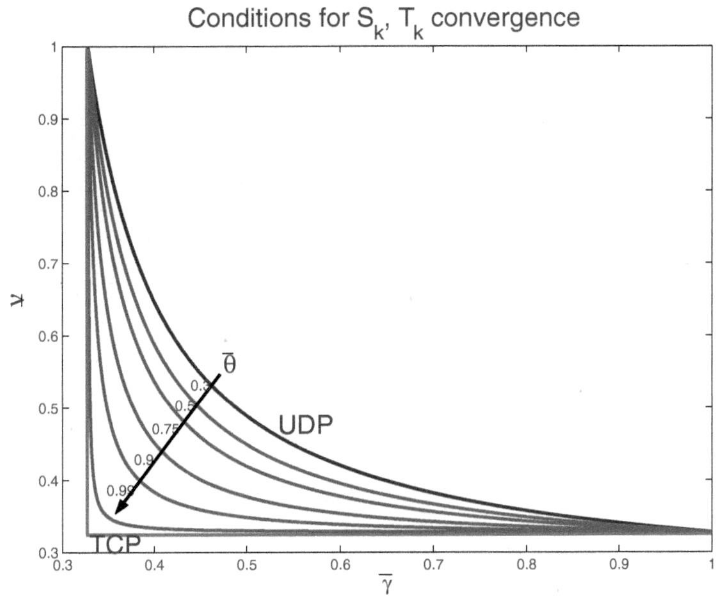


**Fig. 12.4** Region of convergence relative to measurement packet arrival probability $\bar{\gamma}$, and the control packet arrival probability $\bar{\nu}$, parameterized into the acknowledgment packet arrival probability $\bar{\theta}$. Blue and magenta lines depict respectively the bound of the UDP-like and TCP-like cases.

Such a model is discretized with a sampling time $\tau = 0.1s$ and embedded in our networked scheme:

$$x_{k+1} = \begin{pmatrix} 1.1782 & 0.0015 & 0.5116 & -0.4033 \\ -0.0515 & 0.6619 & -0.0110 & 0.0613 \\ 0.0762 & 0.3351 & 0.5607 & 0.3824 \\ -0.0006 & 0.3353 & 0.0893 & 0.8494 \end{pmatrix} x_k + \begin{pmatrix} 0.0045 & -0.0876 \\ 9.4672 & 0.0012 \\ 0.2132 & -0.2353 \\ 0.2131 & -0.0161 \end{pmatrix} \nu_k u_k$$

$$y_k = \gamma_k x_k.$$

Figure 12.4 shows the different necessary stability regions with respect to $\bar{\nu}$ and $\bar{\gamma}$, parameterized by the acknowledgement probability $\bar{\theta}$. In particular, it is possible to show that, as $\bar{\theta} \to 1$ the stability region converges to the one computed for the TCP-like case. Figure 12.5 shows the expected cost $J_N^*$ as a function of $\bar{\theta}$ for $\bar{\gamma} = 0.5, Q = I, U = I, W = I$. It is worth to notice that the influence of packet acknowledgment on performance decreases as $\bar{\nu}$ tends to 1. This matches with the intuition that in case of perfect communication between the control unit and the actuators (i.e. $\bar{\nu} = 1$) the acknowledgment does not carry any additional information.

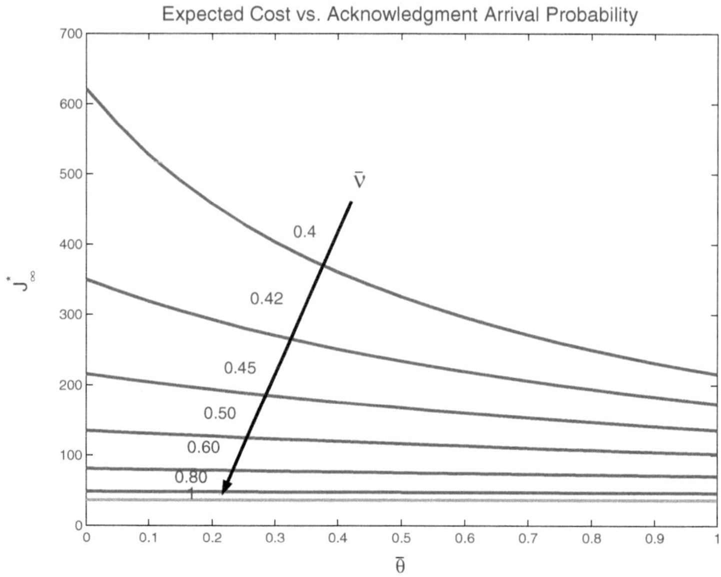


**Fig. 12.5** Expected cost $J_N^*$ as a function of the acknowledgment packet arrival $\bar{\theta}$ for a fixed measurement packet arrival $\bar{\gamma} = 0.5$ and for different control packet arrival $\bar{\nu}$. Green line denotes perfect communication between control unit and actuator.

## 12.7 Conclusions

In this work we analyzed a generalized version of the LQG control problem in the case where both observation and control packets may be lost during transmission over a communication channel. This situation arises frequently in distributed systems where sensors, controllers and actuators reside in different physical locations and have to rely on data networks to exchange information. In this context controller design heavily depends on the communication protocol used. In fact, in TCP-like protocols, acknowledgements of successful transmissions of control packets are provided to the controller, while in UDP-like protocols, no such feedback is provided. In the first case, the separation principle holds and the optimal control is a linear function of the state. As a consequence, controller and estimator design problems are decoupled. UDP-like protocols present a much more complex problem. We have shown that the even partial lack of acknowledgement of control packets results in the failure of the separation principle. Estimation and control are now intimately coupled. In addition, the LQG optimal control is, in general, nonlinear in the estimated state. In the particular case where the observation packet contains full state information the optimal controller is linear in the state. In this particular case we could show how the partial presence of acknowledgement increases both the performance and the stability range of the overall system, which converges to the TCP-like with deterministic acknowledgements as the arrival rate for the acknowledgement packets tends to one.

## References

1. Witsenhausen, H.: A counterexample in stochastic optimum control. SIAM Journal of Control 6 (1968)
2. Murray, R.M., Astrom, K.J., Boyd, S.P., Brockett, R.W., Stein, G.: Control in an information rich world. IEEE Control System Magazine 23 (2003)
3. Hespanha, J., Naghshtabrizi, P., Xu, Y.: A survey of recent results in networked control systems. Proceedings of the IEEE, special issue 95(1) (2007)
4. Yang, T.C.: Networked control system: a brief survey. IEE Proceedings of Control Theory and Applications (2006)
5. Wong, W.S., Brockett, R.E.: Systems with finite communication bandwidth constraints.II: Stabilization with limited information feedback. IEEE Transaction on Automatic Control 44(5) (1999)
6. Nair, G.N., Evans, R.J.: Exponential stabilisability of finite-dimensional linear systems with limited data rates. Automatica 39(4) (2003)
7. Braslavsky, J.H., Middleton, R.H., Freudenberg, J.S.: Feedback stabilization over signal-to-noise ratio constrained channels. In: American Control Conference (2004)
8. Zhang, W., Branicky, M.S., Phillips, S.M.: Stability of networked control systems. IEEE Control System Magazine 21(1) (2001)
9. Luck, R., Ray: An observer-based compensator for distributed delays. Automatica 26(5) (1990)
10. Montestruque, L.A., Antsaklis, P.J.: Stability of model-based networked control systems with time-varying transmission times. IEEE Transaction on Automatic Control 49(9) (2004)
11. Walsh, G.C., Ye, H., Bushnell, L.: Stability analysis of networked control systems. IEEE Transaction on Control System Technology 10(3) (2001)
12. Nesic, D., Teel, A.: Input-output stability properties of networked control systems. IEEE Transaction on Automatic Control 49(10) (2004)
13. Hadjicostis, C.N., Touri, R.: Feedback control utilizing packet dropping network links. In: IEEE Conference on Decision and Control, Las Vegas (2002)
14. Nilsson, J.: Real-time control systems with delays. Ph.D. dissertation, Dept. Automatic Control, Lund Inst. Technology, Lund, Sweden (1998)
15. Schenato, L.: To hold or to zero control inputs with lossy links? Accepted for publication in IEEE Transactions on Automatic Control (2008)
16. Imer, O.C., Yuksel, S., Basar, T.: Optimal Control of dynamical systems over unreliable communicaion links. In: NOLCOS, Stutgart, Germany (2004)
17. Sinopoli, B., Schenato, L., Franceschetti, M., Poolla, K., Jordan, M., Sastry, S.: Optimal Control wih unreliable communication: the TCP case. In: American Control Conference, Portland, OR, USA (2005)
18. Sinopoli, B., Schenato, L., Franceschetti, M., Poolla, K., Sastry, S.: LQG control with missing observation and control packets. In: IFAC World Congress, Prague, Czech Republic (2005)
19. Sinopoli, B., Schenato, L., Franceschetti, M., Poolla, K., Sastry, S.: An LQG optimal linear controller for control system with packet losses. In: Conference on Decision and Control, Sevilla, Spain (2005)
20. Sinopoli, B., Schenato, L., Franceschetti, M., Poolla, K., Sastry, S.: Optimal Linear LQG control over lossy networks wihout packet acknowledgment. In: Conference on Decision and Control, San Diego, CA, USA (2006)

21. Epstein, M., Shi, L., Murray, R.M.: An estimation algorithm for a class of networked control systems using UDP-like communication scheme. In: Conference on Decision and Control, San Diego, CA, USA (2006)
22. Padmasola, P., Elia, N.: Mean Square Stabilization of LTI Systems over Packet-drop Networks with Imperfect Side Information. In: American Control Conference Minneapolis (2006)
23. Tanenbaum, A.: Computer Networks. Prentice-Hall, Englewood Cliffs (1981)
24. Gupta, V., Spanos, D., Hassibi, B., Murray, R.M.: Optimal LQG control across a packet-dropping link. System and Control Letters 56 (2007)
25. Xu, Y., Hespanha, J.: Estimation under controlled and uncontrolled communications in networked control system. In: Conference on Decision and Control, Sevilla, Spain (2007)
26. Smith, S., Seiler, P.: Estimation with lossy measurements: jump estimators for jump systems. IEEE Transaction on Automatic Control 48 (2003)
27. Huang, M., Dey, S.: Stability of Kalman filtering with Markovian packet losses. Automatica 43 (2007)
28. Drew, M., Liu, X., Goldsmith, A., Hedrick, J.: Control system design over a wireless lan. In: Conference on Decision and Control, Sevilla, Spain (2005)
29. Liu, X., Goldsmith, A.: Cross-layer Design of Distributed Control over Wireless Networks. In: Basar, T. (ed.). Birkhauser, Basel (2005)
30. Elia, N., Eisembeis, J.: Limitation of linear control over packet drop networks. In: Conference on Decision and Control, Bahamas (2004)
31. Elia, N.: Remote stabilization over fading channels. System and Control Letters 54 (2005)
32. Schenato, L., Sinopoli, B., Franceschetti, M., Poolla, K., Sastry, S.: Foundations of Control and estimation over lossy Networks. Proceedings of the IEEE 95 (2007)
33. Garone, E., Sinopoli, B., Casavola, A., Goldsmith, A.: LGQ control for distributed systems over TCP-like erasure channels. In: Conference on Decision and Control, New Orleans, LS, USA (2007)
34. Green, M., Limebeer, D.J.N.: Linear Robust Control. Prentice-Hall, Englewood Cliffs (1995)

# Chapter 13
# State Estimation in a Sensor Network under Bandwidth Constraints

Giorgio Battistelli, Alessio Benavoli, and Luigi Chisci

**Abstract.** This paper deals with the problem of estimating the state of a discrete-time linear stochastic dynamical system on the basis of data collected from multiple sensors subject to a limitation on the communication rate from the remote sensor units. More specifically, the attention is devoted to a centralized sensor network consisting of: (1) $N_S$ remote nodes which collect measurements of the given system, compute state estimates at the full measurement rate and transmit them at a reduced communication rate; (2) a fusion node $F$ that, based on received estimates, provides an estimate of the system state at the full rate. A measurement-independent strategy for deciding when transmitting estimates from each sensor to $F$ will be considered. Sufficient conditions for the boundedness of the state covariance at node $F$ will be given. Further, the possibility of determining a communication strategy with optimal performance in terms of minimum mean square estimation error will be investigated.

## 13.1 Introduction

This paper deals with the problem of estimating the state of a discrete-time linear Gaussian stochastic dynamical system subject to a limitation on the communication rate from the remote sensor units to the state estimation unit. More specifically, the attention will be focused on the use of a centralized network (see Fig. 13.1)

Giorgio Battistelli
Dipartimento di Sistemi e Informatica, Università di Firenze, 50139 Firenze, Italy
e-mail: battistelli@dsi.unifi.it

Alessio Benavoli
Istituto "Dalle Molle" di Studi sull'Intelligenza Artificiale, 6928 Manno-Lugano, Switzerland
e-mail: alessio@idsia.ch

Luigi Chisci
Dipartimento di Sistemi e Informatica, Università di Firenze, 50139 Firenze, Italy
e-mail: chisci@dsi.unifi.it

consisting of: $N_S$ remote *sensing nodes* which collect noisy measurements of the given system, can process them to find filtered estimates and transmit such estimates at a reduced communication rate; a local *fusion node* $F$ which receives data from the $N_S$ sensors and, based on such data, should estimate, in the best possible way, the system's state.

The objective is to devise a *transmission strategy* (TS) with fixed rate that guarantees bounded state covariance and possibly optimal estimation performance, in terms of minimum *Mean Square Error* (MSE), at the node $F$.

The above scenario reflects the practical situation in which the sensing units and the monitoring unit are remotely dislocated one with respect to each other and the communication rate between them is severely limited by energy, band and/or security concerns. This happens, for example, in wireless sensor networks wherein every transmission typically reduces the lifetime of the sensor devices (wireless communication being the major source of energy consumption [5]). Further, a reduction in the sensors data transmission rate can be crucial in networked control systems in order to reduce the network traffic and hopefully avoid congestion [17].

State estimation under finite communication bandwidth has been thoroughly investigated, see e.g. [8, 15]. In the above cited references, the emphasis is on the analysis of the quantization effects due to the encoding of transmitted data into a finite alphabet of symbols as well as on the design of efficient, possibly optimal, coding algorithms. Conversely, following [6, 7, 16, 17] the present work tackles the issue of communication bandwidth finiteness from a completely different viewpoint. Specifically, it is assumed that infinite-precision data are transmitted over the communication channels[1] while the bandwidth limitation is accomplished by imposing suitable values of the communication rates. In this context, the focus will be on the choice of a communication strategy for deciding which data transmit from each sensor to $F$. Notice that similar frameworks have been considered in [7, 16, 17], where strategies were proposed for controlling data transmission in a networked control system so as to achieve a trade-off between communication and estimation performance.

In this paper, probabilistic measurement-independent strategies are considered and issues related to the stability and optimality of such strategies are investigated. Specifically, attention is devoted to TS wherein the times between consecutive transmissions are random variables governed by a finite-state Markov chain. This is the same TS analyzed in [10] in the context of a model-based networked control system. Further, in the signal processing literature, the idea of randomly varying the time between consecutive measurements, commonly referred to as *additive random sampling* [3], has been extensively studied in order to overcome the aliasing problem. Finally, additive random sampling is a common practice in Internet flow monitoring to avoid synchronization problems [11].

Two main results will be provided extending to centralized sensor networks the results of [2] obtained in the case of a single remote sensing unit. First it will be shown that boundedness of the MSE can be ensured provided that an upper bound on

[1] This assumption, though incompatible with the finite bandwidth, holds in practice provided that quantization errors are negligible with respect to measurement errors.

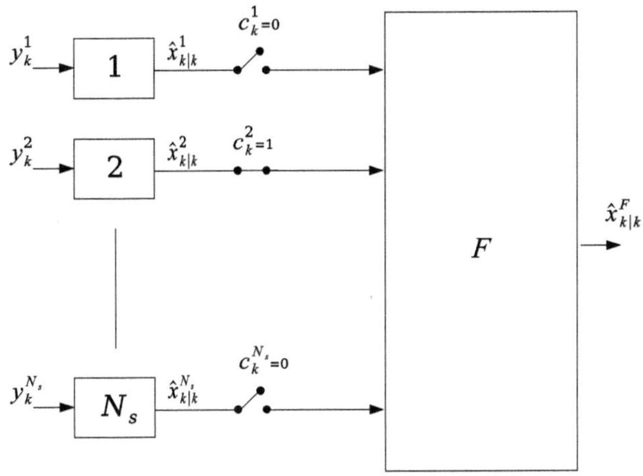


**Fig. 13.1** The considered centralized sensor network.

the time between consecutive transmissions is imposed. Secondly, an optimization-based approach will be proposed to derive the communication strategy with minimum MSE.

The rest of the paper will be organized as follows. Section 13.2 reports the preliminary results of [2] concerning the single-sensor case. Section 13.3 present the theoretical results concerning stability and optimality of the communication strategy in the case of multi-sensor networks. Section 13.4 concludes the paper.

The notations are quite standard: $\mathbb{Z}_+$ is the set of nonnegative integers; given a square matrix $\mathbf{M}$, $\mathrm{tr}(\mathbf{M})$ and $\rho(\mathbf{M})$ denote its trace and, respectively, its spectral radius; finally, $\mathbb{E}\{\cdot\}$ and $\mathbb{P}\{\cdot\}$ denote the expectation and, respectively, probability operators.

## 13.2 State Estimation with a Remote Sensor

Before addressing the multi-sensor case, it is convenient to report the results of [2] concerning the single-sensor case. To this end, consider the discrete-time linear Gaussian stochastic dynamical system

$$\mathbf{x}_{k+1} = \mathbf{A}\mathbf{x}_k + \mathbf{w}_k \tag{13.1}$$

$$\mathbf{y}_k = \mathbf{C}\mathbf{x}_k + \mathbf{v}_k \tag{13.2}$$

### *13.2.1 Transmission Strategy*

The aim of this section is to formalize the concept of TS with fixed rate $\alpha$. To this end, let us introduce binary variables $c_k$ such that

$$c_k = \begin{cases} 1, \text{ if node } S \text{ transmits at time } k \\ 0, \text{ if node } S \text{ does not transmit at time } k \end{cases}$$

The communication strategy is characterized by the following two choices:

- the type of data being transmitted by the node $S$, either measurements $\mathbf{y}_k$ or filtered state estimates $\hat{\mathbf{x}}^S_{k|k}$;
- the mechanism by which the node $S$ decides whether to transmit or not at time $k$, i.e. the mechanism of generating $c_k \in \{0,1\}$.

As far as the first choice is concerned, the following options can for instance be adopted:

- measurement transmission: when $c_k = 1$, node $S$ transmits $\mathbf{y}_k$ to node $F$;
- estimate transmission: when $c_k = 1$, node $S$ transmits $\hat{\mathbf{x}}^S_{k|k}$ to node $F$.

In this work, the attention is restricted to estimate transmission, assuming that node $S$ has enough processing capability to update on-line the optimal state estimate $\hat{\mathbf{x}}^S_{k|k}$. Note that, with such a choice, there is no loss of information due to the finite bandwidth as the estimate $\hat{\mathbf{x}}^S_{k|k}$ represents a sufficient statistics.

As far as the decision mechanism is concerned, this can be formally defined as follows (see also [16] and [7]).

**Definition.** A decision mechanism with rate $\alpha \in (0,1)$ is any, deterministic or stochastic, mechanism of generating $c_k$ such that

$$\lim_{t\to\infty} \frac{1}{t} \sum_{k=1}^{t} \mathbb{E}\{c_k\} = \alpha \tag{13.3}$$

Several decision mechanisms can clearly be devised. In this work, the attention will be restricted to *measurement-independent* strategies that, at time $k$, decide whether to transmit or not independently of the measurement sequence $\mathbf{y}^k \triangleq \{\mathbf{y}_0, \mathbf{y}_1, \ldots, \mathbf{y}_k\}$. More specifically, it is assumed that each $c_k$ is chosen according to some probabilistic criterion, possibly adapted only on the basis of the past transmission pattern, so as to ensure the desired communication rate. With this respect, let $n_k \geq 0$ denote, at a generic time $k$, the number of time instants elapsed from the last transmission, i.e. $c_{k-n_k} = 1$ and $c_k = c_{k-1} = \cdots = c_{k-n_k+1} = 0$. Note that $n_k$ is a function of $c^k \triangleq \{c_0, c_1, \ldots, c_k\}$ and can be recursively computed as

$$n_k = (1 - c_k)(n_{k-1} + 1).$$

It should be evident that, in the considered framework, such a quantity provides a rough estimate of how good is the estimate $\hat{\mathbf{x}}^F_{k|k}$ at node $F$. Then it seems reasonable to take into account the value of $n_{k-1}$ when choosing whether to transmit or not at the generic time $k$. In this connection, the conditional probabilities

$$\mathbb{P}\{c_k = 1 | n_{k-1} = i\}, \quad i = 0, 1, \ldots$$

can be considered as design parameters in the communication strategy that can be suitably tuned to improve performance (e.g., to reduce the MSE). Specifically, in this paper the focus is on transmission strategies wherein, at each time $k$, the variable $c_k$ is chosen to be a Bernoulli random variable with parameter $\varphi(n_{k-1})$, i.e., $c_k$ takes value 1 with probability $\varphi(n_{k-1})$ and value 0 with probability $1-\varphi(n_{k-1})$. Clearly, this corresponds to the assignments

$$\mathbb{P}\{c_k = 1 | n_{k-1} = i\} = \varphi(i),$$

for $i = 0, 1, \ldots$. The function $\varphi : \mathbb{Z}_+ \to [0,1]$ must be chosen so that thc transmission rate constraint is met. The choice of such a function will be the subject of Sections 13.2.3 and 13.2.4.

### *13.2.2 Operations of Nodes S and F*

This section describes the operations that must be performed in the nodes $S$ and $F$ in order to recover at the node $F$ the optimal estimate regardless of the actually employed strategy.

To enable the estimate transmission option, it is assumed that the node $S$ has enough processing capabilities to compute the optimal state estimate. Let $\mathbf{y}^k$ be the set of measurements gathered by node $S$ at time $k$. It is well known that for the linear system (13.1)-(13.2) under Gaussian assumptions on the disturbances $\mathbf{w}_k$ and $\mathbf{v}_k$, the optimal estimates and covariances, fully characterizing the conditional PDF $p(\mathbf{x}_k|\mathbf{y}^k)$, are recursively provided by the Kalman Filter:

$$\begin{cases} \mathbf{P}^S_{k|k-1} = \mathbf{A}\mathbf{P}^S_{k-1|k-1}\mathbf{A}' + \mathbf{Q} \\ \mathbf{K}_k = \mathbf{P}^S_{k|k-1}\mathbf{C}'\left(\mathbf{R} + \mathbf{C}\mathbf{P}^S_{k|k-1}\mathbf{C}'\right)^{-1} \\ \hat{\mathbf{x}}^S_{k|k} = \mathbf{A}\hat{\mathbf{x}}^S_{k-1|k-1} + \mathbf{K}_k(\mathbf{y}_k - \mathbf{C}\mathbf{A}\hat{\mathbf{x}}^S_{k-1|k-1}) \\ \mathbf{P}^S_{k|k} = \mathbf{P}^S_{k|k-1} - \mathbf{P}^S_{k|k-1}\mathbf{C}'\left(\mathbf{R} + \mathbf{C}\mathbf{P}^S_{k|k-1}\mathbf{C}'\right)^{-1}\mathbf{C}\mathbf{P}^S_{k|k-1} \end{cases} \tag{13.4}$$

where $\mathbf{Q}$ and $\mathbf{R}$ are the covariances of $\mathbf{w}_k$ and, respectively, $\mathbf{v}_k$. Notice that the covariance matrices of the Kalman Filter do not depend on the sequence of measurements and, hence, can be autonomously computed by the node $F$ so that only transmission of estimates but no transmission of covariances from $S$ to $F$ is needed.

For the node $F$, the available information at time $k$ is given by $n_k$ and $\hat{\mathbf{x}}^S_{k|k-n_k}$. It is easy to verify that the PDFs at the fusion node $F$ turn out to be Gaussian. Further, the optimal estimates $\hat{\mathbf{x}}^F_{k|k}$ and covariances $\mathbf{P}^F_{k|k}$ can be recursively computed by

$$\begin{cases} \hat{\mathbf{x}}^F_{k|k} = c_k\hat{\mathbf{x}}^S_{k|k} + (1-c_k)\mathbf{A}\hat{\mathbf{x}}^F_{k-1|k-1} \\ \mathbf{P}^F_{k|k} = c_k\mathbf{P}^S_{k|k} + (1-c_k)\left(\mathbf{A}\mathbf{P}^F_{k-1|k-1}\mathbf{A}' + \mathbf{Q}\right) \end{cases} \tag{13.5}$$

where the covariance matrix $\mathbf{P}^S_{k|k}$ is updated in the node $F$ in the same way as in the node $S$; see (13.4). It is important to note that, in this case, also the covariance $\mathbf{P}^F_{k|k}$ is a random variable as it depends on $c^k$.

### *13.2.3 Boundedness of the State Covariance*

In this section, the boundedness of the state covariance at node $F$ is discussed for different choices of the transmission probabilities $\mathbb{P}\{c_k = 1 | n_{k-1} = i\} = \varphi(i)$. This allows one to gain some insights on how the choice of such probabilities may significantly affect the performance of the TS.

In order to ensure the well-posedness of the state estimation problem, the following preliminary assumption is needed.

**A1.** $(\mathbf{A},\mathbf{C})$ is detectable, $(\mathbf{A},\mathbf{Q}^{1/2})$ is stabilizable, $\mathbf{R} > 0$.

As well known, assumption A1 ensures that the a-priori state covariance $\mathbf{P}^S_{k|k-1}$ and the a-posteriori covariance $\mathbf{P}^S_{k|k}$ exponentially converge to the steady-state values $\mathbf{P}_b$ and $\mathbf{P}_a$, respectively, where $\mathbf{P}_b$ is the unique positive definite solution of the algebraic Riccati equation

$$\mathbf{P}_b = \mathbf{A}\left[\mathbf{P}_b - \mathbf{P}_b\mathbf{C}'\left(\mathbf{R}+\mathbf{C}\mathbf{P}_b\mathbf{C}'\right)^{-1}\mathbf{C}\mathbf{P}_b\right]\mathbf{A}' + \mathbf{Q} \tag{13.6}$$

and $\mathbf{P}_a$ can be obtained as

$$\mathbf{P}_a = \mathbf{P}_b - \mathbf{P}_b\mathbf{C}'\left(\mathbf{R}+\mathbf{C}\mathbf{P}_b\mathbf{C}'\right)^{-1}\mathbf{C}\mathbf{P}_b\,. \tag{13.7}$$

A first important observation is that the dependence of the transmission probabilities on the elapsed time $n_k$ can be crucial to ensure that the estimation error is bounded in mean square. With this respect, in [2, 6, 16] it has been shown that, in the case of a single remote sensor, when $\varphi(i) = \alpha$ for $i = 0, 1, \ldots$ the MSE at node $F$ may diverge if the transmission rate $\alpha$ does not exceed a certain critical value. More specifically, the following result can be stated.

**Proposition 13.1.** *Let $c_k$, $k = 0, 1, \ldots,$ be mutually independent Bernoulli random variables with parameter $\alpha$. Under assumption A1, the following facts hold:*

*(i) if $\alpha > 1 - 1/\rho^2(\mathbf{A})$ then*

$$\lim_{k\to\infty} \mathbb{E}\{\mathbf{P}^F_{k|k}\} = \mathbf{X}$$

*where $\mathbf{X}$ is the unique positive definition solution of the Lyapunov equation*

$$(1-\alpha)\mathbf{A}\mathbf{X}\mathbf{A}' - \mathbf{X} = -(1-\alpha)\mathbf{Q} - \alpha\mathbf{P}_a;$$

*(ii) if $\alpha \le 1 - 1/\rho^2(\mathbf{A})$ then*

$$\lim_{k\to\infty} \operatorname{tr}\left(\mathbb{E}\{\mathbf{P}^F_{k|k}\}\right) = +\infty .$$

Proposition 13.1 states that, when the matrix $\mathbf{A}$ is not Schur stable (i.e., $\rho(\mathbf{A}) \geq 1$), the expected covariance is ultimately bounded if and only if the transmission rate exceeds a critical value $\alpha_c \triangleq 1 - 1/\rho^2(\mathbf{A})$. It is worth noting that in [12] the existence of a similar critical value was proved for Kalman filtering with intermittent observations, which in the present framework corresponds to transmitting measurements instead of estimates. Thus, Proposition 13.1 shows that equipping the remote sensor with the processing capability of computing the optimal estimates is not sufficient to overcome such a drawback.

Nevertheless, it turns out that boundedness of the MSE can be ensured provided that an upper bound on the time between consecutive transmissions is imposed. In this connection, consider a strategy wherein $\varphi(N-1) = 1$ for some $N > 1$. This amounts to assuming that the time interval between two consecutive transmissions never exceeds $N$. As will be clear in the sequel, transmission strategies of this kind, satisfying the communication rate constraint (13.3), always exist provided that $N > 1/\alpha$.

As the covariance $\mathbf{P}^F_{k|k}$ is a random variable, in order to analyze the performance of the considered strategy, the expected value $\mathbb{E}\{\mathbf{P}^F_{k|k}\}$ has to be considered. With this respect, since $\varphi(N-1) = 1$ implies that $\sum_{i=0}^{N-1} \mathbb{P}\{n_k = i\} = 1$, the expected covariance can be written as

$$\mathbb{E}\{\mathbf{P}^F_{k|k}\} = \sum_{i=0}^{N-1} \mathbb{P}\{n_k = i\}\, \mathbf{P}^S_{k|k-i}$$

where $\mathbf{P}^S_{k|k-i}$ is the covariance at the node $F$ provided that the last transmission has occurred $i$ time instants ago.

In order to analyze the asymptotic behavior of the expected state covariance, first note that under assumption A1 we have

$$\lim_{k\to\infty} \mathbf{P}^S_{k|k-i} = \mathbf{P}(i), \quad i = 0, \ldots, N-1$$

where the covariances $\mathbf{P}(i)$ can be obtained by repeatedly applying the Lyapunov difference equation:

$$\begin{cases} \mathbf{P}(0) = \mathbf{P}_a \\ \mathbf{P}(i) = \mathbf{A}\mathbf{P}(i-1)\mathbf{A}' + \mathbf{Q}, \quad i = 1, \ldots, N-1 . \end{cases} \tag{13.8}$$

Moreover, note that the sequence $n_0, n_1, \ldots$ can be described by a discrete-time Markov chain characterized by the state set $\{0, 1, \ldots, N-1\}$ and the state transition diagram of Fig. 13.2. Then, by means of straightforward calculations (see [2]), one can see that the unique invariant distribution for the probabilities

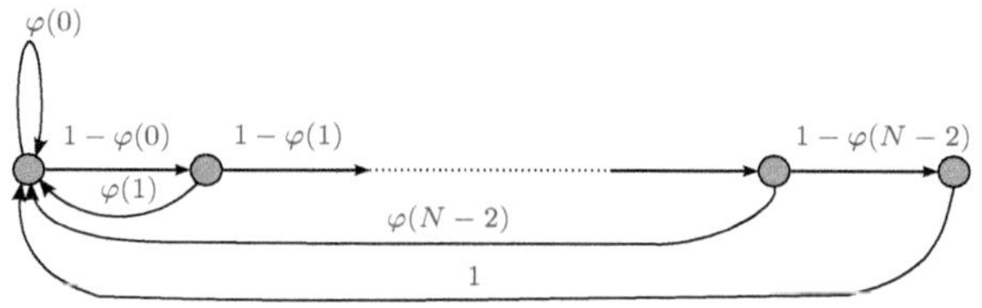


**Fig. 13.2** State transition diagram of the Markov chain $n_0, n_1, \ldots$ for the considered TS.

$\mathbb{P}\{n_k = i\} = p(i), \quad i = 0, \ldots, N-1$ (representing also the long-run average distribution) is given by

$$p(i) = \left( \sum_{l=0}^{N-1} \prod_{j=0}^{l-1} (1 - \varphi(j)) \right)^{-1} \prod_{j=0}^{i-1} (1 - \varphi(j)), \qquad \text{for } i = 0, \ldots, N-1 \tag{13.9}$$

where, for the sake of compactness, we define $\prod_{j=0}^{-1} (\cdot) \triangleq 1$.

The foregoing discussion leads to the following result.

**Proposition 13.2.** *Let $c_k$, $k = 0, 1, \ldots$, be Bernoulli random variables with parameter $\varphi(n_{k-1})$. Moreover, let $\varphi(N-1) = 1$ for some $N > 1$. Then, under assumption A1, we have*

$$\lim_{t \to \infty} \frac{1}{t} \sum_{k=1}^{t} \mathbb{E}\{\mathbf{P}^F_{k|k}\} = \left( \sum_{i=0}^{N-1} \prod_{j=0}^{i-1} (1 - \varphi(j)) \right)^{-1} \sum_{i=0}^{N-1} \prod_{j=0}^{i-1} (1 - \varphi(j)) \mathbf{P}(i).$$

In other words, the proposed TS always provides an ultimately bounded expected state covariance regardless of the transmission rate $\alpha$ and of the stability of the system matrix $\mathbf{A}$.

## *13.2.4 Optimal Single-Sensor Transmission Strategy*

For a given $N$, the TS introduced in the previous section is not uniquely determined in that there are $N-1$ degrees of freedom, i.e., the parameters $\varphi(i)$ for $i = 0, \ldots, N-2$. To be more precise, when the transmission rate $\alpha$ is fixed, the degrees of freedom are only $N-2$ since the parameters have to be chosen so that the equality constraint (13.3) is satisfied. It is quite natural to think of exploiting such degrees of freedom to optimize some performance index. With this respect, a reasonable choice for the performance index is given by the cost

$$J \triangleq \lim_{t \to \infty} \frac{1}{t} \sum_{k=1}^{t} \operatorname{tr} \left( \mathbb{E}\{\mathbf{P}^F_{k|k}\} \mathbf{W} \right)$$

where $\mathbf{W}$ is a positive definite weight matrix.

In the light of Proposition 13.2, it is immediate to write cost $J$ as a function of the parameters $\varphi(i)$ for $i=0,\ldots,N-2$:

$$J=\left(\sum_{i=0}^{N-1}\prod_{j=0}^{i-1}(1-\varphi(j))\right)^{-1}\sum_{i=0}^{N-1}\operatorname{tr}\left(\mathbf{P}(i)\,\mathbf{W}\right)\prod_{j=0}^{i-1}(1-\varphi(j)). \tag{13.10}$$

In the sequel, for the sake of compactness, we shall use the notation $w(i)\triangleq\operatorname{tr}\left(\mathbf{P}(i)\,\mathbf{W}\right)$.

As to the transmission rate constraint (13.3), the following result holds.

**Lemma 13.1.** *For the considered TS, the transmission rate constraint (13.3) is equivalent to*

$$\sum_{i=0}^{N-1}\prod_{j=0}^{i-1}(1-\varphi(j))=1/\alpha. \tag{13.11}$$

Note that one can exploit such a constraint and, instead of $J$, consider the simplified cost

$$J'\triangleq\sum_{i=0}^{N-1}w(i)\prod_{j=0}^{i-1}(1-\varphi(j)).$$

Summing up, the following constrained optimization problem can be stated that leads to the synthesis of an optimized TS.

**Problem 13.1.** *For a given $N>1$, find the optimal parameters $\varphi(i)$, $i=0,\ldots,N-2$, that solve the constrained minimization problem*

$$\begin{aligned}&\textit{minimize} && \sum_{i=0}^{N-1}w(i)\prod_{j=0}^{i-1}(1-\varphi(j))\\ &\textit{subject to} && \sum_{i=0}^{N-1}\prod_{j=0}^{i-1}(1-\varphi(j))=1/\alpha,\\ &&& 0\le\varphi(i)\le 1,\quad i=0,\ldots,N-2.\end{aligned} \tag{13.12}$$

As shown in [2], the above-defined problem admits the following closed-form solution.

**Theorem 13.1.** *Problem 13.1 admits solution if and only if $N\ge 1/\alpha$. Moreover, under assumption A1, for any $\mathbf{W}>0$ the optimal solution is given by*

$$\begin{aligned}&\varphi(i)=0, && i=0,\ldots,M-3,\\ &\varphi(M-2)=M-1/\alpha, &&\\ &\varphi(M-1)=1, &&\end{aligned} \tag{13.13}$$

*where $M$ is the smallest integer such that $M\ge 1/\alpha$.*

A few remarks about Theorem 13.1 are in order. First note that, when $1/\alpha$ is integer, the optimal TS is periodic and consists of transmitting the estimate $\hat{\mathbf{x}}^S_{k|k}$ once every

$1/\alpha$ instants. The same does not hold true if $1/\alpha$ is not integer. However, in the case in which $1/\alpha$ is rational, i.e., $1/\alpha = p/q$ with $p,q \in \mathbb{Z}_+$ coprime, it is still possible to devise a periodic strategy with period $p$ equivalent to the optimal one (i.e., with the same cost). Such a strategy consists of transmitting the estimate $\hat{\mathbf{x}}^S_{k|k}$ $q$ times every $p$ instants in such a way that the maximum time between consecutive transmissions is minimized.

Another important observation is that, since the weights $w(i)$ are monotonically increasing, the TS of Theorem 13.1 turns out to be optimal also in a minimax sense, i.e. it minimizes $\lim_{t\to\infty} \sup_{k\geq t} \left(\mathbb{E}\{\mathbf{P}^F_{k|k}\}\,\mathbf{W}\right)$. In fact, minimizing the maximum time between consecutive transmissions corresponds to minimizing the maximum of the estimation error $\left(\mathbb{E}\{\mathbf{P}^F_{k|k}\}\,\mathbf{W}\right)$ with respect to all the transmission patterns.

## 13.3 Extension to Multi-sensor Networks

In this section, the previous results will be extended to the case of a centralized multi-sensor network. To this end, suppose that $N_S$ remote sensors be available, each one collecting noisy measurements

$$\mathbf{y}^s_k = \mathbf{C}^s \mathbf{x}_k + \mathbf{v}^s_k, \tag{13.14}$$

computing locally optimal filtered estimates $\hat{\mathbf{x}}^s_{k|k}$ by means of the Kalman filter recursions, and transmitting such estimates to the fusion node. Hereafter, the subscript $s\,(s = 1,\ldots,N_S)$ will be used to refer to the $s$-th sensor. The local *fusion node* $F$ receives data from the $N_S$ sensors and, based on such data, should provide, in the best possible way, a fused estimate $\hat{\mathbf{x}}^F_{k|k}$ of the system's state.

Throughout this section, the following notation will be adopted

$$\mathbf{c}_k \stackrel{\triangle}{=} \text{col}\left(c^1_k,\ldots,c^{N_S}_k\right), \quad \mathbf{n}_k \stackrel{\triangle}{=} \text{col}\left(n^1_k,\ldots,n^{N_S}_k\right),$$
$$\mathbf{y}_k \stackrel{\triangle}{=} \text{col}\left(\mathbf{y}^1_k,\ldots,\mathbf{y}^{N_S}_k\right).$$

It is supposed that each sensor transmits its local estimate according to the probabilistic TS defined in Section 13.2.1 Further, in order to ensure boundedness of the error covariance regardless of the system stability, attention will be devoted to strategies satisfying the following assumption.

**A2.** The TS is such that for each sensor $s \in \{1,\ldots,N_S\}$, one has $\varphi^s(N^s - 1) = 1$ for some $N^s > 1$.

This amounts to assuming that the time interval between two consecutive transmissions of sensor $s$ never exceeds $N^s$. Clearly, assumption A2 implies that the vector of elapsed times $\mathbf{n}_k$ always takes value in the set

$$\mathscr{N} \triangleq \{(n^1,\ldots,n^{N_S})' \in \mathbb{Z}_+^S : n^s \le N^s - 1\,,\, s = 1,\ldots,N_S\}\,.$$

In view of Theorem 13.1, one has that transmission strategies of this kind satisfying the communication rate constraint (13.3) always exist provided that $N^s > 1/\alpha^s$.

As well known, different fusion algorithms can be devised to combine data collected from multiple sensors and related information in order to derive a fused estimate $\hat{\mathbf{x}}^F_{k|k}$ of the true state $\mathbf{x}_k$. As to the computation of the MSE corresponding to a generic transmission strategy, some hypotheses about the fusion algorithm are needed. Specifically, it is supposed that the fusion algorithm has the following properties.

**A3.** At each time $k = 0,1,\ldots$, the fused estimate $\hat{\mathbf{x}}^F_{k|k}$ is unbiased.
**A4.** At each time $k = 0,1,\ldots$, the covariance matrix of the estimation error $\mathbf{x}_k - \hat{\mathbf{x}}^F_{k|k}$ is a function of the last transmission instant of each sensor, i.e.,

$$\mathbb{E}\left\{(\mathbf{x}_k - \hat{\mathbf{x}}^F_{k|k})(\mathbf{x}_k - \hat{\mathbf{x}}^F_{k|k})' \,|\, \mathbf{y}_{0:k}\,;\, \mathbf{c}_{0:k}\right\} = \mathbf{P}^F_{k|k}(\mathbf{n}_k)\,.$$

**A5.** For each $\mathbf{n} \in \mathscr{N}$, the covariance matrix $\mathbf{P}^F_{k|k}(\mathbf{n})$ converges exponentially to a steady state value $\mathbf{P}^F(\mathbf{n})$.

Assumptions A3-A5 can be seen as a sort of minimal requirements that a sensible information fusion algorithm should satisfy. For optimal fusion algorithms satisfying all such properties see [9] and the references therein.

By exploiting assumptions A2 and A4, one can write the expected covariance of the estimation error as

$$\mathbb{E}\left\{(\mathbf{x}_k - \hat{\mathbf{x}}^F_{k|k})(\mathbf{x}_k - \hat{\mathbf{x}}^F_{k|k})'\right\} = \sum_{\mathbf{n}\in\mathscr{N}} \mathbb{P}\{\mathbf{n}_k = \mathbf{n}\}\mathbf{P}^F_{k|k}(\mathbf{n})\,.$$

Since in the considered framework the sensors' transmission strategies are independent, then for each vector $\mathbf{n} = (n^1,\ldots,n^{N_s})$ the probability $\mathbb{P}\{\mathbf{n}_k = \mathbf{n}\}$ can be decomposed as

$$\mathbb{P}\{\mathbf{n}_k = \mathbf{n}\} = \prod_{s=1}^{N_S} \mathbb{P}\{n^s_k = n^s\}\,.$$

Let us now consider the asymptotic behavior of the Markov chain $n^s_0, n^s_1, \ldots$. As well known, two possibilities may arise:

(i) the Markov chain is aperiodic and each probability, $\mathbb{P}\{n^s_k = i\}$, converges exponentially to the unique invariant distribution $p^s(i)$ given in (13.9);
(ii) the Markov chain is periodic and its asymptotic behavior depends on the initial probabilities.

In the context of our application, it is desirable that in the long run the probabilities $\mathbb{P}\{n^s_k = i\}$ do not depend on their initial values, otherwise the error covariance at node $F$ would depend on the phase shift among the sensors and a synchronization would be needed to optimize the performance. Then the following constraint is imposed.

**A6.** For each sensor $i = 1, \ldots, S$, the Markov chain $n_0^s, n_1^s, \ldots$ is aperiodic.

The foregoing discussion leads to the following result.

**Theorem 13.2.** *Suppose that assumptions A2-A6 hold. Then the steady-state mean covariance of the estimation error can be obtained as*

$$\lim_{k\to\infty} \mathbb{E}\left\{(\mathbf{x}_k - \hat{\mathbf{x}}_{k|k}^F)(\mathbf{x}_k - \hat{\mathbf{x}}_{k|k}^F)'\right\} = \sum_{(n^1,\ldots,n^S)\in\mathcal{N}} \left(\prod_{s=1}^{N_S} p^s(n^s)\right) \mathbf{P}\left([n^1,\ldots,n^{N_S}]'\right).$$

Note that, such a covariance is always bounded regardless of the transmission rates $\alpha^s$ and of the stability of the system matrix $\mathbf{A}$.

### *13.3.1 Optimized Transmission Strategies*

This section is devoted to the synthesis of an optimized TS. In this connection, note that, for given $N^1, \ldots, N^{N_S}$, the strategy introduced in the previous section is not uniquely determined in that there are $\sum_{s=1}^{N_S}(N^s - 1)$ degrees of freedom, i.e., the parameters $\varphi^s(i)$ for $i = 0, \ldots, N^s - 2$ and $s = 1, \ldots, N_S$. To be more precise, when the transmission rates $\alpha^s$ are fixed, the degrees of freedom are only $\sum_{s=1}^{N_S}(N^s - 2)$ since the parameters have to be chosen so that the transmission rate constraint is satisfied for each sensor. Like in the single-sensor case, it is quite natural to think of exploiting such degrees of freedom in order to optimize some performance index. With this respect, a reasonable choice for the performance index is given by the cost

$$J \triangleq \lim_{t\to\infty} \frac{1}{t} \sum_{k=1}^{t} \operatorname{tr}\left(\mathbb{E}\left\{(\mathbf{x}_k - \hat{\mathbf{x}}_{k|k}^F)(\mathbf{x}_k - \hat{\mathbf{x}}_{k|k}^F)'\right\} \mathbf{W}\right)$$

where $\mathbf{W}$ is a positive definite weight matrix.

In the light of Theorem 13.2, it is immediate to write cost $J$ as a function of the parameters $\varphi^s(i)$ for $i = 0, \ldots, N^s - 2$ and $s = 1, \ldots, N_S$:

$$J = \sum_{(n^1,\ldots,n^{N_S})\in\mathcal{N}} \operatorname{tr}\left(\mathbf{P}([n^1,\ldots,n^{N_S}]')\mathbf{W}\right) \left(\prod_{s=1}^{N_S} p^s(n^s)\right) \tag{13.15}$$

where the steady state probabilities $p^s(n^s)$ can be obtained as in (13.9). In the sequel, for the sake of compactness, we shall use the notation $w(n^1,\ldots,n^{N_S}) \triangleq \operatorname{tr}\left(\mathbf{P}([n^1,\ldots,n^{N_S}]')\mathbf{W}\right)$.

Then, taking into account Lemma 13.1, the following constrained optimization problem can be stated that leads to the synthesis of an optimized strategy.

**Problem 13.2.** *Given $N^1,\ldots,N^{N_S}$, find the optimal parameters $\varphi^s(i)$ for $i = 0,\ldots,N^s-2$ and $s=1,\ldots,N_S$ that minimize cost $J$ under the constraints*

$$\begin{aligned}
&\sum_{i=0}^{N^s-1}\prod_{j=0}^{i-1}(1-\varphi^s(j)) = 1/\alpha^s\,, && s=1,\ldots,N_S \\
&0 \le \varphi^s(i) \le 1\,, && i=0,\ldots,N^s-2, \\
& && s=1,\ldots,N_S\,.
\end{aligned} \tag{13.16}$$

*Remark 13.1.* The formulation of Problem 13.2 does not take into account the fact that, in order to satisfy assumption A6, the Markov chain $n_0^s, n_1^s, \ldots$ has to be aperiodic for each sensor $s$. Then, solution of Problem 13.2 may yield periodic Markov chains. As already pointed out, in the multi-sensor context this does not correspond to the desired behavior and, in this case, the resulting periodic Markov chains should be somehow perturbed (in the respect of the transmission rate constraint) so as to ensure aperiodicity.

In order to address the solution of Problem 13.2, it is convenient to recast it in terms of the steady state probabilities $p^s(n^s)$. To this end, note that by (13.9) each constraint (13.11) corresponds to $p^s(0) = \alpha^s$. Further, if one refers to (13.9), the probabilities $p^s(i)$ can also be obtained recursively as

$$p^s(i) = p^s(i-1)(1-\varphi^s(i-1))\,, \quad i=1,\ldots,N^s-1\,.$$

Then the feasibility constraints $0 \le \varphi^s(i) \le 1$, $i=0,\ldots,N^s-2$, turn out to be equivalent to

$$\begin{aligned}
&p^s(i) \ge 0\,, && i=1,\ldots,N^s-1\,, \\
&p^s(i+1) \le p^s(i)\,, && i=0,\ldots,N^s-2\,.
\end{aligned} \tag{13.17}$$

Finally, notice that $\varphi^s(N^s-1)=1$ implies that the probabilities $p^s(i)$, $i=0,\ldots,N^s-2$, sum up to one, i.e.,

$$\sum_{i=0}^{N^s-1} p^s(i) = 1\,. \tag{13.18}$$

As a consequence, each vector $\mathbf{p}^s$ is constrained to belong to the polytope $\mathscr{P}^s$ defined by (13.17), (13.18) and $p^s(0) = \alpha^s$. Thus, Problem 13.2 admits the equivalent formulation:

$$\begin{aligned}
&\text{minimize} \sum_{(n^1,\ldots,n^{N_S})\in\mathscr{N}} w(n^1,\ldots,n^{N_S})\left(\prod_{s=1}^{N_S} p^s(n^s)\right) \\
&\text{subject to } \mathbf{p}^s \in \mathscr{P}^s\,, \quad s=1,\ldots,N_S\,.
\end{aligned} \tag{13.19}$$

In the light of such a reformulation, the following feasibility result can be readily established.

**Proposition 13.3.** *Problem 13.2 admits solution if and only if $N^s \geq 1/\alpha^s$ for $s = 1, \ldots, N_S$.*

For what concerns the solution of (13.19), it is worth noting that for any $s$ the objective function $J(\mathbf{p}^1, \ldots, \mathbf{p}^{N_S})$ is linear in $\mathbf{p}^s$ when all $\mathbf{p}^q$, $q \neq s$, are fixed. Thus solving (13.19) amounts to finding the minimum of a multilinear function over the Cartesian product of Polytopes $\mathscr{P} \triangleq \mathscr{P}^1 \times \cdots \times \mathscr{P}^{N_S}$. Then, at least in principle, one can exploit the known fact that there always exists an extreme point of $\mathscr{P}$ where the multilinear function is minimized and solve the problem in finite time by means of an extensive search over all the extreme points of $\mathscr{P}$.

The main drawback of such an approach is that the number of extreme points of $\mathscr{P}$ grows exponentially with the number of sensors $N_S$. As a consequence, unless very small instances of the problem are considered, the possibility of solving it exactly is ruled out by the so-called *curse of dimensionality*, i.e., the exponential growth of the computational burden. Then, suboptimal solutions should be sought after, for example by applying some local descent algorithm (see [1] and the references therein).

Of course, an alternative approach consists in seeking a closed form for the optimal transmission probabilities (as it happens in the single-sensor case) for some particular sensor configuration and fusion rule. This issue will be the subject of further investigation.

## 13.4 Conclusions

The paper has addressed the state estimation problem in a centralized sensor network assuming that: (1) estimates are required at a distant location from the sensors connected via a communication link; (2) a limitation on the communication rate of each sensor is imposed; (3) each sensor node has enough processing capability to compute local state estimates. Probabilistic measurement-independent strategies for deciding which data transmit have been investigated. It has been shown that: (1) boundedness of the mean square error is guaranteed provided that an upper bound on the inter-transmission time is enforced; (2) communication strategies with optimal performance in terms of minimum mean square error can be devised. Future work will concern the extension to measurement-dependent transmission strategies [7, 14, 16] and distributed sensor networks [4, 13].

## References

1. Absil, P.A., Sepulchre, R.: Continuous dynamical systems that realize discrete optimization on the hypercube. Systems & Control Letters 52, 297–304 (2004)
2. Battistelli, G., Benavoli, A., Chisci, L.: State estimation with a remote sensor under limited communication rate. In: Proc. of the 3rd International Symposium on Communications, Control and Signal Processing, pp. 654–659 (2008)

3. Bilinskis, I., Mikelsons, A.: Randomized Signal Processing. Prentice Hall International, Englewood Cliffs (1992)
4. Carli, R., Chiuso, A., Schenato, L., Zampieri, S.: Distributed kalman filtering based on consensus strategies. IEEE Journal on Selected Areas in Communications 26, 622–633 (2008)
5. Feeney, L.M., Nilsson, M.: Investigating the energy consumption of a wireless network interface in an ad hoc networking environment. In: Proc. IEEE Infocom, pp. 1548–1557 (2001)
6. Gupta, V., Hassibi, B., Murray, M.: Optimal LQG control across packet-dropping links. Systems & Control Letters 56, 439–446 (2007)
7. Hespanha, J.P., Naghshtabrizi, P., Xu, Y.: A survey of recent results in networked control systems. Proc. of IEEE Special Issue on Technology of Networked Control Systems 95, 138–162 (2007)
8. Li, X., Wong, W.S.: State estimation with communication constraints. Systems & Control Letters 28, 49–54 (1996)
9. Li, X.R., Zhu, Y.M., Wang, J., Han, C.Z.: Optimal linear estimation fusion-part I: unified fusion rules. IEEE Transactions on Information Theory 49, 2192–2208 (2003)
10. Montestruque, L.A., Antsaklis, P.: Stability of model-based networked control systems with time-varying transmission times. IEEE Transactions on Automatic Control 49, 1562–1572 (2004)
11. Paxson, V.: End-to-end routing behavior in the internet. ACM SIGCOMM Computer Communication Review 26, 25–38 (1996)
12. Sinopoli, B., Schenato, L., Franceschetti, M., Poolla, K., Jordan, M., Sastry, S.: Kalman filtering with intermittent observations. IEEE Transactions on Automatic Control 49, 1453–1464 (2004)
13. Spanos, D.P., Murray, R.M.: Distributed sensor fusion using dynamic consensus. In: Proc. of the 16th IFAC World Congress (2005)
14. Suh, Y.S., Nguyen, V.H., Ro, Y.S.: Modified Kalman filter for networked monitoring systems employing a send-on-delta method. Automatica 43, 332–338 (2007)
15. Wong, W.S., Brockett, R.W.: Systems with finite communication bandwidth constraints - Part I: state estimation problem. IEEE Transactions on Automatic Control 42, 1294–1299 (1997)
16. Xu, Y., Hespanha, J.P.: Estimation under uncontrolled and controlled communications in networked control systems. In: Proc. of the 44th IEEE Conference on Decision and Control, pp. 842–847 (2005)
17. Yook, J.K., Tilbury, D.M., Soparkar, N.R.: Trading computation for bandwidth: Reducing communication in distributed control systems using state estimators. IEEE Transactions on Control Systems Technology 10, 503–518 (2002)

# Chapter 14
# Admission Control in Variable Capacity Communication Networks*

A. Pietrabissa and F. Delli Priscoli

**Abstract.** This paper defines a theoretical framework based on Markov Decision Processes (MDP) to deal with call control algorithms in links with variable capacity supporting multiple classes of service. The variable capacity problem, which arises in wireless network scenarios, is addressed by incorporating the link model into the MDP formulation and by introducing, beside the standard call admission policy, a call dropping policy. In this way, the proposed approach is capable of controlling class-level quality of service in terms of both blocking and dropping probabilities.

**Keywords:** Call Control, Markov Decision Processes (MDP), Linear Programming (LP), Communication Networks.

## 14.1 Introduction

We consider a multi-class network, where calls of different classes compete for bandwidth and are regulated by a Call Admission Control (CAC) algorithm, which decides whether a new call can be safely accepted by the network or not. The CAC problem has been successfully modelled as a Markov Decision Process (MDP) ([1]). By formulating it as a Linear Programming (LP) problem ([4], [16]), global constrainits on blocking and dropping probabilities can be enforced.

In the MDP formulations introduced so far (e.g., [6]-[13],[14], [21]-[23]), the link capacity is considered as a given constant value. On the contrary, when we consider wireless networks this is not true, since the link capacity is time-varying, for instance due to weather conditions, interferences, speed of mobile users, adaptive coding and

A. Pietrabissa and F. Delli Priscoli
Università di Roma "La Sapienza" Dipartimento di Informatica e Sistemistica,
via Ariosto 25, 00162, Roma (IT)
e-mail: `{pietrabissa,dellipriscoli}@dis.uniroma1.it`

* This work was supported in part by the SATSIX european project ([18]).

modulation schemes. If the link capacity decreases, the network might have to drop one or more calls. To control dropping probabilities, this paper proposes to incorporate the link model within the MDP framework and to compute simultaneously both an optimal admission policy and an optimal dropping policy. The main drawback of the proposed solution is that it worsens the curse of dimensionality problem of the MDP approach ([1]), since the state space dramatically increases as the number of states of the link model increases; thus, also an approximate MDP approach is introduced.

In the literature, the time-varying link capacity problem has been faced separately from the MDP-based admission control problem. The most common and effective way to model a link is Markov-chain modeling, where each link state is characterized by a different available capacity ([20], [2], [15], [17]). MDP-based admission control algorithms (considering links with constant capacity) are widespread in the literature ([6] and [10] for broadband multi-service networks, [11] for ATM networks, [8] for optical networks, [9], [21], [12], [23], [22] for wireless networks).

## 14.2 MDP for Links with Variable Capacity

### *14.2.1 Time-varying Link Model*

We consider a generic time-varying link modelled as a Markov chain. Each state of the Markov chain is characterized by a different available capacity, denoted with $\xi_l$, $l = 1,\dots,L$, such that $\xi_{l+1} < \xi_l$, $l = 1,\dots,L\text{-}1$. The number $L$ of link states is defined based on the link characteristics. The generic transition frequency $\chi_{lk}$ between two link states and the stationary probabilities $\pi_{link}(m)$ that the system is in state $m$, $m = 1,\dots,L$, can be estimated by link measures (see, for example, [2]).

### *14.2.2 DTMDP Model*

Let us consider a link supporting $C$ classes of service, each one characterized by a bitrate $e_c$, $c = 1,\dots, C$. The network is represented by a discrete-time system, whose generic state $\mathbf{x}(t)$ is given by the number of calls of each class $c$ on-going at time $t$, $n_c(t)$, plus the link capacity at time $t$, $\xi(t)$: $\mathbf{x}(t) = (n_1(t),n_2(t),\dots,n_C(t),\xi(t))$. At time $t$, the load $\eta(t)$ associated to state $\mathbf{x}(t)$ is then $\eta(t) = \eta[\mathbf{x}(t)] = \sum_{c=1,\dots,C} n_c(t)e_c$. Two cases arise: 1) $\eta(t) \le \xi(t)$: in this case, if a call request arrives, the role of the admission controller is to admit or reject it based on the current state $\mathbf{x}(t)$; 2) $\eta(t) > \xi(t)$: in this case, one ore more calls must be dropped until the system reaches a state whose load is less than the link capacity. Hereafter, a given state $\mathbf{x}(t)$ will be denoted as *available* at time $t$ if $\eta[\mathbf{x}(t)] \le \xi(t)$, *unavailable* otherwise.

For each class $c$, call attempts are distributed according to a Poisson process with mean arrival frequency $\lambda_c$, and the call holding time of class $c$ is exponentially distributed with mean termination frequency $\mu_c$. The control $\mathbf{u}(t)$ is relevant either at call attempts or when a state becomes unfeasible due to the variation of the link capacity: in the former case $\mathbf{u}(t)$ is the admission decision, in the latter it is the dropping decision. The objective is to find the optimal control law $\mathbf{u}^*(t)$ with respect to an appropriate reward function.

Hereafter, the state space, the action space, the transition matrix and the reward function of the MDP are defined.

STATE SPACE $S$

The system state at time $t$ is defined by the number of on-going connections and by the link state: $\mathbf{x}(t) = (n_1(t), n_2(t), ..., n_C(t), \xi(t))$. Let us consider a generic link state $l$; there is a finite number of states identified by the following equation: $\sum_{c=1,...,C} n_c(t)e_c \leq \xi_L$ . By denoting with $M$ the number of states, the system states when the link is in state $l$ are denoted as $\mathbf{x}_{i,l} = (n_1(i), n_1(i), ..., n_C(i), \xi_l)$, $l = 1,...,M$. The set of the states whose load is lower than or equal to $\xi_l$ are available, whereas the other states are unavailable. Let $N_l$ be the number of available states when the link state is $l$, and let the states $\mathbf{x}_{i,l}$ be ordered with respect to their load, i.e., $\eta(\mathbf{x}_{i+1,l}) \geq \eta(\mathbf{x}_{i,l})$, $i = 1,...,M$-1. The set of available states is then defined as follows:

$$S_{AV} = \left\{ \mathbf{x}_{i,l} = (n_1(i), n_2(i), ..., n_C(i), \xi_l), | \sum_{c=1,...,C} n_c(i)e_c \leq \xi_l \right. \\ \left. , i = 1, ..., M, \ l = 1, ..., L \right\} = \{\mathbf{x}_{i,l}\}_{l=1,...,L;\ i=1,...,N_l}, \tag{14.1}$$

the set of unavailable states is

$$S_{UN} = \left\{ \mathbf{x}_{i,l} = (n_1(i), ..., n_C(i), \xi_l), | \xi_l < \sum_{c=1,...,C} n_c(i)e_c \leq \xi_L \right. \\ \left. , i = 1, ..., M, \ l = 1, ..., L \right\} = \{\mathbf{x}_{i,l}\}_{l=1,...,L;\ i=N_l+1,...,M}, \tag{14.2}$$

and the whole state space is $S_{AV} \cup S_{UN}$. The total number of states is $LM$.

ACTION SPACE $A$

Let us define $\delta_c$ as a 1x($C$+1) vector of zeros but the element of column $c$ equal to 1. In the generic available state $\mathbf{x}_{i,l} \in S_{AV}$, if a call attempt of class $c$ occurs, the controller might block the call (and the state remains the same) or accept it, provided that $\mathbf{x}_{j,l} + \delta_c \in S_{AV}$. Let us denote such decision as $u_{adm}(\mathbf{x}_{i,l}, c)$, and let us associate the value 1 if the decision is to accept the new call, 0 if it is to reject it; no decision can be taken on call terminations. The action space $A_{AV}$, which associates the admission decisions to the available state space $S_{AV}$, is then defined as follows:

$$A_{AV} = \left\{ \mathbf{u}_{adm}(\mathbf{x}_{i,l}) = (u_{adm}(\mathbf{x}_{i,l}, 1), ..., u_{adm}(\mathbf{x}_{i,l}, C)) \mid \right. \\ u_{adm}(\mathbf{x}_{i,l}, c) \in \{0, 1\} \text{ if } \mathbf{x}_{i,l} + \delta_c \in S_{AV}; \\ \left. u_{adm}(\mathbf{x}_{i,l}, c) = 0 \text{ if } \mathbf{x}_{i,l} + \delta_c \notin S_{AV}; c = 1, ..., C; i = 1, ..., N_l \right\}. \tag{14.3}$$

In the generic unavailable state, if a call attempt of class $c$ occurs, the controller has to block the call; moreover, the controller must drop one call to try to reach an available state. The frequency of the dropping action (which determines the time interval between two consecutive dropping action) is denoted with $f_{drop}$. Considering a generic unavailable state $\mathbf{x}_{i,k} \in S_{UN}$, the controller decision is to select the class $c$ of the call to be dropped among all classes $c$ such that $n_c(i) > 0$. Let us denote such decision as $u_{drop}(\mathbf{x}_{i,k},c)$, and let us associate the value 1 if the decision is to drop a class $c$ call, 0 if it is to drop a call of another class. Note that exactly one dropping decision must be equal to 1; the action space $A_{UN}$ associates the dropping decisions to the unavailable state space $S_{UN}$ is then defined as follows:

$$
\begin{aligned}
&A_{UN} = \Big\{\mathbf{u}_{drop}(\mathbf{x}_{i,k}) = \big(u_{drop}(\mathbf{x}_{i,k},1),\dots,u_{drop}(\mathbf{x}_{i,k},C)\big) \mid \\
&u_{drop}(\mathbf{x}_{i,k},c) \in \{0,1\} \text{ if } \mathbf{x}_{i,k}-\delta_c \in S, u_{drop}(\mathbf{x}_{i,k},c)=0 \text{ if } \mathbf{x}_{i,k}-\delta_c \notin S, \\
&\sum_{c=1}^{C} u_{drop}(\mathbf{x}_{i,k},c) = 1,\, i = N_k+1,\dots,M,\ k=1,\dots,L\Big\}.
\end{aligned}
\tag{14.4}
$$

The action space $A$ is the set of the available and the unavailable action spaces: $A = A_{AV} \cup A_{UN}$.

Summarizing: the admission policy $\mathbf{u}_{adm} = \{\mathbf{u}_{adm}(\mathbf{x}_{i,l})\}_{l=1,\dots,L;\, i=1,\dots,N_l}$ maps each available state $\mathbf{x}_{i,l} \in S_{AV}$ to the admission control action $\mathbf{u}_{adm}(\mathbf{x}_{i,l})$; the dropping policy $\mathbf{u}_{drop} = \{\mathbf{u}_{drop}(\mathbf{x}_{i,l})\}_{l=1,\dots,L;\, i=N_l+1,\dots,M}$ maps each unavailable state $\mathbf{x}_{i,l} \in S_{UN}$ to the dropping control action $\mathbf{u}_{drop}(\mathbf{x}_{i,l})$; the controller policy is $\mathbf{u} = \mathbf{u}_{adm} \cup \mathbf{u}_{drop}$.

The policy space $U$ is defined as the set of all the feasible policies:

$$
\begin{aligned}
U = &\{\mathbf{u}_{adm} \mid \mathbf{u}_{adm}(\mathbf{x}_{i,l}) \in A_{AV}, \forall \mathbf{x}_{i,l} \in S_{AV}\} \\
&\cup \{\mathbf{u}_{drop} \mid \mathbf{u}_{drop}(\mathbf{x}_{i,l}) \in A_{UN}, \forall \mathbf{x}_{i,l} \in S_{UN}\}.
\end{aligned}
\tag{14.5}
$$

TRANSITION MATRIX **T**

The transition frequencies $\phi(\mathbf{x}_{i,l}, \mathbf{x}_{j,k})$ between couples of states $\mathbf{x}_{i,l}$ and $\mathbf{x}_{j,k}$ are inferred from the link model, from the traffic statistics, and from the action space $A$:

a) The transition frequency $\phi(\mathbf{x}_{i,l}, \mathbf{x}_{j,l})$ between an available state $\mathbf{x}_{i,l} \in S_{AV}$ and $\mathbf{x}_{j,l} = \mathbf{x}_{i,l} - \delta_c$ is equal to $n_c(i)\mu_c$ if $\mathbf{x}_{j,l} \in S$, to 0 otherwise;

b) the transition frequency $\phi(\mathbf{x}_{i,l}, \mathbf{x}_{j,l})$ between an available state $\mathbf{x}_{i,l} \in S_{AV}$ and $\mathbf{x}_{j,l} = \mathbf{x}_{i,l} + \delta_c$ is equal to $u_{adm}(\mathbf{x}_{i,l},c)\lambda_c$ if $\mathbf{x}_{j,l} \in S_{AV}$, to 0 otherwise;

c) the transition frequency $\phi(\mathbf{x}_{i,l}, \mathbf{x}_{j,l})$ between an unavailable state $\mathbf{x}_{i,l} \in S_{UN}$ and $\mathbf{x}_{j,l} = \mathbf{x}_{i,l} - \delta_c$ is equal to $f_{drop}u_{drop}(\mathbf{x}_{i,l}) + n_c(i)\mu_c$ if $\mathbf{x}_{j,l} \in S$, to 0 otherwise;

d) the transition frequency $\phi(\mathbf{x}_{i,l}, \mathbf{x}_{i,k})$ between the states $\mathbf{x}_{i,l} \in S$ and $\mathbf{x}_{i,k} \in S$ is equal to $\chi_{lk}$.

All the other transition frequencies are null. A standard procedure to compute the transition probabilities is the *uniformization* method ([7]):

i) The transition probabilities of each state are computed by dividing the corresponding transition frequencies by the constant $\gamma > \max_{l=1,...,L;\ i=1,...,M} \left\{ \sum_{k=1}^{L} \sum_{j=1}^{M} \phi(\mathbf{x}_{i,l}, \mathbf{x}_{j,k}) \right\}$, which computes the maximum total output frequency among all the states:

$$p_{\mathbf{u}}(\mathbf{x}_{i,l}, \mathbf{x}_{j,k}) = \frac{\phi(\mathbf{x}_{i,l}, \mathbf{x}_{i,k})}{\gamma}, i = 1, ..., M, l = 1, ..., L, \tag{14.6}$$

where $p_{\mathbf{u}}(\mathbf{x}_{i,l},\mathbf{x}_{j,k})$ is the transition probability between states $\mathbf{x}_{i,l} \in S$ and $\mathbf{x}_{j,k} \in S$, and where the base $\mathbf{u}$ explicitly highlights that the transition matrix depends on the policy $\mathbf{u}$.

ii) For each state, a self-transition is added to let the sum of the transitions leaving a state be equal to 1:

$$p_{\mathbf{u}}(\mathbf{x}_{i,l}, \mathbf{x}_{i,l}) = 1 - \sum_{\substack{j=1,...,M \\ k=1,...,L \\ (j,k)\neq(i,l)}} p_{\mathbf{u}}(\mathbf{x}_{i,l}, \mathbf{x}_{j,k}), i = 1, ..., M, l = 1, ..., L. \tag{14.7}$$

The elements $p_{\mathbf{u}}(\mathbf{x}_{i,l},\mathbf{x}_{j,k})$ define the transition matrix $\mathbf{T}$.

REWARD FUNCTION $r$

For admission control algorithms, the most common evaluation parameter is the link utilization or *throughput*; the reward associated to the available state $\mathbf{x}_{i,l} \in S_{AV}$ is then defined as the load occupied by the on-going accepted calls:

$$r(\mathbf{x}_{i,l}) = \sum_{c=1}^{C} n_c(i) e_c,\ l = 1, ..., L,\ i = 1, ..., N_l. \tag{14.8}$$

In the unavailable states, where the state load is greater than the link capacity, the actual throughput is equal to the link capacity; the reward associated to the unavailable state $\mathbf{x}_{i,l} \in S_{UN}$ is then the link capacity $\xi_l$:

$$r(\mathbf{x}_{i,l}) = \xi_l,\ l = 1, ..., L,\ i = N_l + 1, ..., M. \tag{14.9}$$

### 14.2.3 Approximated DTMDP Model

The second model will be developed under the same assumptions on the traffic statistics considered for the first model.

STATE SPACE $S$

In the second proposed model, the generic state is given by the number of on-going calls of each class plus the state availability: $\mathbf{x}_{i,a} = (n_1(i),...,n_C(i),a)$ denotes an available state, whereas $\mathbf{x}_{i,u} = (n_1(i),...,n_C(i),u)$ denotes an unavailable state. Similarly with the first model, $M$ is the finite number of available states identified by the equation $\eta(i) = \sum_{c=1,\dots,C} n_c(i)e_c \le \xi_L$, and the available states are ordered with respect to their load, i.e., $\eta(i+1) \ge \eta(i)$, $i = 1,\dots,M-1$. For each available state $\mathbf{x}_{i,a} = (n_1(i),\dots,n_C(i), a)$, the corresponding unavailable state is defined as $\mathbf{x}_{i,u} = (n_1(i),\dots,n_C(i),u)$, unless $\eta(i)$ is less than or equal to the minimum link load $\eta_1$: in this case, the unavailable state does not exist. Let $N$ be the number of unavailable states; obviously, $N < M$ since (at least) the load of the available state $\mathbf{x}_{1,a} = (0,\dots, 0, a)$ is 0. The set of available states is then defined as follows:

$$S_{AV} = \left\{ \mathbf{x}_{i,a} = (n_1(i),\dots,n_C(i),a), \middle| \sum_{c=1,\dots,C} n_c(i)e_c \le \xi_L \right\} = \{\mathbf{x}_{i,a}\}_{i=1,\dots,M}, \tag{14.10}$$

the set of unavailable states is

$$S_{UN} = \left\{ \mathbf{x}_{i,u} = (n_1(i),\dots,n_C(i),u), \middle| \xi_1 < \sum_{c=1,\dots,C} n_c(i)e_c \le \xi_L \right\} = \{\mathbf{x}_{i,u}\}_{i=M-N+1,\dots,M}, \tag{14.11}$$

and the whole state space is $S = S_{AV} \cup S_{UN}$.

The state space dimension is $M+N < 2N$. In the following, for the sake of simplicity, a generic state (either available or unavailable) will be denoted with $\mathbf{x}_{i,l}$.

ACTION SPACE $A$

As in the first DTMDP, an admission decision $u_{adm}(\mathbf{x}_{i,a},c)$ is associated to each available state $\mathbf{x}_{i,a} \in S_{AV}$ at arrival of a class $c$ call: $u_{adm}(\mathbf{x}_{i,a},c) = 1$ if the call is accepted, provided that $\mathbf{x}_{i,a} + \delta_c \in S_{AV}$, 0 otherwise. The action space $A_{AV}$ is then defined as follows:

$$\begin{aligned} A_{AV} = \{ &\mathbf{u}_{adm}(\mathbf{x}_{i,a}) = (u_{adm}(\mathbf{x}_{i,a},1),\dots,u_{adm}(\mathbf{x}_{i,a},C)) \mid \\ &u^{(c)}_{adm}(\mathbf{x}_{i,a},c) \in \{0,1\} \text{ if } \mathbf{x}_{i,a} + \delta_c \in S_{AV}; \\ &u^{(c)}_{adm}(\mathbf{x}_{i,a},c) = 0 \text{ if } \mathbf{x}_{i,a} + \delta_c \notin S_{AV};\ \mathrm{c} = 1,\dots,\mathrm{C};\ i = 1,\dots,M \}. \end{aligned} \tag{14.12}$$

Similarly, each unavailable state $\mathbf{x}_{i,u} \in S_{UN}$ is associated to the dropping decision: $u_{drop}(\mathbf{x}_{i,u},c) = 1$ if the controller decides to drop a class $c$ call, provided that $n_c(i) > 0$, $u_{drop}(\mathbf{x}_{i,u},c) = 0$ otherwise. The dropping frequency is $f_{drop}$, and exactly one dropping decision must be equal to 1 in each unavailable state; the action space $A_{UN}$ is then defined as follows:

$$\begin{aligned} A_{UN} = \{ &\mathbf{u}_{drop}(\mathbf{x}_{i,u}) = (u_{drop}(\mathbf{x}_{i,u},1),\dots,u_{drop}(\mathbf{x}_{i,u},C)) \mid \\ &u_{drop}(\mathbf{x}_{i,u},C) \in \{0,1\} \text{ if } \mathbf{x}_{i,u} - \delta_c \in S,\ u_{drop}(\mathbf{x}_{i,u},c) = 0 \\ &\text{if } \mathbf{x}_{i,u} - \delta_c \notin S, \sum_{c=1}^{C} u_{drop}(\mathbf{x}_{i,u},c) = 1,\ i = M-N+1,\dots,M \}, \end{aligned} \tag{14.13}$$

The action space $A$ is the set of the available and the unavailable action spaces: $A = A_{AV} \cup A_{UN}$.

Summarizing: the admission policy $\mathbf{u}_{adm} = \{\mathbf{u}_{adm}(\mathbf{x}_{i,a})\}_{i=1,...,M}$ maps each available state $\mathbf{x}_{i,a} \in S_{AV}$ to the admission control action $\mathbf{u}_{adm}(\mathbf{x}_{i,a})$; the dropping policy $\mathbf{u}_{drop} = \{\mathbf{u}_{drop}(\mathbf{x}_{i,u})\}_{i=M-N+1,...,M}$ maps each unavailable state $\mathbf{x}_{i,u} \in S_{UN}$ to the dropping control action $\mathbf{u}_{drop}(\mathbf{x}_{i,u})$; the controller policy is $\mathbf{u} = \mathbf{u}_{adm} \cup \mathbf{u}_{drop}$.

The policy space $U$ is defined as the set of all the feasible policies:

$$\begin{aligned} U = &\{\mathbf{u}_{adm} \mid \mathbf{u}_{adm}(\mathbf{x}_{i,a}) \in A_{AV}, \forall \mathbf{x}_{i,a} \in S_{AV}\} \\ &\cup \{\mathbf{u}_{drop} \mid \mathbf{u}_{drop}(\mathbf{x}_{i,u}) \in A_{UN}, \forall \mathbf{x}_{i,u} \in S_{UN}\}. \end{aligned} \tag{14.14}$$

TRANSITION MATRIX **T**

The transition frequencies are inferred from the assumptions on traffic statistics, from the action space $A$ and from the link model. Differently with respect to the first DTMDP, the link model is used to compute the transition probabilities from each available state $\mathbf{x}_{i,a}$ to its corresponding unavailable state $\mathbf{x}_{i,u}$, and vice-versa. Let us consider an available state $\mathbf{x}_{i,a} \in S_{AV}$ such that $\xi_{l-1} < \eta(i) \leq \xi_l$, with $l \in [2,...,L]$. If at time $t$ the system is in $\mathbf{x}_{i,a}$, which is an available state, by definition it means that the channel capacity at time $t$ is equal to or greater than $\eta(i)$. Then, when the system is in state $\mathbf{x}_{i,a}$, the link is in state $m = l,...,L$ with probability:

$$\pi_{link}(m|\mathbf{x}(t) = \mathbf{x}_{i,a}) = \begin{cases} \dfrac{\pi_{link}(m)}{\sum\limits_{m'=l}^{L} \pi_{link}(m')} & \text{if } m \geq l \\ 0 & \text{if } m < l \end{cases} \tag{14.15}$$

(where $\pi_{link}$ are the stationary probabilities of the channel model, see Section 2.1). Thus, the transition frequencies from the available to the unavailable states are the following:

$$\begin{aligned} \phi(\mathbf{x}_{i,a}, \mathbf{x}_{i,u}) &= \sum_{n=1}^{l-1} \sum_{m=l}^{L} \pi_{link}(m|\mathbf{x}(t) = \mathbf{x}_{i,a})\, \chi_{mn} \\ &= \sum_{n=1}^{l-1} \sum_{m=l}^{L} \frac{\pi_{link}(m)}{\sum\limits_{m'=l}^{L} \pi_{link}(m')} \chi_{mn}, \end{aligned} \tag{14.16}$$

where $l$ is such that $\xi_{l-1} < \eta(i) \leq \xi_l, i = M - N + 1, ..., M$.

Similarly, the transition frequencies from the unavailable to the available states are computed as follows:

$$\begin{aligned} \phi(\mathbf{x}_{i,u}, \mathbf{x}_{i,a}) &= \sum_{m=l}^{L} \sum_{n=1}^{l-1} \pi_{link}(n|\mathbf{x}(t) = \mathbf{x}_{i,u})\, \chi_{nm} \\ &= \sum_{m=l}^{L} \sum_{n=1}^{l-1} \frac{\pi_{link}(n)}{\sum\limits_{n'=1}^{l-1} \pi_{link}(n')} \chi_{nm}, \end{aligned} \tag{14.17}$$

where $l$ is such that $\xi_{l-1} < \eta(i) \leq \xi_l, i = M-N+1,...,M$. The other transition frequencies are defined as follows:

a) The transition frequency $\phi(\mathbf{x}_{i,a}, \mathbf{x}_{i,a} - \delta_c)$ is equal to $n_c(i)\mu_c$ if $\mathbf{x}_{i,a} - \delta_c \in S$, to 0 otherwise.

b) Let us consider the transition frequency $\phi(\mathbf{x}_{i,a}, \mathbf{x}_{i,a} + \delta_c)$. If $\mathbf{x}_{i,a} + \delta_c \notin S$, the frequency is 0. Otherwise, the controller should accept a call only if the current link state $k$ is such that $\eta(i) + e_c \leq \xi_k$. Let $l$ denote the link state such that $\xi_{l-1} < \eta(i) + e_c \leq \xi_l$ (note that $\xi_l$ must be greater than or equal to $\xi_k$). When the system is in state $\mathbf{x}_{i,a}$, the probability that $\mathbf{x}_{i,a} + \delta_c$ is available is equal to $\sum_{m=l}^{L} \pi_{link}(m|\mathbf{x}(t) = \mathbf{x}_{i,a}) = \sum_{m=l}^{L} \frac{\pi_{link}(m)}{\sum_{m'=l'}^{L} \pi_{link}(m')}$. Thus, the transition frequency $\phi(\mathbf{x}_{i,1}, \mathbf{x}_{i,1} + \delta_c)$ is $\sum_{m=l}^{L} \frac{\pi_{link}(m)}{\sum_{m'=l'}^{L} \pi_{link}(m')} u_{adm}(\mathbf{x}_{i,a}, c)\lambda_c$.

c) The transition frequency $\phi(\mathbf{x}_{i,u}, \mathbf{x}_{i,u} - \delta_c)$ is equal to $f_{drop}u_{drop}(\mathbf{x}_{i,u}, c) + n_c(i)\mu_c$ if $\mathbf{x}_{i,u} - \delta_c \in S$, to 0 otherwise.

d) The transition frequency $\phi(\mathbf{x}_{i,u}, \mathbf{x}_{i,u} - \delta_c)$ is equal to $f_{drop}u_{drop}(\mathbf{x}_{i,u}, c) + n_c(i)\mu_c$ if $\mathbf{x}_{i,u} - \delta_c \notin S$ and $\mathbf{x}_{i,a} - \delta_c \in S$, to 0 otherwise.

All the other transition frequencies are null.

As with the first DTMDP, we obtain the generic transition probabilities (i) by dividing the frequencies by the uniformization constant $\gamma > \max_{\mathbf{x}_{i,l} \in S} \left\{ \sum_{\mathbf{x}_{j,l'} \in S} \phi(\mathbf{x}_{i,l}, \mathbf{x}_{j,l'}) \right\}$, (ii) by adding the self transitions.

REWARD FUNCTION $r$

As with the first algorithm, we define the reward associated to the available state $\mathbf{x}_{i,a} \in S_{AV}$ as the load occupied by the on-going accepted calls:

$$r(\mathbf{x}_{i,a}) = \sum_{c=1}^{C} n_c(i)e_c, \; i = 1,...,M, \tag{14.18}$$

and the reward associated to the unavailable state $\mathbf{x}_{i,u} \in S_{UN}$ as the expected link capacity when the system is in $\mathbf{x}_{i,u}$:

$$\begin{aligned} r(\mathbf{x}_{i,u}) &= \sum_{n=1}^{l-1} \pi_{link}(n|\mathbf{x}(t) = \mathbf{x}_{i,u})\, \xi_n \\ &= \sum_{n=1}^{l-1} \frac{\pi_{link}(n)}{\sum_{n'=1}^{l-1} \pi_{link}(n')} \xi_n, \; i = M-N+1,...,M, \end{aligned} \tag{14.19}$$

where $l$ is such that $\xi_{l-1} < \eta(i) \leq \xi_l$.

## 14.3 LP formulations of the MDPs

The LP formulation of DTMDP problems is based on the definition of *randomized policies* ([4], [16]): each admission decision $u_{adm}(\mathbf{x}_{i,l},c)$ is a real number between 0 and 1, and represents the probability that a call of class $c$ is accepted when the system is in an available state $\mathbf{x}_{i,l} \in S_{AV}$; similarly, each dropping decision $u_{drop}(\mathbf{x}_{i,l},c)$ is a real number between 0 and 1, and represents the probability that a call of class $c$ is dropped when the system is in an unavailable state $\mathbf{x}_{i,l} \in S_{UN}$.

In a network supporting $C$ classes, each available state is associated to the decision of accepting/rejecting a class $c$ call. Consequently, the following generic acceptance action $\mathbf{a}_k(i)$ can be associated to state $\mathbf{x}_{i,l} \in S_{AV}$:

$$\begin{aligned} \mathbf{a}_k(\mathbf{x}_{i,l}) = \left[a_k^{(1)} \ldots a_k^{(C)}\right], \; a_k^{(c)} \in \{0,1\} \text{ if } \mathbf{x}_{i,l} + \delta_c \in S_{AV}, \\ a_k^{(c)} = 0 \text{ otherwise}, \forall \mathbf{x}_{i,l} \in S_{AV}. \end{aligned} \tag{14.20}$$

Thus, up to $2^C$ actions are associated to a single state $\mathbf{x}_{i,l} \in S_{AV}$. The set of actions in the available state $\mathbf{x}_{i,a}$ is denoted with $A_a$.

Similarly, each state $\mathbf{x}_{i,l} \in S_{UN}$ is associated to the decision of dropping or not a class $c$ call. Recalling the definition of $\delta_c$, since one (and only one) dropping decision must be equal to 1 (see equations (4) and (13)), the following dropping actions $\mathbf{d}_c$ can be associated to each state $\mathbf{x}_{i,l} \in S_{UN}$:

$$\mathbf{d}_c(\mathbf{x}_{i,l}) = \delta_c; c \text{ such that } n_c(i) > 0, \forall \mathbf{x}_{i,l} \in S_{UN}. \tag{14.21}$$

Thus, up to $C$ actions are associated to a single state $\mathbf{x}_{i,l} \in S_{UN}$. The set of actions in the unavailable state $\mathbf{x}_{i,l}$ is denoted with $A_d$.

By introducing the unknown $z(\mathbf{x}_{i,l},\mathbf{a}_k)$, which is the probability that the system is in the available state $\mathbf{x}_{i,l} \in S_{AV}$ and admission action $\mathbf{a}_k \in A_a(i)$ is chosen, and the unknown $z(\mathbf{x}_{i,l},\mathbf{d}_c)$, which is the probability that the system is in the unavailable state $\mathbf{x}_{i,l} \in S_{UN}$ and dropping action $\mathbf{d}_c$ is chosen, the following LP is defined (see [16]):

*Maximize*

$$\begin{aligned} f(\mathbf{z}) = \sum_{\mathbf{x}_{i,l} \in S_{AV}} \sum_{k \in A_a(i)} z(\mathbf{x}_{i,l}, \mathbf{a}_k) r(\mathbf{x}_{i,l}) \\ + \sum_{\mathbf{x}_{i,m} \in S_{UN}} \sum_{c \in A_d(i)} z(\mathbf{x}_{i,m}, \mathbf{d}_c) r(\mathbf{x}_{i,m}), \end{aligned} \tag{14.22}$$

*subject to*

$$\begin{aligned}
&\sum_{k' \in A_a(i')} z(\mathbf{x}_{i',l'}, \mathbf{a}_{k'}) - \sum_{\mathbf{x}_{i,l} \in S_{AV}} \sum_{k \in A_a(i)} z(\mathbf{x}_{i,l}, \mathbf{a}_k) p(\mathbf{x}_{i,l}, \mathbf{x}_{i',l'}; \mathbf{a}_k) \\
&- \sum_{\mathbf{x}_{i,l} \in S_{UN}} \sum_{c \in A_d(i)} z(\mathbf{x}_{i,l}, \mathbf{d}_c) p(\mathbf{x}_{i,l}, \mathbf{x}_{i',l'}; \mathbf{d}_c) = 0, \ \forall \mathbf{x}_{i',l'} \in S_{AV}, \\
&\sum_{c' \in A_d(i')} z(\mathbf{x}_{i',l'}, \mathbf{d}_{c'}) - \sum_{\mathbf{x}_{i,l} \in S_{AV}} \sum_{c \in A_a(i)} z(\mathbf{x}_{i,l}, \mathbf{a}_k) p(\mathbf{x}_{i,l}, \mathbf{x}_{i',l'}; \mathbf{a}_k) \\
&- \sum_{\mathbf{x}_{i,l} \in S_{UN}} \sum_{c \in A_d(i)} z(\mathbf{x}_{i,l}, \mathbf{d}_c) p(\mathbf{x}_{i,l}, \mathbf{x}_{j',l'}; \mathbf{d}_c) = 0, \ \forall \mathbf{x}_{i',l'} \in S_{UN},
\end{aligned} \tag{14.23}$$

$$\sum_{\mathbf{x}_i, l \in S_{AV}} \sum_{k \in A_a(\mathbf{x}_{i,l})} z(\mathbf{x}_{i,l}, \mathbf{a}_k) + \sum_{\mathbf{x}_{j,l} \in S_{UN}} \sum_{c \in A_d(\mathbf{x}_{j,l})} z(\mathbf{x}_{j,l}, \mathbf{d}_c) = 1, \tag{14.24}$$

*and*

$$\begin{aligned}
z(\mathbf{x}_{i,l}, \mathbf{a}_k) &\geq 0, \ k \in A_a(i), \ \mathbf{x}_{i,l} \in S_{AV}, \\
z(\mathbf{x}_{i,l}, \mathbf{d}_c) &\geq 0, \ c \in A_d(i), \ \mathbf{x}_{i,l} \in S_{UN}.
\end{aligned} \tag{14.25}$$

Since the stationary probabilities that the system is in state $\mathbf{x}_{i,l} \in S$ can be computed as

$$\begin{aligned}
&\sum_{k \in A_a(i)} z(\mathbf{x}_{i,l}, \mathbf{a}_k) = \pi(\mathbf{x}_{i,l}), \mathbf{x}_{i,l} \in S_{AV}, \\
&\sum_{c \in A_d(i)} z(\mathbf{x}_{i,l}, \mathbf{d}_c) = \pi(\mathbf{x}_{i,l}), \mathbf{x}_{i,l} \in S_{UN},
\end{aligned} \tag{14.26}$$

it follows (i) that the reward function (22) computes the expected throughput, and (ii) that constraints (23)-(25) express the balance equations of the DTMDP. Note that the number of constraints (23) is *LM* for the first DTMDP, *M*+*N* for the second one. As demonstrated in [16], since the described DTMDPs are unichain (i.e., all stationary policies have a single recurrent class and possibly a non-empty set of transient states), an optimal solution $\mathbf{z}^*$ generates an optimal policy $\mathbf{u}^*$ (which is retrieved form $\mathbf{z}^*$ by standard computations [16]).

The LP formulation allows the explicit control of both blocking and dropping probabilities. Different policies can be pursued by telecom operators; in this paper, we consider three objectives: 1) the throughput must be maximized; 2) the fairness among classes in terms of blocking probabilities must be maximized; 3) the class dropping probabilities must be below a give threshold. The first two objectives can be achieved by properly defining the cost function in a multi-objective fashion, whereas the third objective requires to add class-level constraints.

By defining the set $B_c$ as the set of indexes $k$ corresponding to the actions which block the admission of a class $c$ call, and recalling that the action $\mathbf{d}_c$ corresponds to the dropping of class $c$ calls, we can interpret the quantities

$$P^{(c)}_{block}(\mathbf{z}) = \sum_{\mathbf{x}_{i,l}\in S_{AV}} \sum_{k\in A_a(i)\cup B_c} z(\mathbf{x}_{i,l},\mathbf{a}_k) + \sum_{\mathbf{x}_{i,l}\in S_{UN}} \sum_{c\in A_d(i)} z(\mathbf{x}_{i,l},\mathbf{d}_c),$$
$$P^{(c)}_{drop}(\mathbf{z}) = \sum_{\mathbf{x}_{i,l}\in S_{UN}} z(\mathbf{x}_{i,l},\mathbf{d}_c)\frac{f_{drop}}{\lambda(c)\,[1-P_{block}(c)]}, \tag{14.27}$$

$c = 1,...,C$, as the blocking probabilities and dropping probabilities of class $c$. Note that (i) in the computation of the blocking probabilities we consider that if a call attempt arrives when the system is in an unavailable state it must be blocked, and that (ii) the expression of the dropping probabilities takes into account that the call dropping rate has to be evaluated over the admission rate of class $c$ calls, computed as $\lambda(c)[1\text{-}P_{block}(c)]$, and not over the dropping frequency $f_{drop}$.

Thus, the proposed reward function aimed at objectives 1. and 2. is defined as follows:

$$f(\mathbf{z}) = \sum_{\mathbf{x}_{i,l}\in S_{AV}} \sum_{k\in A_a(i)} z(\mathbf{x}_{i,l},\mathbf{a}_k)r(\mathbf{x}_{i,l})$$
$$+ \sum_{\mathbf{x}_{i,l}\in S_{UN}} \sum_{c\in A_d(i)} z(\mathbf{x}_{i,l},\mathbf{d}_c)r(\mathbf{x}_{i,l}) - w_{fair}\max\left[P^{(c)}_{block}(\mathbf{z})\right], \tag{14.28}$$

where $w_{fair}$ is a weight specifying the relevance of the third term, aimed at obtaining a fair solution, with respect to the first two terms, aimed at maximizing the throughput. Cost function (28) can be expressed by a linear cost function by introducing a new variable t and $C$ constraints. The LP becomes:

*Maximize*

$$f(\mathbf{z},t) = \sum_{\mathbf{x}_{i,l}\in S_{AV}} \sum_{k\in A_a(i)} z(\mathbf{x}_{i,l},\mathbf{a}_k)r(\mathbf{x}_{i,l})$$
$$+ \sum_{\mathbf{x}_{i,l}\in S_{UN}} \sum_{c\in A_d(i)} z(\mathbf{x}_{i,l},\mathbf{d}_c)r(\mathbf{x}_{i,l}) - w_{fair}\, t, \tag{14.29}$$

*subject to constraints (23), (24), (25) and to the following additional C constraints*:

$$t \geq P^{(c)}_{block}(z),\ c = 1,...,C. \tag{14.30}$$

Finally, to enforce a pre-defined maximum dropping probability $\rho_c$ tolerated by the network for class $c$, the following constraints on per-class dropping probabilities must be added to the LP:

$$P^{(c)}_{drop}(z) \leq \rho_c\ ,\ c = 1,...,C, \tag{14.31}$$

which are linear in the unknowns (see equations (27))

In the following Section, we will compare the results obtained by two different LPs: LP1, defined by the cost function (29) with $w_{fair} = 0$ and constraints (23), (24), (25), aimed at maximizing the total throughput; LP2, defined by the cost function

(29) with $w_{fair} > 0$ and constraints (23), (24), (25), (30), (32), aimed at maximizing the total throughput and at minimizing the maximum blocking probability among the classes, while keeping the dropping probabilities below given thresholds.

## 14.4 Comparison of the Two Models

The state space dimension of the second model is $M$+N; since N $< M$, the state space is smaller than the state space of the first model, whose size is $LM$, whenever the channel model has more than one state. Thus, the second model is scalable with respect the number $L$ of channel states, since its size remains the same regardless of $L$, whereas the state space of the first model grows linearly with $L$. Moreover, considering the LP formulations, also the number of constraints (23) is $M+N$ and $LM$, respectively. On the other hand, the policy space of the second model is smaller than the policy space of the first model: the former defines one admission policy for each available state and one dropping policy for each unavailable state, regardless of the current link state, whereas the latter defines different policies depending on the current channel state. Thus, the scalability of the second model is obtained at the price of a worse effectiveness of the computed policy. In the following, the two models are compared and evaluated by numerical simulations.

Tables 1-4 summarize link model, class, simulation and LP parameters.

The policies were computed by using the LP solver provided by the PCx software ( 1996, University of Chicago, USA, [3]). Two evaluation metrics were used to evaluate the LP complexity ([19]): the problem size and the CPU time required to solve it. To evaluate the obtained fairness we use the Jains fairness index ([5]), which rates the fairness of the set of the $C$ blocking probability values (the greater the index, the fairer the policy):

$$fairness\_index = \frac{\left(\sum_{c=1}^{C} P_{block}^{(c)}\right)^2}{C\left(\sum_{c=1}^{C} P_{block}^{(c)}\right)^2} \tag{14.32}$$

**Table 14.1** Values of the channel model

| Channel state load [Mbps] | Transition frequency matrix [$\min^{-1}$] |
|---|---|
| $\xi_1 = 0.25$<br>$\xi_2 = 0.40$<br>$\xi_3 = 0.55$<br>$\xi_4 = 0.70$<br>$\xi_5 = 0.85$<br>$\xi_6 = \xi_L = 1$ | $\chi = \begin{bmatrix} 0 & 0.2641 & 0.1208 & 0.0302 & 0 & 0 \\ 0.0058 & 0 & 0.1968 & 0.2290 & 0.0526 & 0 \\ 0.0029 & 0.0049 & 0 & 0.5427 & 0.1715 & 0 \\ 0 & 0.0019 & 0.0107 & 0 & 0.5349 & 0.1647 \\ 0 & 0.0010 & 0.0049 & 0.0175 & 0 & 0.29331 \\ 0 & 0 & 0.0010 & 0.0117 & 0.0370 & 0 \end{bmatrix}$ |

**Table 14.2** Class parameters (the parameter $\sigma$ is used to vary the offered traffic)

| Subject | Class 1 | Class 2 | Class 3 |
|---|---|---|---|
| $e_c$ [kbps] | 64 | 96 | 128 |
| $\lambda_c$ [min$^{-1}$] | $2.125\sigma$ | $1.25\sigma$ | $0.5\sigma$ |
| $\mu_c$ [min$^{-1}$] | 0.333 | 0.2 | 0.2 |
| $\varepsilon_d$ [%] | 1 | 1 | 1 |

**Table 14.3** Simulation parameters ($\eta_{OFF} = \sum_{c=1}^{C} \frac{\lambda_c}{\mu_c} e_c$: offered load)

| Simulation set number | 1 | 2 | 3 | 4 | 5 |
|---|---|---|---|---|---|
| $\eta$ [Mbps] | 1 | 1 | 1 | 1 | 1 |
| $\sigma$ | 0.433 | 0.471 | 0.508 | 0.546 | 0.584 |
| $\eta_{OFF}$ [Mbps] | 0.575 | 0.625 | 0.675 | 0.725 | 0.775 |
| $f_{drop}$ [min-1] | 600 | 600 | 600 | 600 | 600 |
| Length of each sim. run [s] | 14400 | 14400 | 14400 | 14400 | 14400 |
| Number of sim. runs per set | 10 | 10 | 10 | 10 | 10 |

**Table 14.4** Problem parameters

| Problem | LP1 | LP2 |
|---|---|---|
| $w_{fair}$ | 0 | 2 |
| Dropping probability constraints | No | Yes |

The fairness index is defined only if at least one of the $P_{block}^{(c)}$ is not null, and its range is (0,1].

Simulation results, summarized by the Tables 5 and 6, shows that for both developed models (*i*) by adding the fairness term to the reward function (LP2) the fairness indexes are dramatically increased with respect to the basic reward function (LP1); (ii) while the dropping probability constraint is violated in the unconstrained LP (LP1), by enforcing the dropping probability constraint (LP2), the dropping policy manages to redistribute the dropping probabilities among the supported classes, so that the dropping probability constraints are met, without affecting the overall throughput and the blocking probabilities. By using the approximated approach, the load utilization is only slightly worsened with respect to the one obtained by the first MDP: the worsening is between 1% and 3%.

Finally, Table 7 evaluates the scalability of the two models, by using channel models with increasing number of channels $L$. The link model parameters were set as follows: $\chi_{i,i+1}$ = 0.15 min$^{-1}$, $\chi_{i+1,i}$ = 0.15 min$^{-1}$, $i$ = 1,..., $L$ 1; $\chi_{i,i+2}$ = 0.05 min$^{-1}$, $\chi_{i+2,i}$ = 0.05 min$^{-1}$, $i = 1,\ldots,L-2$; $\xi_i = \frac{\xi_L}{L} i, i = 1,\ldots,L$. Simulation parameters were set as follows: $\xi_L$ = 0.95 Mbps, $\eta_{OFF}$ = 0.675, Mbps, $f_{drop}$ = 600

**Table 14.5** Simulation results with Model 1: average link load $\eta_{AVG,i}$, average fairness index $fair\_index_i$ and average maximum dropping probability $max\_drop\_prob_i$ for $LP_i$, $i = 1,2$

| Simulation set number | 1 | 2 | 3 | 4 | 5 |
|---|---|---|---|---|---|
| $\eta_{AVG,1}$ [%] | 48.33 | 51.89 | 54.84 | 57.33 | 59.35 |
| $\eta_{AVG,2}$ [%] | 47.69 | 51.21 | 54.07 | 56.30 | 58.40 |
| $fair\_index_1$ [%] | 91.72 | 83.02 | 76.67 | 90.77 | 93.15 |
| $fair\_index_2$ [%] | 98.29 | 99.50 | 98.69 | 99.16 | 99.69 |
| $max\_drop\_prob_1$ [%] | 2.65 | 2.93 | 3.79 | 3.74 | 4.36 |
| $max\_drop\_prob_2$ [%] | 0.92 | 1.01 | 1.28 | 1.04 | 1.00 |

**Table 14.6** Simulation results with Model 2: average link load $\eta_{AVG,i}$, average fairness index $fair\_index_i$ and average maximum dropping probability $max\_drop\_prob_i$ for $LP_i$, $i = 1,2$

| Simulation set number | 1 | 2 | 3 | 4 | 5 |
|---|---|---|---|---|---|
| $\eta_{AVG,1}$ [%] | 47.76 | 51.19 | 53.94 | 55.88 | 57.50 |
| $\eta_{AVG,2}$ [%] | 46.60 | 49.95 | 51.77 | 53.45 | 55.13 |
| $fair\_index_1$ [%] | 88.95 | 92.07 | 90.49 | 91.04 | 91.93 |
| $fair\_index_2$ [%] | 99.96 | 99.92 | 99.77 | 99.96 | 99.99 |
| $max\_drop\_prob_1$ [%] | 2.06 | 2.40 | 3.48 | 3.17 | 3.90 |
| $max\_drop\_prob_2$ [%] | 0.96 | 1.25 | 1.13 | 1.04 | 1.22 |

**Table 14.7** Simulation results with Model 1 and Model 2: number of variables *Var*, number of constraints *Const*, LP size *Size*, CPU time *CPU*

| *L* | Model 1 | | | | Model 2 | | | |
|---|---|---|---|---|---|---|---|---|
| | *Var* | *Const* | *Size* | *CPU* [s] | *Var* | *Const* | *Size* | *CPU* [s] |
| 3 | 3086 | 816 | 2.52E+06 | 4.67 | 2262 | 276 | 6.24E+05 | 0.29 |
| 4 | 3944 | 1086 | 4.28E+06 | 8.42 | 2262 | 276 | 6.24E+05 | 0.29 |
| 5 | 4903 | 1356 | 6.65E+06 | 9.59 | 2262 | 276 | 6.24E+05 | 0.29 |
| 6 | 5802 | 1626 | 9.43E+06 | 13.57 | 2262 | 276 | 6.24E+05 | 0.29 |
| 7 | 6703 | 1896 | 1.27E+07 | 24.02 | 2262 | 276 | 6.24E+05 | 0.29 |
| 8 | 7625 | 2166 | 1.65E+07 | 27.1 | 2262 | 276 | 6.24E+05 | 0.29 |
| 9 | 8566 | 2436 | 2.09E+07 | 32.46 | 2262 | 276 | 6.24E+05 | 0.29 |
| 10 | 9370 | 2706 | 2.54E+07 | 41.56 | 2262 | 276 | 6.24E+05 | 0.29 |
| 100 | - | - | - | - | 2262 | 276 | 6.24E+05 | 3.06 |

min-1. The table clearly shows how, with the first DTMDP, the state space explodes $L$, while with the first DTMDP it remains constant. Consequently, with the second model it was possibile to compute the optimal policy even considering a link model with 100 states.

## 14.5 Conclusions and Future Work

The theoretical relevance of this paper is that it extends the MDP framework currently available for the admission control problem in communication networks to the case of networks with time-varying link capacity; this is typically the case of wireless networks such as CDMA networks, DVB-S2 satellite networks and WiMAX networks. The innovative approach consists in i) integrating a Markov Chain-based link model within the MDP framework and ii) computing a class dropping policy beside the standard admission policy to control the call dropping due to the variable link capacity. Numerical simulations validate the effectiveness of the proposed algorithms.

## References

1. Altman, E.: Applications of Markov decision processes in communication networks- A survey. Tech. Rep., INRIA (2000), http://www.inria.fr/RRRT/RR-3984.html
2. Castanet, L., Deloues, T., Lemorton, J.: Channel modelling based on N-state Markov chains for satcom systems simulation. In: Twelfth International Conference on Antennas and Propagation (ICAP 2003), vol. 1, pp. 119–122 (2003)
3. Czyzyk, J., Mehrotra, S., Wagner, M., Wright, S.J.: PCx User Guide (Version 1.1). Optimization Technology Center, Technical Report OTC 96/01 (1997)
4. Hillier, F.S., Lieberman, G.J.: Introduction to Operations Research, 6th edn., ch. 21. McGraw Hill, New York (1995)
5. Jain, R.K., Chiu, D.-M.W., Hawe, W.R.: A quantitative measure of fairness and discrimination for resource allocation in shared computer systems. Digital Equipment Corporation, Tech. Rep. (1984)
6. Kalyanasundaram, S., Chong, E.K.P., Shroff, N.B.: Admission control schemes to provide class-level QoS in multiservice networks. Computer Networks 35, 307–326 (2001)
7. Kalyanasundaram, S., Chong, E.K.P., Shroff, N.B.: Optimal resource allocation in multiclass networks with user-specified utility functions. Computer Networks 38, 613–630 (2002)
8. Mosharaf, K., Talim, J., Lambadaris, I.: Call admission and fairness control in WDM networks with grooming capabilities. In: IEEE Proc. 43rd Conference on Decision and Control, Nassau (Bahamas), pp. 3738–3743 (2004)
9. Nasser, N., Hassanein, H.: An Optimal and Fair Call Admission Control Policy for Seamless Handoff in Multimedia Wireless Networks with QoS Guarantees. In: IEEE Proc. Globecom, pp. 3926–3930 (2004)
10. Ni, J., Tsang, D.H.K., Tatikonda, S., Bensaou, B.: Optimal and structured call admission control policies for resource-sharing systems. IEEE transactions on Communications 55(1), 158–170 (2007)
11. Nordstrom, E., Carlstrom, J.: Call admission control and routing for integrated CBR/VBR and ABR services: a Markov decision approach. In: IEEE Proc. ATM Workshop 1999, pp. 71–76 (1999)
12. Park, J.-S., Huang, L., Lee, D.C., Kuo, C.-C.J.: Optimal Code Assignment and Call Admission Control for OVSF-CDMA Systems Constrained by Blocking Probabilities. In: Proc. of IEEE Globecom 2004, pp. 3290–3294 (2004)
13. Pietrabissa, A.: Admission Control in UMTS Networks based on Approximate Dynamic Programming. European Journal of Control 14(1), 62–75 (2008)

14. Pietrabissa, A.: An Alternative LP Formulation of the Admission Control Problem in Multi-Class Networks. IEEE Transaction on Automatic Control 53(3), 839–844 (2008)
15. Pimentel, C., Falk, T., Lisbo, L.: Finite-State Markov Modeling of Correlated Rician-Fading Channels. IEEE Transactions on Vehicular Technology 53(5), 1491–1501 (2004)
16. Puterman, M.L.: Markov Decision Processes. Jhon Wiley & Sons, New Jersey (1994)
17. Sanchez-Salas, D.A., Cuevas-Ruiz, J.L.: N-states Channel Model using Markov Chains. In: Electronics, Robotics and Automotive Mechanics Conference (CERMA 2007), pp. 342–347 (2007)
18. SATSIX (Satellite-based communications systems within IPv6 networks) project (contract IST-2006-26950), `http://www.ist-satsix.org/`
19. Vanderbei, R.J.: Linear Programming: Foundation and Extension, 2nd edn. Kluwer Academic Publishers, Boston (2001)
20. Vucetic, B., Du, J.: Channel Modeling and Simulation in Satellite Mobile Communication Systems. IEEE Journal On Selected Areas In Communications 10(8), 1209–1218 (1992)
21. Xiao, Y., Chen, C., Wang, Y.: Optimal admission control for multi-class of wireless adaptive multimedia services. IEICE Transactions on Communications 2001 E84-B(4), 795–804 (2001)
22. Yang, X., Feng, G.: Optimizing Admission Control for Multiservice Wireless Networks With Bandwidth Asymmetry Between Uplink and Downlink. IEEE Transactions on Vehicular Technology 56(2), 907–917 (2007)
23. Yu, F., Krishnamurthy, V., Leung, V.C.M.: Cross-Layer Optimal Connection Admission Control for Variable Bit Rate Multimedia Traffic in Packet Wireless CDMA Networks. IEEE Transactions on Signal Processing 54(2), 542–555 (2006)

FSC
www.fsc.org
MIX
Papier | Fördert gute Waldnutzung
FSC® C083411

Kolibri 360 GmbH
Ferdinand-Jühlke-Straße 7
99095 Erfurt, Deutschland
produktsicherheit@kolibri360.de